ATHEROSCLEROSIS

Cellular and Molecular Interactions in the Artery Wall

ALTSCHUL SYMPOSIA SERIES

Series Editors: Sergey Fedoroff and Gary D. Burkholder

Volume 1 • ATHEROSCLEROSIS: Cellular and Molecular Interactions in the
Artery Wall
Edited by Avrum I. Gotlieb, B. Lowell Langille, and Sergey Fedoroff

ATHEROSCLEROSIS

Cellular and Molecular Interactions in the Artery Wall

Edited by

Avrum I. Gotlieb
B. Lowell Langille

University of Toronto
Toronto, Ontario, Canada

and

Sergey Fedoroff

University of Saskatchewan
Saskatoon, Saskatchewan, Canada

SPRINGER SCIENCE+BUSINESS MEDIA, LLC

Library of Congress Cataloging in Publication Data

Altschul Symposium (1st: 1990: Saskatoon, Sask.)
  Atherosclerosis: cellular and molecular interactions in the artery wall / edited by
Avrum I. Gotlieb . . . [et al.].
    p.   cm. — (Altschul symposia series; v. 1)
  "Proceedings of the First Altschul Symposium . . . held Apr. 29–May 2, 1990, in
Saskatoon, Saskatchewan, Canada" — T.p. verso.
  Includes bibliographical references and index.
  ISBN 978-1-4613-6672-0   ISBN 978-1-4615-3754-0 (eBook)
  DOI 10.1007/978-1-4615-3754-0
  1. Atherosclerosis — Pathogenesis — Congresses. 2. Atherosclerosis — Molecular
aspects — Congresses. I. Gotlieb, Avrum I. II. Title. III. Series.
  [DNLM: 1. Atherosclerosis — etiology — congresses. 2. Atherosclerosis — pathology —
congresses. 3. Atherosclerosis — physiopathology — congresses. 4. Endothelium,
Vascular — injuries — congresses. 5. Endothelium, Vascular — pathology — congresses. 6.
Endothelium, Vascular — physiopathology — congresses. WG 500 A469a 1990]
  RC692.A38   1991
  616.1'36 — dc20
  DNLM/DLC   91-3800
  for Library of Congress   CIP

Proceedings of the First Altschul Symposium, Atherosclerosis:
Cellular and Molecular Interactions in the Artery Wall,
held April 29 — May 2, 1990, in Saskatoon, Saskatchewan, Canada

ISBN 978-1-4613-6672-0

© 1991 by Springer Science+Business Media New York
Originally published by Plenum Press New York in 1991
Softcover reprint of the hardcover 1st edition 1991

RUDOLF ALTSCHUL
1901 – 1963

PREFACE

This volume contains the papers which were presented at the First Altschul Symposium, *Atherosclerosis: Cellular and Molecular Interactions in the Artery Wall.* The symposium was held in Saskatoon, at the University of Saskatchewan in May 1990 in memory of Dr. Rudolph Altschul, a pioneer in the field of vascular biology and the prevention of atherosclerosis. Dr. Altschul was Professor and Head of the Department of Anatomy at the University of Saskatchewan from 1955 to 1963.

The challenge for biomedical scientists is to unravel the multifactorial etiology of atherosclerosis. For the last hundred and sixty years, anatomical pathologists have carefully studied the morphological changes of the human vascular wall during the initiation and evolution of the fibrofatty atherosclerotic plaque. Based on these elegant morphological observations, theories on atherogenesis were put forth by pathologists in the 1840's. Rudolf Virchow suggested that the movement of substances from the blood into the vessel wall was important for atherogenesis while Carl von Rokitansky felt that the deposition of substances on the lumenal surface of the artery resulted in the formation of atherosclerotic plaque. Since these original theories, it has become apparent that the pathogenesis of atherosclerosis is multifactorial and the disease evolves in stages. It is also likely that not all plaques arise through the same sequence of events and that many steps are involved in the development of each plaque.

Today, our understanding of the complicated processes of atherogenesis is still incomplete. Histopathology data have been used to develop animal models to study pathogenesis. Tissue culture and cell biology have enhanced our understanding of those cells which we believe to be the seat of disease. New powerful morphological techniques have appeared, including the molecular biology technique of *in situ* hybridization and the application of monoclonal antibodies in immunohistochemistry, which are providing sensitive methods to study the composition of lesions. Based on a solid foundation of traditional morphology, modern cell and molecular biology studies are now providing us with a new understanding of the pathogenesis of the atherosclerotic plaque.

The purpose of this book is to review the field of atherosclerosis in depth and to present state of the art information on the pathogenesis of atherosclerosis in its broadest sense. The focus is on cell and molecular biology, highlighting interactions between vascular and blood cells, their products, serum constituents, blood flow and vessel wall constituents. In the chapters of this book, detailed studies of intricate and complex interactions of the cells of the plaque are described. The discovery and understanding of these interactions are now providing new areas of possible prevention and therapy of atherosclerosis and its clinical complications.

The sponsors of the symposium include the University of Saskatchewan, Medical Research Council of Canada, Canadian Atherosclerosis Society, Pharmaceutical Manufacturers Association of Canada, Heart and Stroke Foundation of Saskatchewan, and the Toronto Hospital. We are also indebted to the following companies for their generous support: Astra Pharma Inc., Carl Zeiss Canada Ltd., Glaxo Canada Inc., Medical Arts Laboratories in Saskatoon, Merck Frosst Canada Inc., and Parke-Davis.

A.I. Gotlieb
B.L. Langille
S. Fedoroff

CONTENTS

Rudolph Altschul: Man and Scientist... 1
 S. Fedoroff

1 THE DYNAMIC VESSEL WALL

Natural History of Atherosclerosis.. 7
 S.N. Schwartz, D. Gordon and J.N. Wilcox

Adaptations of Mature and Developing Arteries
 to Local Hemodynamics.. 33
 B.L. Langille

The Cytoskeleton of Arterial Smooth Muscle Cells during
 Development, Atheromatosis and Tissue Culture.. 53
 G. Gabbiani and L. Rubbia

2 VASCULAR WALL INJURY AND REPAIR

Repair of Arterial Endothelium... 59
 M.A. Reidy and V. Lindner

The Role of the Cytoskeleton in Endothelial Repair....................................... 67
 A.I. Gotlieb

Thrombin-Stimulated Endothelial Cell Functions:
 Monocyte Adhesion and PDGF Production... 89
 P.E. DiCorleto, C. de la Motte and
 R. Shankar

3 EXTRACELLULAR MATRIX INTERACTIONS IN ATHEROGENESIS

Dynamic Responses of Collagen and Elastin to Vessel
 Wall Perturbation... 101
 F.W. Keeley

Dynamic Interaction of Proteoglycans.. 115
 T.N. Wight

Endothelial Cell - Extracellular Matrix Interactions:
 Modulation of Vascular Cell Phenotype by Matrix
 Components and Soluble Factors.. 127
 J.A. Madri

4   AUTOCRINE AND PARACRINE CONTROL OF VASCULAR TISSUE

Regulation of PDGF Expression in Vascular Cells.. 139
T. Collins, R. Young, A.E. Mendoza,
J.W.U. Fries, A.J. Williams, P. Sultan
and D.T. Bonthron

Endothelial Regulation of Vasomotor Tone in
  Atherosclerosis... 153
D.G. Harrison

Production of Cytokines by Vascular Wall Cells:
  An Update and Implications for Atherogenesis.. 161
P. Libby, H. Loppnow, J.C. Fleet,
H. Palmer, H.M. Li, S.J.C. Warner,
R.N. Salomon and S.K. Clinton

5   THROMBOSIS AND FIBRINOLYSIS

Hypoxia and Endothelial Cell Function: Alterations
  in Barrier and Coagulant Properties... 173
S. Ogawa, M. Matsumoto, J. Brett,
M. Clauss and D.M. Stern

Regulation of Type One Plasminogen Activator Inhibitor
  Gene Expression in Cultured Endothelial Cells
  and the Vessel Wall.. 187
M. Sawdey and D.J. Loskutoff

Platelet Reactions in Thrombosis... 209
M.A. Packham

Eicosanoid Metabolism and Endothelial Cell Adhesion
  Molecule Expression: Effects on Platelet/Vessel
  Wall Interactions.. 227
M.R. Buchanan, M.C. Bertomeu, S.J. Brister
and T.A. Haas

6   LIPIDS IN ATHEROGENESIS

Receptor-Mediated Low Density Lipoprotein Metabolism............................... 237
W.J. Schneider

The Molecular Basis for Lipoprotein Interaction with
  Vascular Tissue... 247
G. Bondjers, E.H. Camejo, O. Wiklund, U. Olsson,
S.-O. Oloffson and G. Camejo

Cholesterol: Is there a Consensus?.. 255
L. Horlick

Contributors... 263

Index... 265

RUDOLPH ALTSCHUL: MAN AND SCIENTIST

S. Fedoroff

Department of Anatomy
University of Saskatchewan
Saskatoon, Saskatchewan, Canada

Professor Rudolf Altschul was a colorful, highly-cultured individual who left an enduring legacy through his friends, his students and his writings. For those who were not privileged to know him, I have a difficult task to convey a sense of his spirit, trends of thought, and love of science, art and music. To those who did know him, perhaps I can jog the memory.

Rudolf Altschul was born in Prague, in what was then Bohemia, at the turn of the century and graduated in 1925 from the University of Prague with the degree of Doctor of Universal Medicine, which encompassed both Medicine and Dentistry. He did postdoctoral work in neurology and neuropathology in the Charcot Clinic in the Hôpital de la Salpétrière in Paris under Professor Georges Guilain and in the University of Rome under Professor Giovanni Mingazzini. In Rome he met and married Anni Fischer. They were a devoted couple, a truly complementary pair.

Scientifically, the two years in Rome were most productive. Altschul demonstrated originality, a penchant to search for the unusual and keen powers of observation. For example, during his stay in Rome there was a fire in a theatre in which four female dancers lost their lives. Seizing this unique opportunity, Altschul obtained their brains at postmortem and looked for common lesions that could be related to carbon monoxide poisoning. He did not find any such lesions but thought it curious that there should not be specific lesions linked to the effects of the carbon monoxide. He returned to this problem later on.

Altschul described a unique case of myeloschisis with triplication of the spinal cord and also a condition in which there is one undeveloped breast in females with thalamic syndrome. He was also interested in histology. By that time the fad of metallic impregnation of sections of nervous tissue had already died down, and the only methods surviving were those of Cajal's gold chloride and del Rio Hortega's silver carbonate. However, results obtained by the gold chloride method were difficult to reproduce. He persistently investigated all the variables and finally concluded that temperature was the key.

With this background, he returned to Prague in 1929 and entered the private practice of neuropsychiatry and at the same time began research in the Department of Histology at the German University. The department was chaired by the renowned Professor Alfred Kohn, who was one of the pioneers of modern endocrinology; he contributed significantly to understanding of the parathyroid and the development and function of chromaffin cells, a term he coined. During his 10 years as assistant to Professor Kohn, Altschul published widely. Remembering the dancers in Rome, he reinvestigated carbon monoxide poisoning using animal experiments and concluded that the effects result from anoxia and that variability of lesions in the CNS depend on the amount of carbon monoxide inhaled over a period of time. He also studied the phylogenetic development of the hippocampus and continued his experiments with metal impregnation methods. He had just begun to develop an interest in the vascular system of the brain when in 1939, he and Anni were forced to flee from the Nazi occupation of Czechoslovakia.

Altschul had somehow heard that a medical school was being established in far-off Saskatoon, Canada and he wrote to President J.S. Thompson, a well known theologian. Dr. Thompson contacted a theological colleague in Prague and on receiving a favorable recommendation, offered Dr. Altschul a position as instructor in Anatomy.

Rudolf Altschul and Anni eagerly began their journey to Canada but unfortunately, they were aboard the ill-fated S.S. Athenia, the first Allied ship to be torpedoed in the Second World War. During the rescue operation they were separated, then later reunited in Ireland. Eventually they arrived in New York, only to be interned as aliens on Ellis Island. A telegram from President Thompson expedited their release and they proceeded to Saskatoon to begin a new life which for Altschul meant the rebuilding of his scientific career.

Until then, Altschul's primary interest had been the normal and pathological morphology of the nervous system. In Saskatoon, he was asked to assist a local pathologist with the routine examination of brains. Among these was the brain of a 45-year-old man in which he found unusually clear cellular reactions in some atherosclerotic vessels. He could see gradual transition from elements that were unmistakably endothelial cells, through less characteristic cells, to typical foam cells. He concluded that foam cells are not histiocytes. It appeared to him that endothelial cells lining arteries of all sizes and even capillaries are the progenitors of foam cells as well as cells that morphologically resemble mesenchymal or reticular cells and that they migrate into subjacent layers. These views met with considerable criticism. To prove his point, he began animal experiments to investigate response to atherogenic agents.

At that time, at the end of the war, there was little money for research. Rabbits for the project were caught just across College Drive on the vacant land where people from Saskatoon will know Brunskill School is now situated, and gophers were trapped on the prairie where the University Hospital now stands. Moreover, people trained to work with animals were scarce and precise administration of cholesterol to animals was a problem. Altschul thought of the dried powdered eggs that had been shipped to Britain during the war as part of the food supply, and after a few phone calls he obtained some powdered egg yolk. The rabbits, however, were uncooperative and spit out the dry yolk, or spilled it; the yolk powder mixed with water or milk became sour. There was no way to determine how much cholesterol the rabbits ingested. He came up with the great idea that Anni should bake the yolk into a cake - 4 parts yolk powder, one part flour, plus water. The resulting cake looked delicious and luckily, the rabbits loved it. The egg yolk cake produced more intensive vascular lesions and the distribution of the lesions in the vascular bed was more widespread than had been reported by other investigators using pure cholesterol. This led to a series of experiments in which pure cholesterol was heated and given in capsules. Sure enough, heated cholesterol was more strongly atherogenic than unheated cholesterol. One explanation was oxidation of the cholesterol.

Experiments were undertaken to feed rabbits cholesterol treated with peroxide, irradiated with ultraviolet light, or irradiated in the wet state by X-rays. Contrary to expectations, atherogenic properties decreased or were wanting as the length of treatment of the cholesterol increased. Altschul then began experiments by which he hoped to increase the oxidative processes *in vivo*. Animals were fed cholesterol for 3 months and during this time they were irradiated periodically with ultraviolet light. Others were exposed three times weekly to increased oxygen inhalation. In both series, the atherogenic lesions were less than in animals given cholesterol without additional treatment.

These experiments were later extended to humans, and serum cholesterol was lowered by increased oxygen tension and ultraviolet irradiation. These treatments, however, were cumbersome and Altschul looked for other means to accomplish the results of UV and oxygen treatments. About that time in Saskatchewan, Dr. Abram Hoffer and Dr. Humphrey Osmond were investigating the effect of large doses of nicotinic acid and its amide in schizophrenia. Both substances raise the levels of nicotinamide adenine dinucleotide (NAD), an important component in cell respiration. Altschul reasoned that this might increase oxidative processes, thereby influencing cholesterol metabolism. He tested relatively large doses of nicotinic acid and in short term experiments found a significant decrease of serum cholesterol both in rabbits and in human volunteers. Correspondingly large doses of the amide did not lower the serum cholesterol. This led eventually to the proposal of niacin therapy in hypercholesterolemia and hyperlipemia. Altschul also continued work on the normal and pathological morphology of the central nervous system, metallic impregnation of sections of nervous tissue, and on denervated muscle.

In 1950, Altschul published his first book, "Selected Studies on Arteriosclerosis"[1]. In the preface he wrote: "Looking over the book after it was written and recognizing its incompleteness and the paucity of its conclusions, I was tempted to label it 'Volume I' thus stressing that it is an unfinished work." Typical of his humility and honesty, he did not call it Volume I, just in case he never did get around to writing Volume II.

A few years later he published his second book, "Endothelium"[2], now considered a classic in the field. In the preface he said, "It has been said that one is as old as one's arteries. In view of the supreme importance of endothelium in arterial function I should like to modify, or rather simplify, the statement by saying that one is as old as one's endothelium."

His final book, "Niacin in Vascular Disorders and Hyperlipemia"[3], was a compilation of a number of papers on the subject. He described the book as "the proceedings of a symposium which did not take place". While in the final stages of the preparation of the book Rudolf Altschul died peacefully during one of his regular afternoon naps, at his home, in 1963. I had the privilege of completing the book.

A man of many interests and broad culture, Altschul was an enthusiastic photographer, a linguist, and a connoisseur of the arts. He loved conversation and had a fund of anecdotes gained from his wide experience, which he told with great wit and charm. Always happiest in the laboratory and library, he could be found there any day of the week; holidays were a burden to him.

At that time, the President of the University had fairly regular afternoon tea parties in his beautiful residence, situated on the University campus. Anni and "Rudl" as his friends called him, were often invited, and he would go along through the reception line, then make a beeline out the back door straight to his laboratory. He would spend a happy couple of hours, whistling operatic arias as he worked over his microtome or peered down the microscope. Then he would slip back to the reception, rejoin Anni, and make his good-byes, thanking his host for a lovely afternoon. The President knew very well what he had been up to but wisely just thanked Rudl for coming.

He was nevertheless gregarious and had regular Saturday evening gatherings at his home. Anni was a perfect hostess with her natural warmth and humor. Usually they invited people from various faculties and added some from outside the academic circle. He made the evenings exciting and encouraged debate. He always carried in his wallet a few pictures of endothelial cells or foam cells or some other cell he happened to be studying at the moment. He could not resist showing the photos to a startled Engineering or English professor, enthusiastically explaining the beauty and significance of the cells.

In the classroom he quickly won the trust and deep affection of his students, to whom he taught not only neuroanatomy and histology, but also the wisdom of life. Our first Dean of Medicine, Dr. Stewart Lindsay, recounted how Altschul had replied to a student who had given a plausible but incorrect answer to a question. "If that is not true, it is nicely invented", he said. Dean Lindsay commented that "underlying Dr. Altschul's interest in students was a keen critical mind which possessed a tolerance for almost anything but shoddy work. This he never forgave."[4]

Altschul was ahead of his time. He already understood the complexity of tissues, their plasticity and the dynamics of cell interactions. He expressed this in his book on endothelium and in his essays. In the introduction to "Endothelium"[2] he wrote: "While working on problems of arteriosclerosis, I realized not only how little I know about endothelium but also how much I ought to know for the proper understanding of arteriosclerosis".

In one of his essays Altschul wrote: "Their distribution [blood vessels] is studied in the first year of Anatomy and the average student is soon familiar with it. Much simpler still appears the microscopic structure: intima, media and adventitia, minus the media for capillaries. Histologically proper, the wall consists of endothelium, smooth muscle and connective tissue, with some elastic tissue thrown in for good weight. Armed with this knowledge, many investigators try to tackle the problem of arteriosclerosis. However, one very quickly finds that things are not so simple, that endothelial cells can migrate and either form plaques or regress to cells of the mesenchymal type and later redifferentiate either into fibroblasts or smooth muscle cells. Some of these cells may become phagocytic. The regrouping of the cellular pattern of the vessel wall will most likely be followed by functional disturbance, affecting permeability of endothelium and contraction of the media of the vessel wall. The bottom line is that not only function but the structure of the blood vessel depends to a large part on the state of the endothelium. Its dysfunction will significantly alter the structural pattern of the vessel."

Considering the primitive methodology at his disposal, Altschul, through intuition, keen observation and his unorthodox way of doing research, came remarkably close to present day thinking. He never pretended to have final answers. For example, in his review on the effect of nicotinic acid on hypercholesterolemia and hyperlipemia[3], he said that before or soon after publication of the review a better way of dealing with hypercholesterolemia or hyperlipemia than by treatment with large doses of nicotinic acid might be found but that the work on nicotinic acid would not lose its scientific value.

Altschul was always surprised to receive the acclaim of his peers. I remember when he returned once from a scientific meeting and I asked, "How was the meeting?", he replied, "I met a very famous person. His name was Rudolf Altschul." In 1961, he was elected a Fellow of the Royal Society of Canada.

The members of the Department of Anatomy of the University of Saskatchewan are proud to be organizing the Altschul Symposia, made possible by the endowment left by Anni Altschul. The symposia will be held biennially and will alternate in their emphasis on the cardiovascular and nervous systems.

REFERENCES

1.   R. Altschul, "Selected Studies on Atherosclerosis," Charles C. Thomas, Springfield, Ill. (1950).
2.   R. Altschul, "Endothelium.   Its Development, Morphology, Function and Pathology," The MacMillan Company, New York (1954).
3.   R. Altschul, ed., "Niacin in Vascular Disorders and Hyperlipemia", Charles C. Thomas, Springfield, Ill. (1964).
4.   D.J. Buchan, "Greenhouse to Medical Centre. Saskatchewan's Medical School, 1926-78." College of Medicine, University of Saskatchewan, Saskatoon, Canada, p.157 (1983).

SECTION 1

THE DYNAMIC VESSEL WALL

NATURAL HISTORY OF ATHEROSCLEROSIS

Stephen M. Schwartz, David Gordon
and Josiah N. Wilcox

Department of Pathology
University of Washington
Seattle, WA 98185

Perhaps the three features most characteristic of atherosclerosis are lipid accumulation, abnormal growth of smooth muscle, and thrombotic occlusion. This review will focus on the role of smooth muscle proliferation in the atherosclerotic process and will attempt to identify key questions relevant to the ultimate conversion of a benign, smooth muscle lesion into a progressive and fatal, vaso-occlusive lesion.

IDENTIFICATION OF SMOOTH MUSCLE CELLS AS KEY CELLS IN EARLY LESIONS

As late as the 1960's the composition of the atherosclerotic lesion was covered under such vague terms as "fibromuscular proliferation". The identification of many of the lesion cells as smooth muscle cells, resulted from the application of electron microscopy to the lesions. Cells in the atherosclerotic plaque were characterized as smooth muscle on the basis of their content of dense accumulations of actin bundles, dense membrane-associated bands, caveolae in the plasmalemma, and the presence of a surrounding layer of amorphous material called basement membrane[1-4].

The source of these smooth muscle cells was only discussed in general terms. Investigators were, however, aware of three facts that still form the basis for most discussions of the process of smooth muscle accumulation in atherosclerosis[5,6]. First, the lesion is intimal; any hypothesis needs to explain the selective accumulation of smooth muscle cells in the intima at sites of injury[6]. Second, the normal intima also contains smooth muscle cells. Function and properties of these intimal smooth muscle cells were unknown, although most investigators believed that "diffuse intimal thickening" was a normal, not a pathological process[6,7]. Third, the smooth muscle cells were accompanied by macrophages[6].

SMOOTH MUSCLE GROWTH FACTORS

During the time that these smooth muscle cells were being identified, relatively little thought was given to the molecules responsible for stimulating smooth muscle accumulation, although investigators were aware of the possible roles of proliferation and migration of cells from the media[6,8]. Proliferation was seen as a nonspecific reaction to injury, either the result of toxic products accumulating along with the lipid at a site of atherogenesis, or the result of cell injury at sites where hypercholesterolemia had caused necrosis of the vessel wall. These rather general ideas have been replaced over the last two decades by more specific hypotheses based in large part on studies *in vitro* identifying specific molecules, primarily polypeptides, able to stimulate smooth muscle cell replication.

Emphasis on polypeptide mitogens originated with the hypothesis that pathologic smooth muscle replication was initiated by actions of growth factors derived from platelets or leukocytes. The focus of this hypothesis was platelet-derived growth factor (PDGF)[9]. The idea that a platelet factor controlled smooth muscle replication grew out of the observation that smooth muscle cells would not replicate in serum prepared without platelet release[9]. This led to the suggestion that PDGF was required for smooth muscle replication. The hypothesis was supported by the observation by Stemerman and others that denudation of arteries by a balloon embolectomy catheter produces intimal proliferation[10]. Immediately after balloon catheterization, the denuded surface shows a prominent platelet carpet[10,11]. Anti-platelet serum, given for two weeks, inhibited the final extent of smooth muscle accumulation[12].

The observation that whole blood serum (WBS) contained a platelet-derived mitogen led to the purification and cloning of two chains: PDGF-A and PDGF-B. The PDGF molecule is a dimer of these two chains and can exist in either homodimeric form or as the heterodimer[13-15]. Perhaps most startling was the identification of the oncogene c-sis as the gene for the PDGF-B chain[16,17]. This oncogene is, itself, capable of transforming cells, raising the possibility that atherosclerosis is the result of the secretion of an oncogenic peptide within the vessel wall. Recently, however, attention has begun to be focused on the possibility that regulation of the PDGF receptor may play a critical role in growth control *in vivo*. The PDGF receptor consists of combinations of two subunits[18,19,20]. The "a" subunit is capable of recognizing both PDGF monomers A and B. The "b" subunit only recognizes PDGF-B subunits. Thus bb receptors cannot recognize PDGF-AA dimer and only aa receptors can recognize PDGF-AA dimer. The relative roles of the mix-and-match receptor-agonist system remain unclear, although both receptors appear to have mitogenic capacity[13,14,19,21]. While probes for the "a" subunit are only now becoming available, there is evidence for differential regulation of receptor transcription[22]. Moreover, data on the "b" subunit already show the surprising finding that "b" subunit is low in the normal wall and high in sites prone to smooth muscle proliferation such as the intima of the atherosclerotic plaque or the wall of vessels in inflammatory sites[23,24].

While most attention has been given to growth factors released from platelets, there is no reason to believe that PDGF or any platelet mitogen are the major controls of smooth muscle growth in responses to injury *in vivo*. Recent studies by Fingerle *et al.*[25] using anti-platelet antibodies show that the initial proliferative response occurs even in the absence of platelets. The platelet could, nonetheless play a role in proliferation occurring at later times. Again, as noted above, chronic administration of antiplatelet serum over two weeks greatly diminished the extent of intimal thickening[12]. Antiplatelet serum, however, can only be used effectively for a few days; in the original experiments, most of the tested animals died. Thus the anti-platelet serum approach is mainly useful for looking at the early response to denudation, as in Fingerle's study. We know, however, that proliferation lasts weeks after a balloon injury, while platelet release lasts only hours to days[26,27]. Several possibilities exist for this prolonged effect. The balloon injury is traumatic and, as just noted, it is quite likely that such trauma releases heparin-binding growth factors (HBGFs) from the injured cells. Interestingly, Imai *et al.*[28] found a high correlation of incidence of cell death and stimulation of smooth muscle proliferation in fat-fed swine. They attributed this correlation to the toxic effects secondary to oxidation in cholesterol preparations used to feed the animals. Cell death and platelet release in the plaque may be synergistic in the release of mitogens and hydrolysis of heparin-like growth inhibitory molecules. Platelet released PDGF might also play a local role in control of growth even when platelets are no longer apparent at the site of injury. Smith *et al.*[29] found that residual PDGF on the culture dish could maintain smooth muscle cell growth for days even in the presence of medium containing no PDGF. FGF, possibly released from dying cells, might also persist in areas of cell death for very long times without having an obvious cellular source[30].

In vitro studies give us a large number of molecules as candidate mitogens to control smooth muscle growth (Table 1). In general, all of the polypeptide factors as well as eicosanoids shown to stimulate growth of fibroblasts *in vitro* also stimulate smooth muscle cell replication in culture. The length of the list in Table 1 is confusing. For clarification, it is reasonable to ask which mitogens are able to initiate replication of quiescent smooth muscle rather than simply accelerating the growth of cells that are already replicating. Numerous studies with PDGF, EGF, and FGF show that these polypeptides stimulate growth of quiescent smooth muscle cells in culture. Catecholamines, in contrast, increase the rate of replication of growing smooth muscle cells but neither initiate nor are required for growth *in vitro*[31]. An interesting example of such an effect comes from recent studies of the serotonin receptor transfected into 3T3 cells. Under these conditions, with an alien receptor showing inappropriate function, serotonin is, in effect, a transforming growth factor[32]. Since the metabolic context of cells *in vitro* is clearly different from the cells' status *in vivo*, it may not be surprising that Table 1 shows mitogenic effects of several potent molecules, many of which have not usually been thought of as mitogens *in vivo*. The ability of any growth factors to stimulate replication in culture may depend on alterations in the cells' metabolic balance that are irrelevant to initiation of replication *in vivo*. The heparin-binding growth factors (HBGFs), i.e., acidic and basic FGF, are also good candidates for smooth muscle mitogens *in vivo*. About ten years ago we and others observed that detergent insoluble cell layers as well as lysates of endothelial cells were able to stimulate smooth muscle replication in the absence of defined growth factors[30,33,34]. In retrospect, it seems possible that these effects were due to fibroblast growth factors synthesized by the endothelium[35-37]. Similarly, smooth muscle cells appear able to make both acidic and basic FGF[38,39]. Thus, it seems likely that HBGFs will play an important role in control of smooth muscle replication wherever cell injury has occurred.

Table 1
Smooth Muscle Cell Growth Factors and Inhibitors

Growth Factor	Reference
Platelet-Derived Growth Factor (PDGF)	15,20
Fibroblast Growth Factor (FGF)	39,51,124,204
Epidermal Growth Factor (EGF)	41,42,51,149,205
Transforming Growth Factor beta (TGF-ß*)	43,205,206
Insulin-like Growth Factor-1 (IGF-1)	51,56,207
Catecholamines	31
Angiotensin II**	208
Low Density Lipoprotein (LDL)	46,47,48,52,54,55
Neuraminidase	209
Fibronectin	65
Nicotine	210
Leukotrienes	211
Thrombospondin	61
Fibrin	212
Interleukin-1 (IL-1)***	213,214
Endothelin	215
Serotonin	216
Neurokinin A	217
Substance K	218
Substance P	219

*Depending on the source of cell and culture conditions, growth inhibition or stimulation may occur.
**Stimulates mass change, not replication.
***Appears to work by eliciting PDGF expression.

Much less attention has been given to EGF as a potential mitogen for smooth muscle. In part this reflects the absence of EGF from sources likely to deliver this factor to smooth muscle cells, e.g., platelets. Platelets do, however, contain an EGF-like molecule[40] and, at least *in vitro*, smooth muscle cells can respond to EGFs[41]. Studies of vascular contraction in response to EGF imply that EGF receptors are present in the intact wall as well[42].

TGF-ß is usually seen as a growth inhibitor. However, as pointed out in the original studies by Assoian and Sporn[43], TGF-ß can inhibit or stimulate smooth muscle growth. The difference seems to depend on the need for growing cells to maintain certain kinds of cell contact and the ability of TGF-ß to stimulate formation of the appropriate kinds of extracellular matrix and cell adhesion molecules[44]. Another issue is activation of a TGF-ß precursor. While immunocytological studies have shown plentiful TGF-ß in most tissues, the mechanisms of activation and the presence of active material remain unclear. Recent studies suggest that TGF-ß precursor may become activated as a result of the action of an as yet undefined protein released at sites of smooth muscle endothelial interaction *in vitro*[45].

There are several reports suggesting that lipoproteins stimulate smooth muscle replication[46-54]. Unfortunately, most of these reports have been complicated by the possible contamination by low concentrations of growth factors. A recent report, however, suggests a differential mitogenic effect by LDL from hyperlipemic serum, even in the presence of excess platelet-derived factors[55]. Another concern is that lipoproteins in culture are almost always present in concentrations well below their, presumably non-mitogenic, concentrations *in vivo*. It is unclear how one should interpret "growth factors" when the active molecule is plentiful *in vivo* under normal conditions. We need to distinguish between molecules required for replication *in vitro* and molecules able to stimulate replication *in vivo*.

As with the lipoproteins, the abundance of IGF-1 in plasma poses problems for any putative role in smooth muscle growth control. IGF-1 is a co-factor that smooth muscle cells require for completion of the cell cycle following stimulation with PDGF[56-59]. IGF-1 itself can be synthesized by smooth muscle cells, and antibodies to IGF-1 inhibit cell proliferation[60]. Whether additional IGF-1 is needed *in vivo*, with high nascent concentrations in plasma, is unclear. Thrombospondin is a third example of a potential smooth muscle mitogen that is plentifully available *in vivo*. In this case, however, the critical issue could be presentation of thrombospondin as part of the locally synthesized environment about proliferating cells[61]. In summary, the list of possible mitogens and growth inhibitors is quite long (Table 1). For the most part these molecules have only been demonstrated to have a role *in vitro*. We need methods to determine whether these molecules play a real role *in vivo*.

SMOOTH MUSCLE GROWTH *VERSUS* SMOOTH MUSCLE DIFFERENTIATION

An emphasis on differentiation of smooth muscle cells is also relevant because of the suggestion that differentiation is linked to proliferation in this cell type[62]. Thyberg and the Campbells have published extensively on the conditions required for smooth muscle cells to "modulate" in cell culture. All cells lose differentiated properties as they adapt to culture conditions. Indeed, there is an extensive literature on attempts to define culture media, cell substrates or growth factors able to induce differentiation *in vitro*. The proponents of smooth muscle phenotypic modulation, however, suggest that loss of expression of contractile proteins and appearance of large amounts of endoplasmic reticulum are, in some way, linked to the cells' acquisition of the ability to respond to mitogens[63,64]. Thus, Hedin and Thyberg have suggested that serum fibronectin plays a necessary role in modulation but proliferation itself is not initiated unless the smooth muscle cells are exposed to mitogens[65]. Thyberg and Nilsson claim to be able to control loss of the differentiated phenotype without stimulating replication[66,67]. The reverse also appears to be true. Blank and Owens showed that PDGF was able to stimulate replication of smooth muscle cells *in vitro* without these cells losing expression of smooth muscle α-actin[68]. However, when replication was stimulated by serum, α-actin was lost. In more molecular terms, one might imagine that the nucleus of the smooth muscle cell normally contains trans-acting factors that stimulate transcription of the set of contractile proteins required for contraction and do not permit transcription of genes required for cell-cycle transition.

The idea of modulation, however, becomes more confused when we realize that the normal, non-replicating vessel wall contains "undifferentiated" cells, that is, mesenchymal cells with few of the features of the smooth muscle cell. According to the modulation hypothesis, such cells might be expected to show an increased response to proliferative stimuli. At least *in vitro* this does not appear to be true since Gabbiani and his colleagues were unable to detect any difference between growth of smooth muscle cells that do and those that do not contain desmin, a marker of the contractile phenotype[69]. Nonetheless, the possibility exists that de-differentiation is a necessary step before replication *in vivo*. This important idea has been reviewed elsewhere[62,63].

ORIGINS OF SMOOTH MUSCLE CELLS IN THE ATHEROSCLEROTIC PLAQUE

At this point we can return to the objective of describing a model sequence for the natural history of formation of the lesion. We have attempted to divide the disease into 4 stages, reflecting both the chronology of the lesion and likely differences in etiological mechanisms.

Stage 1. Developmental Origins

The simplest characteristic feature of atherosclerosis is the origin of the lesion in the tunica intima[6]. The issue of localization, however, goes further than just localization to the intima. Rheologists in particular have attempted to explain the localization of lesions to particular sites on the vessel wall by the presence of local disturbances of flow. They have suggested that flow initiates lesions by local endothelial injury or alterations in permeability[70]. This model, at least as applied to the origins of the lesion, is weak. There is evidence that lesions begin with focal accumulations of smooth muscle cells and there is no evidence that these focal cell masses begin with the sort of endothelial injury usually required to implicate flow[71,72].

Substantial portions of vascular structure appear to be determined, though not fully developed, by birth. For example, at least in large vessels, the numbers of layers of smooth muscle appear to be predetermined. At parturition, most arteries appear to have developed their adult number of layers of smooth muscle[73]. After this stage, the number of layers in the wall does not increase, and medial thickening is due to the production of connective tissue, increase of cell number, or increase of cell mass[74,75]. When mice are made transgenic for growth hormone there is a 70% increase in aortic wall thickness, consistent with predicted values for cardiac output, and there is a proportionate increase in cell number. There is, however, no increase in the number of cell layers. This implies that the number of cell layers is fixed either genetically or, at least, by events that are completed during intrauterine development[76]. Genetic control of smooth muscle mass is also suggested by studies comparing fetal development of spontaneously hypertensive and normotensive rats. Sarah Gray and her colleagues have found that the vessel walls of SHR rats are thicker than the walls of WKY rats. They also found, however, that the aortas have an increased number of layers of elastin. Thus, it is conceivable that some genetic tendency to form excessive layers of vessel wall or recruit excessive numbers of cells to the vessel wall is a primary event in genetic hypertension[77,78].

We know something about the origins of vascular smooth muscle cells, though little is known about how layers form. The initial endothelial tubes become surrounded by locally derived, irregularly shaped mesenchymal cells that include the precursors of smooth muscle cells and adventitial fibroblasts[79-82]. Studies in the chick embryo, using a smooth muscle-specific anti-α actin cDNA probe, show the appearance of cell type-specific protein shortly after the endothelial tubes become invested by these poorly differentiated precursors[83]. As organ primordia are invaded by endothelium, the primordia themselves contribute the smooth muscle coats to the developing vessels[84,85]. The local origins of smooth muscle cells about forming vessels require that we consider the possibility that different smooth muscle cells may have quite distinct functional properties due to unique embryological origins.

From the above discussion, it seems likely that endothelial cells recruit or even initiate the differentiation of smooth muscle cells. Studies from this laboratory attempted to explore the ability of adult bovine endothelial cells to induce changes in the developing chick embryo. While there was no evidence of induction of smooth muscle cells, the bovine endothelial cells elicited a marked angiogenic response. Bovine fibroblasts and smooth muscle cells were not similarly angiogenic[86]. The molecule controlling this angiogenic signal is unknown. PDGF has been shown to be released from endothelial cells in a vectorial fashion[87] from the basal surface of the cells, perhaps providing an organizing gradient for smooth muscle cells. A similar mechanism has been proposed for nerve growth factor (NGF) production by smooth muscle cells as an organizer of vascular innervation[88-92].

The observation that smooth muscle cells are locally derived raises important questions of smooth muscle diversity. How "locally derived" can a smooth muscle cell be? Do pericytes, the smooth muscle cells lining very small vessels, have different properties than smooth muscle cells lining larger vessels[93]? Are coronary artery smooth muscle cells distinct from smooth muscle cells of the aorta? Even within a single artery, the uniformity of the smooth muscle phenotype is not clear. At least two distinct phenotypes can be seen in the avian aorta. One of these, called the interlaminar cell, lacks the distinctive morphological features of a smooth muscle cell and would probably be identified as a fibroblast in other tissues[94-96]. Similarly, poorly differentiated cells are seen scattered in the normal vessel wall, where attempts to identify smooth muscle cells with antibodies directed at cell type-specific cytoskeletal proteins always reveal a few percent of unlabelled cells[97-99]. Perhaps more relevant to the theme of this review, undifferentiated mesenchymal cells appear postnatally in the tunica intima as part of normal development and as a prominent feature of atherosclerotic lesions[24,99-102].

Three pieces of data support the hypothesis that the lesion begins early in development with the focal proliferation of these undifferentiated smooth muscle cells. First, studies of smooth muscle replication in the mature plaque show little difference from normal wall[103], implying that most of the replication must be earlier. Second, the cells making up the proliferative mass are monoclonal, implying some local event early in the pathogenesis of the lesion rather than a chronic, diffuse proliferation involving smooth muscle cells in a general region of injury or flow disturbance[104,105]. Third, the distribution of focal accumulations of smooth muscle cells in the intima is a good predictor of the later distribution of classical, recognizable lesions[106].

Since accumulation of smooth muscle cells in the intima is part of the aging process[7,107,108], many investigators would like to distinguish "normal" intimal thickening from the focal, pathological events leading to atherosclerosis[109]. However, careful cell kinetic studies imply that the proliferative lesion in fat-fed pigs also begins in a pre-existing intimal cell mass[110]. The unique, "abnormal" feature of the lesion of atherosclerosis, as opposed to normal intima, may be the focal extent of proliferation within the intima. Thus the distinction between "normal" and "pathologic" may be semantic.

Stage 2. Focal Proliferation, Monoclonality

Monoclonality was the first direct evidence for smooth muscle proliferation in the early stages of human atherosclerotic lesions. Benditt utilized the fact that the gene for glucose-6 phosphate dehydrogenase (G-6-PD) is located on the X chromosome. Each female cell turns off one or the other X chromosome early during differentiation. Tissues in a heterozygotic individual consist of a mosaic of cells, some expressing one allele and other cells expressing the other allele of G-6-PD. The mixture of alleles in a cell mass, when compared with the random mixture in the underlying tissue, is a measure of the clonality of proliferation. Benditt found that a large proportion of atherosclerotic plaques of vessels in black females was monoclonal; whereas the smallest samplings of the normal aorta were polyclonal[104].

Benditt interpreted monoclonality of plaques as evidence that the atherosclerotic plaque is a benign leiomyoma[111]. Leiomyomas of the uterus are monoclonal by the G-6-PD criteria. Pearson and his colleagues confirmed Benditt's observations in human plaques, comparing the plaques to cutaneous scars[112]. Particularly interesting is their finding that monoclonality of atherosclerotic plaques at different stages of development was more definitive in advanced lesions, that is, in lesions with well formed fibrous caps, than in fatty streaks[113]. This might have been expected since the smooth muscle cells of the early fatty streak are diluted by large numbers of monocytes[7]. One would assume that the monocytes were polyclonal. In contrast to advanced plaques, cutaneous scars were polyclonal, that is, both allotypes of G-6-PD were represented equally. This would be expected if the response to a wound was a diffuse process involving stimulation of proliferation in many different cells as proposed above.

Lee, Thomas, and their collaborators have offered a different interpretation of monoclonality[110,114-116]. This group has attempted to reconcile the data in humans with their own experimental work in animal models. Extensive cell kinetic analysis of fat-fed swine has been interpreted as showing that smooth muscle proliferation occurs by repetitive waves of replication and cell death. On the basis of a mathematical model, they argued that monoclonality might arise by a repeated resampling of a proliferative population in which several clones have become expanded. While this model might appear attractive mathematically, it is important to point out that the group has been unsuccessful in producing monoclonal lesions in fat-fed animals. This experiment was performed in hybrid hares bred between two species differing in the G-6-PD allotype[117]. The failure to demonstrate monoclonality in the animal model must be seen as a serious reservation in our interpretation of smooth muscle cell proliferation in animal model systems[110].

It is difficult to reconcile the observation of monoclonality with hypotheses for control of smooth muscle replication that depend on a generalized loss of growth control in an area of injury such as the growth factor or growth inhibitor mechanisms already discussed as responses to injury. An alternative possibility is that monoclonality develops during embryogenesis, with accelerated growth of preexisting, intimal cell masses accounting for the formation of characteristic lesions in later life[106]. The original monoclonal masses might develop by repeated replications of very rare smooth muscle cells that arrive in the intima either by being trapped as the internal elastic lamina is formed or by migrating from the tunica media after formation of the internal elastic lamina[106]. There is good evidence that arterial injury can elicit migration of smooth muscle cells into the intima after the vessel is formed[8,26,118]. In the most detailed study, rats with balloon-denuded carotid arteries were labelled by continuous infusion with tritiated thymidine. All cells entering the cell cycle over 2 weeks were detected by autoradiography. During this interval, the increase in DNA of the media could be accounted for by about three cell doublings. About 90% of the cells in the intima were labeled and, therefore, must represent cells that moved into the intima and replicated on one or both sides of the internal elastic lamina. Since the average labeled cells must have divided three times, the 10% of the cells remaining unlabelled represents about 50% of the total migration into the intima (one cell with three divisions = eight labeled cells; one cell with migration alone = one unlabelled cell). Thus, substantial amounts of migration occur even without replication[26].

The possibility that monoclonality develops as a result of migration of smooth muscle cells and trapping of these isolated cells in the intima is also supported by a reinterpretation of other studies on fat-fed swine. Thomas and his collaborators described focal clusters of smooth muscle cells existing in the intima before the beginning of fat feeding[110]. When animals were fat-fed, these preexisting cells formed lesions by undergoing a relatively small number of cell divisions. In a similar manner, Stary and the Velicans have described focal intimal masses or eccentric intimal thickenings as "normal" stages that precede lesion formation in human atherosclerosis[7,108]. Intimal thickening and lipid accumulation occur over the first two decades of life before the characteristic changes of smooth muscle proliferation, i.e., formation of a smooth muscle "cap" over the fatty deposit, become clearly defined[7]. This long duration would provide ample time for growth factors, released from injured cells or secreted by monocytes or platelets, to act on the preexisting intimal cells.

While the intimal cell mass may explain monoclonality, we still need to explain why cells in this site proliferate. If the plaque is a neoplasm, then perhaps we should consider the possibility that it generates its own mitogens, as has been proposed for transforming growth factors found in neoplasms[119]. Two pieces of *in vivo* evidence provide direct support for the possibility that smooth muscle growth is controlled by locally released mitogens derived from smooth muscle cells. Guyton and Karnovsky found intimal smooth muscle proliferation in an occluded artery, with no platelets in the vascular lumen[120]. Fingerle took this one step further and showed smooth muscle proliferation in organ culture. The extent of cell replication was not dependent on the presence of platelet-released factors[121]. Thus, it is at least possible for smooth muscle cells to initiate DNA synthesis without interacting with platelets or leukocytes.

Both endothelial cells and smooth muscle cells are potential sources of endogenous mitogens. As we have discussed, HBGFs are synthesized by endothelial cells; however, the protein either is not secreted or remains bound to the extracellular matrix[122-124]. Cultured endothelial cells secrete soluble polypeptide mitogens that can stimulate smooth muscle replication. Part of this activity is due to a material which competes with the PDGF receptor and is neutralized by the PDGF antibody[125]. The identity of the non-PDGF soluble mitogens remains unknown[30,126]. The PDGF gene is expressed at high levels by cultured endothelial cells while only low levels of expression of the PDGF gene are seen in endothelium *in vivo*[127]. At least in culture, production of PDGF by endothelium appears to be controlled by several factors including the state of differentiation as well as by exposure to thrombin, catecholamines, TGF-ß or to lipoproteins[128-133]. We do not know whether levels of PDGF protein production can be altered *in vivo*.

Smooth muscle cells, at least *in vitro*, can also synthesize PDGF. The discovery that cultured smooth muscle cells from fetal rats spontaneously produce PDGF-like material grew out of the observation that growth of smooth muscle cells derived from newborn rat aortas cannot be arrested by putting cells in PDGF-deficient serum[134]. The culture medium from these fetal cells stimulates 3T3 cell and smooth muscle cell replication. The mitogen responsible for this activity is inactivated by anti-PDGF antibody and the conditioned medium competes for the PDGF receptor. Thus, it appears likely that fetal smooth muscle cells, at least in one species, can make this important mitogen[134]. There is also evidence that adult smooth muscle cells can be induced to produce similar material. Primary cultures of smooth muscle from rat aorta as well as from human atherosclerotic plaque[135-137] express A-chain message and synthesize a PDGF-like protein. The A chain can also be detected in fresh aortic tissue by *in situ* hybridization, and appears to be localized in intimal smooth muscle cells[24,136]. With passage, however, cultured rat arterial smooth muscle cells produce low or undetectable levels of PDGF-like material unless those cells are derived from balloon injured vessels. Cells cultured from neointima formed after balloon injury to the rat aorta produced levels similar to the levels seen in cultured cells from the immature aorta[138].

It is important to point out that since protein production can only be observed using cells placed in culture, there is no direct evidence for comparable levels of production *in vivo*. Nonetheless, the appearance of PDGF activity in the medium of cells derived from the balloon injured vessel and from the human atherosclerotic plaque implies that some dramatic change has occurred in the capabilities of the smooth muscle cells making up the injured vessel wall[135,136,138]. It is intriguing to consider the possibility that the same mechanism responsible for formation of PDGF secreting cells in culture could be responsible for both the commitment to replication and the maintenance of replication in the neointima during the days and weeks after injury[139]. As with observations of endothelial cell-derived mitogens, interpretation depends on whether the results reflect some property *in vivo*. As already noted, Wilcox and his colleagues began to address this question using *in situ* hybridization. These studies have shown localization of PDGF message in two cell types within human carotid lesions. One cell type, already noted, is the endothelial lining of the capillaries found inside plaques, and the other cell type is the intimal mesenchymal cell, presumably a de-differentiated smooth muscle cell[24]. These results are somewhat surprising, given the high level of PDGF production by macrophages *in vitro*[9]. It will be important to ask whether protein, as well as message, is being produced in the plaques. In this regard, Ross has recently reported the presence of PDGF B-chain antigen in numerous atherosclerotic plaque macrophages[140]. However, given the very low levels of cell proliferation in human plaques[103], such growth factors may be exhibiting properties other than growth stimulation in the intima. Studies correlating the presence of growth factor mRNA, protein, and proliferation in the same tissue are clearly needed.

We need to extend our consideration of smooth muscle growth to the possible role of growth inhibitors. For example, one might imagine that cell contact mechanism would act during development to inhibit smooth muscle replication once a certain density was obtained. Conversely, relatively sparse intimal smooth muscle cells might escape density-dependent inhibition by neighboring cells. As early as 1974, Nam showed that an aqueous extract of pig aorta injected intraperitoneally into 8-week-old living swine drastically reduced the rate of entry of arterial smooth muscle cells into mitosis, with no demonstrable affect on epidermal mitosis[141]. Eisenstein *et al.*[142] reported two classes of molecules derived from bovine aortic extracts that are able to inhibit cell growth *in vitro*. One class, comprised of sulphated polyanions, including heparins and proteoglycans, inhibited the growth of both smooth muscle and endothelial cells. Alcohol precipitates from aqueous extracts of the pig intima plus media were fractionated to produce both stimulators and inhibitors of synthetic state smooth muscle in culture[143]. The most direct experiments, however, were done by Clowes and his collaborators. These investigators discovered that exogenous heparin could inhibit proliferation in response to removal of the endothelium with a balloon catheter[144,145].

One requirement for an inhibitor is that it be available constitutively at the site of action. Possible cellular sources of heparin-like growth inhibitors are suggested by studies showing that cultured vessel wall cells, including both endothelium and smooth muscle, can synthesize heparin sulfates capable of inhibiting smooth muscle growth[146-149]. Fritze *et al.*, continuing the work of Castellot, have shown that smooth muscle cells also make a growth inhibitory heparin sulfate. This material is most prominent in quiescent cultures[150]. Maintenance of the normal wall in a quiescent state could reflect heparin sulfate synthesis by both endothelial cells and smooth muscle cells.

Balloon denudation results in a release of enzymes that digest the extracellular matrix, possibly removing its inhibitory effect. Platelets contain a heparin sulfate degrading endoglycosidase that could act in this way[151-153]. Incubation of endothelial cells with a crude platelet enzyme preparation or with a partially purified preparation of heparin sulfate degrading endoglycosidase from the same source causes a marked release of cell surface heparin sulfate which appears in the incubation medium as oligosaccharides. Castellot *et al.*[154] proposed that following endothelial denudation or injury, platelets are deposited on the subendothelium releasing heparitinase, platelet factor 4 and PDGF. The heparitinase cleaves heparin-like substances on the endothelium and smooth muscle into fragments which diffuse away, with additional inactivation of the substance by platelet factor 4[155]. Thus, it is possible that balloon injury allows growth by initiating breakdown of heparin sulfate.

Platelet-released factors are not very attractive as mitogens in the initial stages of atherosclerosis since, at least in animal models, we know that denudation and thrombosis do not appear until after initial events in the development of atherosclerotic lesions[72,156-159].

Until this point we have focused on the early stages of lesion formation with the assumption that the earliest events involve smooth muscle proliferation. Much of what is actually known about the existence of growth factors in the vessel wall of our species depends on studies of pathological tissue that is already advanced. Two recent studies suggest that endothelial cells in atherosclerotic plaques can synthesize PDGF. Barrett and Benditt[160] used Northern hybridization of human carotid artery plaques to detect messages in atherosclerotic plaques. They found mRNA for von Willebrand factor (an endothelial protein), fms (a leukocyte marker), smooth muscle actin, and both PDGF genes. Correlations between the concentrations of these messages suggested that macrophages were the major source of PDGF B-chain, with endothelium being the second major source. Wilcox and his collaborators used *in situ* hybridization on the same tissue and found a high level of B-chain localization, and some A-chain, over small vessels penetrating the plaques. The large masses of macrophages were negative, suggesting that Barrett's Northern blot hybridizations may have been detecting fms in macrophage associated with these small vessels and suggesting that plaque endothelium is the major source of PDGF B-chain message[24].

Stage 3. Formation of the Classical Lesion, Fatty Change

The classical lesion of atherosclerosis is intimal and consists of a central fatty, necrotic mass called the atheroma covered by a fibrous cap. Until recently, the identity of the cells associated with the dense connective tissue of the fibrous cap was controversial. The identification of smooth muscle cells as the characteristic cells of the fibrous cap arose with the description of fat-filled smooth muscle cells in atherosclerotic lesions of humans, rabbits, and primates[1,2,4]. Presumably, this structure is formed by the interactions of mechanisms involved in lipid accumulation and mechanisms involved in smooth muscle proliferation. The fact that there is dramatic accumulation of smooth muscle cells at sites where lipid accumulates in animals fed fat is one of the cornerstones of current models of this disease.

The traditional view has been that the classical lesion arises in response to the focal accumulation of lipid within the intima. This lipid can be seen from the surface when the vessel is stained with oil red O in humans and in normal animals following a few weeks on a high lipid diet[161,162]. While there is some controversy, most of the evidence suggests that the lipid of the fatty streak appears first in fat-filled macrophages[156,158,163-165]. In the traditional view[9], smooth muscle accumulation occurs as a secondary event, resulting from mitogens either released from the macrophages, from platelets adherent to sites of denudation that form as lesions progress[157], or from mitogens released from cells dying in the lesion as a result of toxic lipid peroxidation products formed by the macrophages.

As a result, much of the recent attention has been paid to the role of monocytes. *In vitro*, monocytes or macrophages produce a PDGF mitogen and possibly other mitogens[166-168]. Recent studies on fat-fed animals show a rapid initial monocytosis of the vessel wall with invasion and formation of intimal foam cell masses. At this point, endothelial continuity is still present[156,157,169]. Although there may be transient breaks as the lining of the vessel is penetrated by the monocytes, it is unlikely that any substantial platelet accumulation occurs at this early stage. Platelets thrombose at later stages, possibly at sites where the endothelium has broken down over accumulations of macrophages[157,158]. The relationship of this disruption to smooth muscle proliferation is an important unknown. Although two studies show that smooth muscle proliferation occurs prior to denudation[72,170], it is likely that the bulk of the proliferation in fat-fed animals occurs at later times[157].

Stage 4. Malignant Conversion of the Classical Lesion

Finally, the processes described above may eventuate in the complicated atherosclerotic plaque which is prone to produce clinical symptoms either by limiting blood flow to distal tissues in a non-occlusive fashion, or by total occlusion with resultant infarction. Considering the latter, atherosclerotic lesions may kill in several ways, including vasospasm, occlusive thrombosis, and plaque rupture.

Vasospasm is the only terminal event for which animal models exist[171]. Perhaps this is because altered contractility is one of the earliest events in lesion progression when animals or people are fed lipid-rich diets. Heistad and his colleagues have studied this phenomenon extensively in the fat fed primate[172,173]. They found that the atherosclerotic plaque is hypercontractile to thromboxane A2, and PGE2. Vasodilator responses to leukotriene D4 were decreased. While some authors have suggested that vasospasm is the result of increased probability of platelet interactions as the endothelial surface degenerates over the advancing plaque, evidence that this actually occurs is limited[172]. Moreover, in studies of the denuded subendothelium following balloon injury, the period of hyperactivity with platelets is less than a day, implying that denudation can exist without thrombosis. On the other hand, there is good evidence for platelet aggregates on the surface of the classical lesion as seen in fat fed monkeys[157] and rabbits[159]. Furthermore, studies by Edgington[174,175] suggest that the macrophage may play a special role in the formation of fibrin and, thus, in stimulation of thrombosis. So platelet thrombi could play a role. Heistad's data, however, argue for a more direct change in the responsiveness of the wall itself since these vasoactive substances were directly active. Moreover, fMLP, an activator of monocytes, was also a selective vasoconstrictor for the atherosclerotic vessel[173].

Plaque rupture and occlusive thrombosis appear to be linked, and result in acutely life threatening clinical sequelae[176-178]. An occlusive mural thrombus accompanies most cases of acute myocardial infarction[179-182]. Plaque rupture or cracking that exposes the necrotic core to the lumen is usually found to underlie such thrombi in both the coronary[176,183-188] and cerebral arteries[189].

Certainly collagen and subendothelial matrix proteins stimulate platelet binding[190-194] and Factor XII activation[195,196]. Additionally, tissue factor appears to play a role. Tissue factor (TF) is a membrane-bound glycoprotein that facilitates both intrinsic and extrinsic pathways of coagulation, and is a key protein in the activation of the coagulation cascade[197-200]. Tissue factor is overexpressed in human atherosclerotic plaques compared to normal vessels[102]. TF protein and mRNA were localized in macrophages adjacent to the cholesterol clefts in the necrotic core of the plaque and in foam cell macrophages in the plaque intima. TF mRNA was also found in cells morphologically similar to the PDGF-expressing mesenchymal intimal cells[24] in the fibrous cap and in areas of organizing thrombi within the plaque[102]. Interestingly, since TF is also an immediate-early gene whose synthesis may be linked to the cell cycle[201,202]; it may play a role in cellular growth as well as in coagulation. We have also found similar mesenchymal plaque cells (presumably smooth muscle-derived) expressing mRNA for plasminogen activator inhibitor[203]. Thus the presence of TF and plasminogen activator inhibitor in human plaques may contribute to the initiation and/or propagation of thrombi associated with plaque rupture. Tissue plasminogen activator mRNA has also been found in the plaque[203] and this could possibly help to modulate the procoagulant factors described above.

In summary, the smooth muscle cell remains the crucial cell type in the development of the atherosclerotic plaque. It is the critical cell which is present in the young intima, and upon which various other artery wall cell types and factors act to modulate smooth muscle proliferation, extracellular matrix production, and cell contractility, as well as possibly its procoagulant activity. A better understanding of the interplay of these factors on the smooth muscle cell will hopefully allow us to eventually inhibit the progression of the plaque to its ultimate occlusive state.

REFERENCES

1. Parker, F. and Odland, G.F. 1966. A light microscopic, histochemical and electron microscopic study of experimental atherosclerosis in rabbit coronary artery and a comparison with rabbit aorta atherosclerosis. Am. J. Pathol. 48(3): 451-481.

2.   Geer, J.C. 1965. Fine structure of human aortic intimal thickening in fatty streaks. Lab. Invest. 14: 1764-1783.

3.   Movat, H.Z., More, R.H. and Haust, M.D. 1959. The morphologic elements in the early lesions of arteriosclerosis. Am. J. Pathol. 35: 93-101.

4.   Haust, M.D., More, R.H. and Movat, H.Z. 1960. The role of smooth muscle cells in the fibrogenesis of arteriosclerosis. Am. J. Pathol. 37: 377-389.

5.   French, J.E., Jennings, M.A., Poole, J.C.F., Robinson, D.S. and Florey, H. 1962. Intimal changes in the arteries of aging swine. Proc. Roy. Soc. Biol. 158: 24-42.

6.   French, J.E. 1966. Atherosclerosis in relation to the structure and function of the arterial intima, with special reference to the endothelium. Int. Rev. Exp. Pathol. 5: 253-354.

7.   Stary, H.C. 1989. Evolution and progression of atherosclerotic lesions in coronary arteries of children and young adults. Arteriosclerosis 9(1 Suppl): 119-132.

8.   Hassler, O. 1970. The origin of the cells constituting arterial intimal thickening. An experimental autoradiographic study. Lab. Invest. 22: 286-295.

9.   Ross, R. 1986. The pathogenesis of atherosclerosis - an update. N. Engl. J. Med. 314: 488-500.

10. Stemerman M.B. and Ross, R. 1972. Fibrous plaque formation in primates, an electron microscope study. J. Exp. Med. 136: 769-789.

11. Schwartz, S.M., Stemerman, M.B. and Benditt, E.P. 1975. The aortic intima. II. Repair of the aortic lining after mechanical denudation. Am. J. Pathol. 81: 15-42.

12. Friedman, R.J., Stemerman, M.B., Wenz, B., Moore, S., Gauldie, J., Gent, M., Tiell, M.L. and Spaet, T.H. 1977. The effect of thrombocytopenia on experimental atherosclerotic lesion formation in rabbits. Smooth muscle cell proliferation and re-endothelialization. J. Clin. Invest. 60: 1191-1201.

13. Hammacher, A., Hellman, U., Johnsson, A., Ostman, A., Gunnarsson, K., Westermark, B., Wasteson, A. and Heldin, C.H. 1988. A major part of platelet-derived growth factor purified from human platelets is a heterodimer of one A and one B chain. J. Biol. Chem. 263: 16493-16498.

14. Nister, M., Hammacher, A., Mellström, K., Seigbahn, A., Rönnstrand, L., Westermark, B. and Heldin, C.H. 1985. Arterial smooth muscle cells in primary culture produce a platelet-derived growth factor-like protein. Proc. Natl. Acad. Sci. USA 82: 4418-4422.

15. Ross, R., Raines, E.W. and Bowen-Pope, D.F. 1986. The biology of platelet-derived growth factor. Cell 46: 155-169.

16. Doolittle, R.F., Hunkapiller, M.W., Hood, L.E., Aaronson, S.A. and Antoniades, H.N. 1983. Simian sarcoma virus onc gene, v-sis, is derived from the gene (or genes encoding a platelet-derived growth factor). Science 221: 275-277.

17. Waterfield, M.D., Scrace, G.T., Whittle, N., Stroobant, P., Johnsson, A., Wasteson, A., Westermark, B., Heldin, C.H., Huang, J.S. and Deuel, T.F. 1983. Platelet-derived growth factor is structurally related to the putative transforming protein p28sis of simian sarcoma virus. Nature 304: 35-39.

18. Matsui, T., Heidaran, M., Miki, T., Popescu, N., La Rochelle, W., Kraus, M., Pierce, J. and Aaronson, S. 1989. Isolation of a novel receptor cDNA establishes the existence of two PDGF receptor genes. Science 243: 800-804.

19.   Seifert, R.A., Hart, C.E., Phillips, P.E., Forstrom, J.W., Ross, R., Murray, M.J. and Bowen-Pope, D.F. 1989. Two different subunits associate to create isoform-specific platelet-derived growth factor receptors. J. Biol. Chem. 264: 8771-8778.

20.   Williams, L.T. 1989. Signal transduction by the platelet-derived growth factor receptor. Science 243: 1564-1570.

21.   Kazlauskas, A., Bowen-Pope, D., Seifert, R. and Hart, C.E. 1988. Different effects of homo- and heterodimers of platelet-derived growth factor A and B chains on human and mouse fibroblasts. Embo. J. 7: 3727-3735.

22.   Grönwald, R.G.K., Seifert, R.A. and Bowen-Pope, D.F. 1989. Differential regulation of expression of two platelet-derived growth factor receptor subunits by transforming growth factor-ß. J. Biol. Chem. 264: 8120-8125.

23.   Rubin, K., Hansson, G.K., Rönnstrand, L., Claesson-Welsh, L., Fellström, B., Tingström, A., Larsson, E., Kareskog, L., Heldin, C.H, and Terracio, L. 1988. Induction of B-type receptors for platelet-derived growth factor in vascular inflammation: possible implications for development of vascular proliferative lesions. Lancet 1: 1353-1356.

24.   Wilcox, J.N., Smith, K.M., Williams, L.T., Schwartz, S.M. and Gordon, D. 1988. Platelet-derived growth factor mRNA detection in human atherosclerotic plaques by in situ hybridization. J. Clin. Invest. 82: 1134-1143.

25.   Fingerle, J., Johnson, R., Clowes, A.W., Majesky, M.W. and Reidy, M.A. 1989. Role of platelets in smooth muscle cell proliferation and migration after vascular injury in rat carotid artery. Proc. Natl. Acad. Sci. USA 86: 8412-8416.

26.   Clowes, A.W. and Schwartz, S.M. 1985. Significance of quiescent smooth muscle migration in the injured rat carotid artery. Circ. Res. 56: 139-145.

27.   Reidy, M.A., Yoshida, K., Harker, L.A. and Schwartz, S.M. 1986. Vascular injury: quantification of experimental focal endothelial denudation in rats using indium-111-labeled platelets. Arteriosclerosis 6: 305-311.

28.   Imai, H., Werthessen, N.T., Taylor, C.B. and Lee, K.T. 1976. Angiotoxicity and arteriosclerosis due to contaminants of USP-grade cholesterol. Arch. Pathol. Lab. Med. 100: 565-572.

29.   Smith, J.C., Singh, J.P., Lillquist, J.S., Goon, D.S. and Stiles, C.D. 1982. Growth factors adherent to cell substrate are mitogenically active in situ. Nature 296: 154-156.

30.   Gajdusek, C.M. and Schwartz, S.M. 1984. Comparison of intracellular and extracellular mitogen activity. J. Cell Physiol. 121: 316-322.

31.   Blaes, N. and Boissel, J.P. 1983. Growth-stimulating effect of catecholamines on rat aortic smooth muscle cells in culture. J. Cell Physiol. 116: 167-172.

32.   Julius, D., Livelli, T.J., Jessell, T.M. and Axel, R. 1989. Ectopic expression of the serotonin 1c receptor and the triggering of malignant transformation. Science 224: 1057-1062.

33.   Gospodarowicz, D., Vlodavsky, I. and Savion, N. 1980. The extracellular matrix and the control of proliferation of vascular endothelial and vascular smooth muscle cells. J. Supramol. Struct. 13: 339-372.

34.   Stavnow, L. and Berg, A.L. 1987. Effects of hypoxia and other injurious stimuli on collagen secretion and intracellular growth stimulating activity of bovine aortic smooth muscle cells in culture. Artery 14: 198-208.

35. Burgess, W.H., Mehlman, T., Marshak, D.R., Fraser, B.A. and Maciag, T. 1986. Structural evidence that endothelial cell growth factor ß is the precursor of both endothelial cell growth factor α and acidic fibroblast growth factor. Proc. Natl. Acad. Sci. USA 83: 7216-7220.

36. Lobb, R.R., Harper, J.W. and Fett, J.W. 1986. Purification of heparin-binding growth factors. Analytical Biochemistry 154: 1-14.

37. Schreiber, A.B., Kenney, J., Kowalski, J., Thomas, K.A., Gimenez-Gallego, G., Rios-Candelore, M., Di Salvo, J., Barritault, D., Courty, J., Courtois, Y., et al. 1985. A unique family of endothelial cell polypeptide mitogens: the antigenic and receptor cross-reactivity of bovine endothelial cell growth factor, brain-derived acidic fibroblast growth factor, and eye-derived growth factor-II. J. Cell Biol. 101: 1623-1626.

38. Gospodarowicz, D., Ferrara, N., Haaparanta, T. and Neufeld, G. 1988. Basic fibroblast growth factor: expression in cultured bovine vascular smooth muscle cells. Eur. J. Cell Biol. 46: 144-51.

39. Winkles, J.A., Friesel, R., Burgess, W.H., Howk, R., Mehlman, T., Weinstein, R. and Maciag, T. 1989. Human vascular smooth muscle cells both express and respond to heparin-binding growth factor I (endothelial cell growth factor). Proc. Natl. Acad. Sci. USA 84: 7124-7128.

40. Oka, Y. and Orth, D.N. 1983. Human plasma epidermal growth factor/beta-urogastrone is associated with blood platelets. J. Clin. Invest. 72: 249-259.

41. Bhargava, G., Rifas, L. and Makman, M.H. 1979. Presence of epidermal growth factor receptors and influence of epidermal growth factor on proliferation and aging in cultured smooth muscle cells. J. Cell Physiol. 100: 365-374.

42. Berk, B.C., Brock, T.A., Webb, R.C., Taubman, M.B., Atkinson, W.J., Gimbrone, M.A. and Alexander, R.W. 1985. Epidermal growth factor, a vascular smooth muscle mitogen, induces rat aortic contraction. J. Clin. Invest. 75: 1083-1086.

43. Assoian, R.K. and Sporn, M.B. 1986. Type beta transforming growth factor in human platelets: release during platelet degranulation and action on vascular smooth muscle cells. J. Cell Biol. 102: 1217-1223.

44. Ignotz, R.A. and Massague, J. 1987. Cell adhesion protein receptors as targets for transforming growth factor-beta action. Cell 51: 189-197.

45. Antonelli-Orlidge, A., Saunders, K.B., Smith, S.R. and D'Amore, P.A. 1989. An activated form of transforming growth factor-beta is produced by cocultures of endothelial cells and pericytes. Proc. Natl. Acad. Sci. USA 86: 4544-4548.

46. Brown, B.G., Mahley, R. and Assmann, G. 1976. Swine aortic smooth muscle in tissue culture. Some effects of purified swine lipoproteins on cell growth and morphology. Circ. Res. 39: 415-424.

47. Chen, J.K., Hoshi, H., McClure, D.B. and McKeehan, W.L. 1986. Role of lipoproteins in growth of human adult arterial endothelial and smooth muscle cells in low lipoprotein-deficient serum. J. Cell Physiol. 129: 207-214.

48. Cox, D.C., Comai, K. and Goldstein, A.L. 1988. Effects of cholesterol and 25-hydroxycholesterol on smooth muscle cell and endothelial cell growth. Lipids 23: 85-88.

49. Daoud, A.S., Fritz, K.E. and Jarmolych, J. 1970. Increased DNA synthesis in aortic explants from swine fed a high-cholesterol diet. Exp. Molec. Pathol. 13: 377-384.

50. Florentin, R.A., Choi, B.H., Lee, K.T. and Thomas, W.A. 1969. Stimulation of DNA synthesis and cell division *in vitro* by serum from cholesterol fed swine. J. Cell Biol. 41: 641-645.

51. Gospodarowicz, D., Hirabayashi, K., Giguere, L. and Tauber, J.P. 1981. Factors controlling the proliferative rate, final cell density, and life span of bovine vascular smooth muscle cells in culture. J. Cell Biol. 89: 568-578.

52. Libby, P., Miao, P., Ordovas, J.M. and Schaefer, E.J. 1985. Lipoproteins increase growth of mitogen-stimulated arterial smooth muscle cells. J. Cell Physiol. 124: 1-8.

53. Pietila, K. 1982. Long-term effect of hyperlipidemic serum on the synthesis of glycosaminoglycans and on the rate of growth of rabbit aortic smooth muscle cells in culture. Atherosclerosis 42: 67-75.

54. Saito, Y., Bujo, H., Morisaki, N., Shirai, K. and Yoshida, S. 1988. Proliferation and LDL binding of cultured intimal smooth muscle cells from rabbits. Atherosclerosis 69: 161-164.

55. Mitsumata, M., Fischer-Dzoga, K., Getz, G.S. and Wissler, R.W. 1988. Sequential change of DNA synthesis in cultured aortic smooth muscle cells stimulated by hyperlipidemic serum. Exp. Mol. Pathol. 48: 24-36.

56. Clemmons, D.R. 1985. Exposure to platelet-derived growth factor modulates the porcine aortic smooth muscle cell response to somatomedin-C. Endocrinology 117:77-83.

57. Okeefe, E.J. and Pledger, W.J. 1983. A model of cell cycle control: sequential events regulated by growth factors. Mol. Cell Endocrinol. 31: 167-186.

58. Russell, W.E., Van Wyk, J.J. and Pledger, W.J. 1984. Inhibition of the mitogenic effects of plasma by a monoclonal antibody to somatomedin C. Proc. Natl. Acad. Sci. USA 81: 2389-2392.

59. Singh, J.P., Chaikin, M.A., Pledger, W.J., Scher, C.D., Stiles, C.D. 1983. Persistence of the mitogenic response to platelet-derived growth factor (competence) does not reflect a long-term interaction between the growth factor and the target cell. J Cell Biol 96: 1497-1502.

60. Clemmons, D.R., Van Wyk and J.J. 1985. Evidence for a functional role of endogenously produced somatomedinlike peptides in the regulation of DNA synthesis in cultured human fibroblasts and porcine smooth muscle cells. J. Clin. Invest. 75: 1914-1918.

61. Majack, R.A., Goodman, L.V. and Dixit, V.M. 1988. Cell surface thrombospondin is functionally essential for vascular smooth muscle cell proliferation. J. Cell Biol. 106: 415-422.

62. Schwartz, S.M., Campbell, G.R. and Campbell, J.H. 1986. Replication of smooth muscle cells in vascular disease. Circ. Res. 58: 427-444.

63. Campbell, G.R., Campbell, J.H., Maderson, J.A., Horrigan, S., Rennick, R.E. 1988. Arterial smooth muscle. A multifunctional mesenchymal cell. Arch. Pathol. Lab. Med. 112: 977-986.

64. Thyberg, J. and Fredholm, B.B. 1987. Modulation of arterial smooth muscle cells from contractile to synthetic phenotype requires induction of ornithine decarboxylase activity and polyamine synthesis. Experimental Cell Research 170: 153-159.

65. Hedin, U. and Thyberg, J. 1987. Plasma fibronectin promotes modulation of arterial smooth-muscle cells from contractile to synthetic phenotype. Differentiation 33: 239-346.

66. Thyberg, J., Nilsson, J., Palmberg, L. and Sjolund, M. 1985. Adult human arterial smooth muscle cells in primary culture. Modulation from contractile to synthetic phenotype. Cell Tissue Res. 239(1): 69-74.

67. Nilsson, J. 1987. Smooth muscle cells in the atherosclerotic process. Acta Med. Scand. Suppl. 715: 25-31.

68. Blank, R.S., Thompson, M.M. and Owens, G.K. 1988. Cell cycle versus density dependence of smooth muscle alpha actin expression in cultured rat aortic smooth muscle cells. J. Cell Biol. 107(1): 299-306.

69. Skalli, O., Bloom, W.S., Ropraz, P., Azzarone, B. and Gabbiani, G. 1986. Cytoskeletal remodeling of rat aortic smooth muscle cells in vitro: relationships to culture conditions and analogies to in vivo situations. J. Submicrosc. Cytol. 18(3): 481-93.

70. Fry, D.L. 1972. Responses of the arterial wall to certain physical factors. Ciba Found. Symp. 12: 93-120.

71. Nerem, R.M. and Cornhill, J.F. 1980. The role of fluid mechanics in atherogenesis. J. Biomechanical Engineering 102: 181-189.

72. Walker, L.N., Reidy, M.A. and Bowyer, D.E. 1986. Morphology and cell kinetics of fatty streak lesion formation in the hypercholesterolemic rabbit. Am. J. Pathol. 125: 450-459.

73. Wolinsky, H. and Glagov, S. 1967. A lamellar unit of aortic medial structure and function in mammals. Circ. Res. 20: 99-101.

74. Looker, T. and Berry, C.L. 1972. The growth and development of the rat aorta. II. Changes in nucleic acid and scleroprotein content. J. Anat. 113: 17-34.

75. Olivetti, G., Anversa, P., Melissari, M. and Loud, A. 1980. Morphometry of medial hypertrophy in the rat thoracic aorta. Lab. Invest. 42: 559-565.

76. Dilley, R.J. and Schwartz, S.M. 1989. Vascular remodeling in the growth hormone transgenic mouse. Circ. Res. 65: 1233-1240.

77. Eccleston-Joyner, C.A. and Gray, S.D. 1988. Arterial hypertrophy in the fetal and neonatal spontaneously hypertensive rat. Hypertension 12: 513-518.

78. Gray, S.D. 1982. Anatomical and physiological aspects of cardiovascular function in Wistar-Kyoto and spontaneously hypertensive rats at birth. Clin. Sci. 63(Suppl. 8): 383s-385s.

79. Blatt, H.J. 1973. Uber die Entwicklung der Coronararterien bei der Ratte Licht- und elektronenmikroskopische Untersuchungen. Z Anat Entwicklungsgesch 142: 53-64.

80. Gonzalez-Crussi, F. 1971. Vasculogenesis in the chick embryo. An ultrasound study. Am. J. Anat. 130: 441-460.

81. Manasek, F.J. 1971. The ultrastructure of embryonic myocardial blood vessels. Dev. Biol. 26: 42-54.

82. Nakamura, H. 1988. Electron microscopic study of the prenatal development of the thoracic aorta in the rat. Am. J. Anat. 181: 406-418.

83. Ruzicka, D and Schwartz, R.J. 1988. Sequential activation of alpha actin genes during avian cardiogenesis: Vascular smooth muscle α-actin gene transcripts mark the onset of cardiac myocyte differentiation. J. Cell Biol. 107: 2575-2586.

84.   Ekblom, P., Sariola, H., Karkinen-Jääskeläinen, M. and Saxén, L. 1982. The origin of the glomerular endothelium. Cell Differentiation 11: 35-39.

85.   Pardanaud, L., Yassine, F. and Dieterlen-Lievre, F. 1989. Relationship between vasculogenesis, angiogenesis and haematopoiesis during avian ontogeny. Development 105: 473-485.

86.   Harris-Hooker, S.A., Gajdusek, C.M., Wight, T.N. and Schwartz, S.M. 1983. Neovascular responses induced by cultured aortic endothelial cells. J. Cell Physiol. 114: 302-310.

87.   Zerwes, H.G. and Risau, W. 1987. Polarized secretion of a platelet-derived growth factor like chemotactic factor by endothelial cells in vitro. J. Cell Biol. 105: 2037-2041.

88.   Davies, A.M., Bandtlow, C., Heumann, R., Korsching, S., Rohrer, H. and Thoenen, H. 1987. Timing and site of nerve growth factor synthesis in developing skin in relation to innervation and expression of the receptor. Nature 326: 353-357.

89.   Hayashi, Y. and Miki, N. 1985. Purification and characterization of a neurite outgrowth factor from chicken gizzard smooth muscle. J. Biol. Chem. 15: 14269-14278.

90.   Rawdon, B.B. and Dockray, G.J. 1983. Directional growth of sympathetic nerve fibres in vitro towards enteric smooth muscle and heart from mice with congenital aganglionic colon and their normal littermates. Developmental Brain Res. 7: 53-59.

91.   Rush, R.A., Abrahamson, I.K., Murdoch, S.Y., Renton, F.J. and Wilson, P.A. 1986. Increase in neuronotrophic activity during the period of smooth muscle innervation. Int. J. Dev. Neurosci. 4: 483-492.

92.   Southwell, B.R., Chamley-Campbell, J.H. and Campbell, G.R. 1985. Tropic interactions between sympathetic nerves and vascular smooth muscle. J. Auton. Nerv. Syst. 13: 342-354.

93.   Herman, I.M. and D'Amore, P.A. 1985. Microvascular pericytes contain muscle and nonmuscle actins. J. Cell Biol. 101: 43-52.

94.   Moss, N.S. and Benditt, E.P. 1970. Spontaneous and experimentally induced arterial lesions. I. An ultrastructural survey of the normal chicken aorta. Lab. Invest. 22: 166-183.

95.   Moss, N.S. and Benditt, E.P. 1970. The ultrastructure of spontaneous and experimentally induced arterial lesions. II. The spontaneous plaque in the chicken. Lab. Invest. 23: 231-245.

96.   Wight, T.N., Cooke, P.H. and Smith, S.C. 1977. An electron microscopic study of pigeon aorta cell cultures. Cytodifferentiation and intracellular lipid accumulation. Exp. Molec. Pathol. 27: 1-18.

97.   Gabbiani, G., Gabbiani, F., Heimark, R.L. and Schwartz, S.M. 1984. Organization of actin cytoskeleton during early endothelial regeneration in vitro. J. Cell Sci. 66:39-50.

98.   Kocher, O. and Gabbiani, G. 1984. Cytoskeletal features of normal and atheromatous human arterial smooth muscle cells. Hum. Pathol. 17: 875-880.

99.   Kocher, O. and Gabbiani, G. 1986. Expression of actin mRNAs in rat aortic smooth muscle cells during development, experimental intimal thickening, and culture. Differentiation 32: 245-251.

100. Jonasson, L., Holm, J., Skalli, O., Gabbiani, G., Bondjers, G. and Hansson, G.K. 1986. Regional accumulations of T cells, macrophages, and and smooth muscle cells in the human atherosclerotic plaque. Arteriosclerosis 6: 131-138.

101. Orekhov, A.N., Ankarpova, I.I., Tertov, V.V., Rudchenko, S.A., Addreeva, E.R., Krushinsky, A.V. and Smirnov, R.N. 1984. Cellular composition of atherosclerotic and uninvolved human aortic subendothelial intima: Light-microscopic study of dissociated aortic cells. Am. J. Pathol. 115: 17-24.

102. Wilcox, J.N., Smith, K.M., Schwartz, S.M. and Gordon, D. 1989. Localization of tissue factor in the normal vessel wall and in the atherosclerotic plaque. Proc. Natl. Acad. Sci. 86: 2839-2843.

103. Gordon, D., Reidy, M.A., Benditt, E.P. and Schwartz, S.M. 1990. Cell proliferation in human coronary arteries. Proc. Natl. Acad. Sci. USA 87: 4600-4604.

104. Benditt, E.P. and Benditt, J.M. 1973. Evidence for a monoclonal origin of human atherosclerotic plaques. Proc. Natl. Acad. Sci. USA 70(6): 1753-1756.

105. Pearson, T.A., Wang, A., Solez, K. and Heptinstall, R.H. 1975. Clonal characteristics of fibrous plaques and fatty streaks from human aortas. Am. J. Pathol. 81(2): 379-388.

106. Schwartz, S.M., Reidy, M.A. and Clowes, A. 1985. Kinetics of atherosclerosis, a stem cell model. Ann. N.Y. Acad. Sci. 454: 292-304.

107. Stary, J.C. and Strong, J.P. 1976. The fine structure of non-atherosclerotic intimal thickening of developing and of regressing atherosclerotic lesions at the bifurcation of the left coronary artery. Adv. Exp. Med. Biol. 67: 89-108.

108. Velican, D. and Velican, C. 1976. Intimal thickening in developing coronary arteries and its relevance to atherosclerotic involvement. Atherosclerosis 23: 345-355.

109. McGill, Jr. H.C. 1984. Persistent problems in the pathogenesis of atherosclerosis. Arteriosclerosis 4: 443-451.

110. Thomas, W.A., Lee, K.T. and Kim, D.N. 1985. Cell population kinetics in atherogenesis. Cell births and losses in intimal cell mass-derived lesions in the abdominal aorta of swine. Ann. N.Y. Acad. Sci. 454: 305-315.

111. Benditt, E.P. and Gown, A.M. 1980. Atheroma: the artery wall and the environment. Int. Rev. Exp. 21: 55-118.

112. Pearson, T.A., Dillman, J.M., Solez, K. and Heptinstall, R.H. 1981. Clonal characteristics of cutaneous scars and implications for atherogenesis. Am. J. Pathol. 102: 49-54.

113. Pearson, T.A., Dillman, J.M. and Heptinstall, R.H. 1987. Clonal mapping of the human aorta. Relationship of monoclonal characteristics, lesion thickness, and age in normal intima and atherosclerotic lesions. Am. J. Pathol. 126: 33-39.

114. Lee, K.T., Thomas, W.A., Florentin, R.A., Reiner, J.M. and Lee, W.M. 1976. Evidence for a polyclonal origin and proliferative heterogeneity of atherosclerotic lesions induced by dietary cholesterol in swine. Ann. N.Y. Acad. Sci. 275: 336-347.

115. Thomas, W.A., Florentin, R.A., Reiner, J.M., Lee, W.M. and Lee, K.T. 1976. Alterations in population dynamics of arterial smooth muscle cell during atherogenesis. IV. Evidence for a polyclonal origin of hypercholesterolemia diet-induced atherosclerotic lesions in young swine. Exp. Mol. Pathol. 24: 244-260.

116. Thomas, W.A., Reiner, J.M., Florentin, R.A. and Scott, R.F. 1979. Population dynamics of arterial cells during atherogenesis. VIII. Separation of the roles of injury and growth stimulation in early aortic atherogenesis in swine originating in pre-existing intimal smooth muscle cell masses. Exp. Mol. Pathol. 31: 124-144.

117. Lee, K.T., Thomas, W.A., Janakidevi, K., Kroms, M., Reiner, J.M. and Borg, K.Y. 1981. Mosaicism in female hybrid hares heterozygous for glucose-6-phosphate dehydrogenase (G-6-PD): I. General properties of a hybrid hare model with special reference to atherogenesis. Exp. Molec. Path. 34: 191-201.

118. Webster, W.S., Bishop, S.P. and Geer, J.C. 1974. Experimental aortic intimal thickening. I. Morphology and source of intimal cells. Am. J. Pathol. 76: 245-284.

119. Keski-Oja, J., Raghow, R., Sawdey, M., Loskutoff, D.J., Postlethwaite, A.E., Kang, A.H. and Moses, H.L. 1988. Regulation of mRNAs for type-1 plasminogen activator inhibitor, fibronectin, and type I procollagen by transforming growth factor-beta. Divergent responses in lung fibroblasts and carcinoma cells. J. Biol. Chem. 263: 3111-3115.

120. Guyton, J.R. and Karnovsky, M.J. 1979. Smooth muscle cell proliferation in the occluded rat carotid artery: lack of requirement for luminal platelets. Am. J. Pathol. 94: 585-602.

121. Fingerle, J. and Kraft, T. 1987. The induction of smooth muscle cell proliferation in vitro using an organ culture system. Int-Angiol. 6: 65-72.

122. Schweigerer, L., Neufeld, G., Friedman, J., Abraham, J.A., Fiddes, J.C. and Gospodarowicz, D. 1987. Capillary endothelial cells express basic fibroblast growth factor, a mitogen that promotes their own growth. Nature 325: 257-259.

123. Vlodavsky, I., Folkman, J., Sullivan, R., Fridman, R., Ishai-Michaeli, R., Sasse, J. and Klagsbrun, M. 1987. Endothelial cell-derived basic fibroblast growth factor: synthesis and deposition into subendothelial extracellular matrix. Proc. Natl. Acad. Sci. USA 84: 2292-2296.

124. Vlodavsky, I., Fridman, R., Sullivan, R., Sasse, J. and Klagsbrun, M. 1987. Aortic endothelial cells synthesize basic fibroblast growth factor which remains cell associated and platelet-derived growth factor-like protein which is secreted. J. Cell Physiol. 131: 402-408.

125. Dicorleto, P.E. and Bowen-Pope, D.F. 1983. Cultured endothelial cells produce a platelet-derived growth factor-like protein. Proc. Natl. Acad. Sci. USA 80: 1919-1923.

126. Dicorleto, P.E. 1984. Cultured endothelial cells produce multiple growth factors for connective tissue cells. Exp. Cell Res. 153: 167-172.

127. Barrett, T.B., Gajdusek, C.M., Schwartz, S.M., McDougall, J.K. and Benditt, E.P. 1984. Expression of the sis gene by endothelial cells in culture and in vivo. Proc. Natl. Acad. Sci. USA 81: 6772-6774.

128. Daniel, T.O., Gibbs, V.C., Milfay, D.F., Garovoy, M.R. and Williams, L.T. 1986. Thrombin stimulates c-sis gene expression in microvascular endothelial cells. J. Biol. Chem. 261: 9579-95826.

129. Daniel, T.O., Gibbs, V.C., Milfay, D.F. and Williams, L.T. 1987. Agents that increase cAMP accumulation block endothelial c-sis induction by thrombin and transforming growth factor-beta. J. Biol. Chem. 262: 11893-11896.

130. Fox, P.L. and Dicorleto, P.E. 1986. Modified low density lipoproteins suppress production of a platelet-derived growth factor-like protein by cultured endothelial cells. Proc. Natl. Acad. Sci. USA 83: 4774-4778.

131. Gajdusek, C., Carbon, S., Ross, R., Nawroth, P. and Stern, D. 1986. Activation of coagulation releases endothelial cell mitogens. J. Cell Biol. 103: 419-428.

132. Jaye, M., McConathy, E., Drohan, W., Tong, B., Deuel, T. and Maciag, T. 1985. Modulation of the sis gene transcript during endothelial cell differentiation *in vitro*. Science 228: 882-885.

133. Starksen, N.F., Harsh, 4th G.R., Gibbs, V.C. and Williams, L.T. 1987. Regulated expression of the platelet-derived growth factor A-chain gene in microvascular endothelial cells. J. Biol. Chem. 262: 14381-14384.

134. Seifert, R.A., Schwartz, S.M. and Bowen-Pope, D.F. 1984. Developmentally regulated production of platelet-derived growth factor-like molecules. Nature 311: 669-671.

135. Libby, P., Warner, S.J., Salmon, R.N. and Birinyi, L.K. 1988. Production of platelet-derived growth factor-like mitogen by smooth-muscle cells from human atheroma. N. Engl. J. Med. 318: 1493-1498.

136. Majesky, M.W., Benditt, E.P. and Schwartz, S.M. 1988. Expression and developmental control of platelet-derived growth factor A-chain and B-chain/Sis genes in rat aortic smooth muscle cells. Proc. Natl. Acad. Sci. 85: 1524-1528.

137. Nilsson, J., Sjolund, M., Palmberg, L., Thyberg, J. and Heldin, C.H. 1985. Arterial smooth muscle cells in primary culture produce a platelet-derived growth factor-like protein. Proc. Natl. Acad. Sci. USA 82: 4418-4422.

138. Walker, L.N., Bowen-Pope, D.F., Ross, R. and Reidy, M.A. 1986. Production of platelet-derived growth factor-like molecules by cultured arterial smooth muscle cells accompanies proliferation after arterial injury. Proc. Natl. Acad. Sci. USA 83: 7311-7315.

139. Clowes, A.W., Clowes, M.M. and Reidy, M.A. 1986. Kinetics of cellular proliferation after arterial injury. III. Endothelial and smooth muscle growth in chronically denuded vessels. Lab. Invest. 54: 295-303.

140. Ross, R., Masuda, J., Raines, E.W., Gown, A.M., Katsuda, S., Sasahara, M., Malden, L.T., Masuko, H. and Sato, H. 1990. Localization of PDGF-B protein in macrophages in all phases of atherogenesis. Science 248: 1009-1012.

141. Nam, S.C., Florentin, R.A., Janakidevi, K., Lee, K.T., Reiner, J. and Thomas, W.A. 1974. Population dynamics of arterial smooth muscle cells. III. Inhibition by aortic tissue extracts of proliferative response to intimal injury in hypercholesterolemic swine. Exp. Mol. Pathol. 21: 259-267.

142. Eisenstein, R., Harper, E., Kuettner, K.E., Schumacher, B. and Matijevitch, B. 1979. Growth regulators in connective tissue. II. Evidence for the presence of several growth inhibitors in aortic extracts. Paroi Arterielle 163-169.

143. Chamley-Campbell, J.H. and Campbell, G.R. 1981. What controls smooth muscle phenotype? Atherosclerosis 40: 347-357.

144. Clowes, A.W. and Karnovsky, M.J. 1977. Suppression by heparin of smooth muscle cell proliferation in injured arteries. Nature 265: 625-626.

145. Majesky, M.W., Schwartz, S.M., Clowes, M.M. and Clowes, A.W. 1987. Heparin regulates smooth muscle S phase entry in the injured rat carotid artery. Circ. Res. 61: 296-300.

146. Campbell, J.H. and Campbell, G.R. 1986. Endothelial cell influences on vascular smooth muscle phenotype. Ann. Rev. Physiol. 48: 295-306.

147. Castellot, Jr. J.J., Addonizio, M.L., Rosenberg, R. and Karnovsky, M.J. 1981. Cultured endothelial cells produce a heparin-like inhibitor of smooth muscle cell growth. J. Cell Biol. 90: 372-379.

148. Furcht, L.T. 1986. Editorial. Critical factors controlling angiogenesis: cell products, cell matrix, and growth factors. Lab. Invest. 55: 505-508.

149. Reilly, C.F., Fritze, L.M.S. and Rosenberg, R.D. 1987. Antiproliferative effects of heparin on vascular smooth muscle cells are reversed by epidermal growth factor. J. Cell Physiology 131: 149-157.

150. Fritz, L.M.S., Reilly, C.F. and Rosenberg, R.D. 1985. An antiproliferative heparan sulfate species produced by postconfluent smooth muscle cells. J. Cell Biol. 100: 1041-1049.

151. Oosta, G.M., Favreau, L.V., Beeler, D.L. and Rosenberg, R.D. 1982. Purification and properties of human platelet heparitinase. J. Biol. Chem. 257: 11249-11255.

152. Wasteson, A., Hook, M. and Westermark, B. 1976. Demonstration of a platelet enzyme, degrading heparan sulphate. FEBS Letters 64: 218-221.

153. Wasteson, A., Glimelius, B., Busch, C., Westermark, B., Heldin, C.H. and Norling, B. 1977. Effect of a platelet endoglycosidase on cell surface associated heparan sulphate of human cultured endothelial and glial cells. Thrombosis Res. 11: 309-321.

154. Castellot, Jr. J.J., Favreau, L.V., Karnovsky, M.J. and Rosenberg, R.D. 1982. Inhibition of vascular smooth muscle cell growth by endothelial cell-derived heparin. Possible role of a platelet endoglycosidase. J. Biol. Chem. 257: 11256-11260.

155. Handin, R.I. and Cohen, J.H. 1976. Purification and binding properties of human platelet factor four. J. Biol. Chem. 251: 4273-4282.

156. Faggiotto, A., Ross, R. and Harker, L. 1984. Studies of hypercholesterolemia in the nonhuman primate. I. Changes that lead to fatty streak formation. Arteriosclerosis 4: 323-340.

157. Faggiotto, A. and Ross, R. 1984. Studies of hypercholesterolemia in the nonhuman primate. II. Fatty streak conversion to fibrous plaque. Arteriosclerosis 4: 341-356.

158. Gerrity, R.G. 1981. The role of the monocyte in atherogenesis. I. Transition of blood-borne monocytes into foam cells in fatty lesions. Am. J. Pathol. 103: 181-190.

159. Rosenfeld, M.E., Tsukada, T., Chait, A., Bierman, E.L., Gown, A.M. and Ross, R. 1987. Fatty streak expansion and maturation in Watanabe heritable hyperlipemic and comparably hypercholesterolemic fat-fed rabbits. Arteriosclerosis 7: 24-34.

160. Barrett, T.B. and Benditt, E.P. 1988. Platelet-derived growth factor gene expression in human atherosclerotic plaques and normal artery wall. Proc. Natl. Acad. Sci. USA 85: 2810-2814.

161. Cornhill, J.F., Barrett, W.A., Herderick, E.E., Mahley, R.W. and Fry, D.L. 1985. Topographic study of sudanophilic lesions in cholesterol-fed minipigs by image analysis. Arteriosclerosis 5: 415-426.

162. Grottum, P., Svindland, A. and Walloe, L. 1983. Localization of early atherosclerotic lesions in the right carotid bifurcation in humans. Acta Path. Microbiol. Immunol. Scand. Sect A 91: 65-70.

163. Stary, H.C. 1980. The intimal macrophage in atherosclerosis. Artery 8(3): 205-207.

164. Fowler, S., Shio, H. and Faley, N.J. 1979. Characterization of lipid-laden aortic cells from cholesterol-fed rabbits. IV. Investigation of macrophage-like properties of aortic cell populations. Lab. Invest. 41(4): 372-378.

165. Rosenfeld, M.E., Tsukada, T., Gown, A.M. and Ross, R. 1987. Fatty streak initiation in Watanabe heritable hyperlipemic and comparably hypercholesterolemic fat-fed rabbits. Arteriosclerosis 7: 9-23.

166. Leibovich, S.J. and Ross, R. 1976. A macrophage-dependent factor that stimulates the proliferation of fibroblasts in vitro. Am. J. Pathol. 84: 501-513.

167. Shimokado, K., Raines, E.W., Madtes, D.K., Barrett, T.B., Benditt, E.P., and Ross, R. 1985. A significant part of macrophage-derived growth factor consists of at least two forms of PDGF. Cell 43: 277-286.

168. Martinet, Y., Bitterman, P.B., Mornex, J., Grotendorst, G.R., Martin, G.R. and Crystal, R.G. 1986. Activated human monocytes express the c-sis proto-oncogene and release a ediator showing PDGF-like activity. Nature 319: 158-160.

169. Goode, T.B., Davies, P.F., Reidy, M.A. and Bowyer, D.E. 1977. Aortic endothelial cell morphology observed in situ by scanning electron microscopy during atherogenesis in rabbits. Atherosclerosis 27: 235-251.

170. Scott, R.F., Thomas, W.A., Lee, W.M., Reiner, J.M. and Florentin, R.A. 1979. Distribution of intimal smooth muscle cell mass and their relationship to early atherosclerosis in the abdominal aortas of young swine. Atherosclerosis 34: 291-301.

171. Vanhoutte, P.M. and Houston, D.S. 1985. Platelets, endothelium and vasospasm. Circulation 72: 728-734.

172. Heistad, D.D., Armstrong, M.L., Marcus, M.L., Piegors, D.J. and Mark, A. 1984. Augmented responses to vasoconstrictor stimuli in hypercholesterolemic monkeys. Circ. Res. 54: 711-718.

173. Lopez, J.A.G., Armstrong, M.L., Harrison, D.G., Piegors, D.J. and Heistad, D.D. 1989. Vascular responses to leukocytic products in atherosclerotic primates. Circ. Res. 65: 1078-1086.

174. Levy, G.A., Schwartz, B.S., Curtiss, L.K. and Edgington, T.S. 1981. Plasma lipoprotein induction and suppression of the generation of cellular procoagulant activity in vitro. J. Clin. Invest. 67: 1614-1622.

175. Schwartz, B.S., Levy, G.A., Curtiss, L.K., Fair, D.S. and Edgington, T.S. 1981. Plasma lipoprotein induction and suppression of the generation of cellular procoagulant activity in vitro. J. Clin. Invest. 67: 1650-1658.

176. Falk, E. 1983. Plaque rupture with severe pre-existing stenosis precipitating coronary thrombosis. Characteristics of coronary atherosclerotic plaques underlying fatal occlusive thrombi. Br. Heart J. 50: 127.

177. Sherman, C.T., Litvack, F., Grundfest, W., Lee, M., Hickey, A., Chaux, A., Kass, R., Blanche, C., Matloff, J., Mogenstern, L., et al. 1986. Coronary angioscopy in patients with unstable angina pectoris. N. Engl. J. Med. 315: 913.

178. Imparato, A.M., Riles, T.S., Mintzer, R. and Baumann, F.G. 1983. The importance of hemorrhage in the relationship between gross morphologic characteristics and cerebral symptoms in 376 carotid artery plaques. Ann. Surg. 197: 195.

179. DeWood, M.A., Spores, J., Notske, R., Mouser, L.T., Burroughs, R., Golden, M.S. and Lang, H.T. 1980. Prevalence of total coronary occlusion during the early hours of transmural myocardial infarction. N. Engl. J. Med. 303: 897.

180. Davies, M.J., Woolf, N. and Robertson, W.B. 1976. Pathology of acute myocardial infarction with particular reference to occlusive coronary thrombi. Br. Heart J. 38: 659.

181. Buja, L.M. and Willerson, J.T. 1981. Clinicopathic correlates of acute ischemic heart disease syndromes. Am. J. Cardiol. 47: 343.

182. Horie, T., Sekiguchi, M. and Hirosawa, K. 1978. Coronary thrombosis in pathogenesis of acute myocardial infarction. Histopathological study of coronary arteries in 108 necropsied cases using serial section. Br. Heart J. 40: 153.

183. Forrester, J.S., Litvack, F., Grundfest, W. and Hickey, A. 1987. A perspective of coronary disease seen through the arteries of living man. Circulation 75: 505.

184. Constantinides, P. 1966. Plaque fissures in human coronary thrombosis. J. Athero Res. 6: 1.

185. Friedman, M. and van den Bovenkamp, G.J. 1966. The pathogenesis of a coronary thrombus. Am. J. Pathol. 48: 19.

186. Friedman, M. 1971. The coronary thrombus: its origin and fate. Hum. Pathol. 2: 81.

187. Chapman, I. 1965. Morphogenesis of occluding coronary artery thrombosis. Arch. Pathol. 80: 256-261.

188. Drury, R.A.B. 1954. The role of intimal haemorrhage in coronary occlusion. J. Path. Bact. 67: 207-215.

189. Constantinides, P. 1967. Pathogenesis of cerebral artery thrombosis in man. Arch. Pathol. 83: 422.

190. Kinlough-Rathbone, R.L., Packham, M.A. and Mustard, J.F. 1983. Vessel injury, platelet adherence, and platelet survival. Arteriosclerosis 3: 529.

191. Groves, H.M., Kinlough-Rathbone, R.L., Richardson, M., Moore, S. and Mustard, J.F. 1979. Platelet interaction with damaged rabbit aorta. Lab. Invest. 40: 194.

192. Parsons, T.J., Haycraft, D.L., Hoak, J.C. and Sage, H. 1986. Interaction of platelets and purified collagens in a laminar flow model. Thromb. Res. 43: 435.

193. Wilner, G.D., Nossel, H.L. and LeRoy, E.C. 1968. Aggregation of platelets by collagen. J. Clin. Invest. 47: 2616.

194. Badimon, L., Badimon, J.J., Turitto, V.T., Vallabhajosula, S. and Fuster, V. 1988. Platelet thrombus formation on collagen type I: A model of deep vessel injury. Circulation 78: 1431-1442.

195. Wilner, G.D., Nossel, H.L. and LeRoy, E.C. 1968. Activation of Hageman factor by collagen. J. Clin. Invest. 47: 2608.

196. Niesiarowski, S., Stuart, R.K. and Thomas, D.P. 1966. Activation of intravascular coagulation by collagen. Proc. Soc. Exp. Biol. Med. 123: 196.

197. Nemerson, Y. 1966. The reaction between bovine brain tissue factor and factors VII and X. Biochemistry 5: 601.

198. Osterud, B. and Rapaport, S.I. 1977. Activation of factor IX by the reaction product of tissue factor and factor VII: additional pathway for initiating blood coagulation. Proc. Natl. Acad. Sci. USA 74: 5260.

199. Nemerson, Y. and Bach, R. 1982. Tissue factor revisited. Prog. Hemostasis Thromb. 6: 237.

200. Nemerson, Y. 1988. Tissue factor and hemostasis. Blood 71: 1.

201. Hartzell, S., Ryder, K., Lanahan, A., Lau, L.F. and Nathans, D. 1989. A growth factor-responsive gene of murine BALB/c 3T3 cells encodes a protein homologous to human tissue factor. Mol. Cell Biol. 9: 2567.

202. Bloem, L.J., Chen, L., Konigsberg, W.H. and Bach, R. 1989. Serum stimulation of quiescent human fibroblasts induces the synthesis of tissue factor mRNA followed by the appearance of tissue factor antigen and procoagulant activity. J. Cell Physiol. 139: 418-423.

203. Wilcox, J.N., Augustine, A.J., Smith, K.M., Schwartz, S.M. and Gordon, D. 1989. Localization of cells expressing tPA, PAI1, and urokinase by in situ hybridization in human atherosclerotic plaques and in the normal rhesus monkey. Thromb. Haemostasis 62: 131,419A.

204. Longenecker, J.P., Kilty, L.A. and Johnson, L.K. 1984. Glucocorticoid inhibition of vascular smooth muscle cell proliferation: influence of homologous extracellular matrix and serum mitogens. J. Cell Biol. 98: 534-540.

205. Ouchi, Y., Hirosumi, J., Watanabe, M., Hattori, A., Nakamura, T. and Orimo, H. 1988. Inhibitory effect of transforming growth factor-beta on epidermal growth factor-induced proliferation of cultured rat aortic smooth muscle cells. Biochem. Biophys. Res. Commun. 157:301-307.

206. Owens, G.K., Geisterfer, A.A., Yang, Y.W. and Komoriya, A. 1988. Transforming growth factor-beta-induced growth inhibition and cellular hypertrophy in cultured vascular smooth muscle cells. J. Cell Biol. 107: 771-780.

207. Clemons, D.R. and Van Wyk, J.J. 1985. Evidence for a functional role of endogenously produced somatomedinlike peptides in the regulation of DNA synthesis in cultured human fibroblasts and porcine smooth muscle cells. J. Clin. Invest. 75: 1914-1918.

208. Geisterfer, A.A., Peach, M.J. and Owens, G.K. 1988. Angiotensin II induces hypertrophy, not hyperplasia, of cultured rat aortic smooth muscle cells. Circ. Res. 62(4): 749-756.

209. Nilsson, J., Ksiazek, T., Thyberg, J. and Wasteson, A. 1983. Cell surface components and growth regulation in cultivated arterial smooth muscle cells. J. Cell Sci. 64: 107-121.

210. Thyberg, J. 1986. Effects of nicotine on phenotypic modulation and initiation of DNA synthesis in cultured arterial smooth muscle cells. Virchows Arch. 52: 25-32.

211. Palmberg, L., Claesson, H.E. and Thyberg, J. 1987. Leukotrienes stimulate initiation of DNA synthesis in cultured arterial smooth muscle cells. J. Cell Sci. 88: 151-159.

212. Ishida, T. and Tanaka, K. 1982. Effects of fibrin and fibrinogen-degradation products on the growth of rabbit aortic smooth muscle cells in culture. Atherosclerosis 44: 161-174.

213. Libby, P., Wyler, D.J., Janicka, M.W. and Dinarello, C.A. 1985. Differential effects of human interleukin-1 on growth of human fibroblasts and vascular smooth muscle cells. Atherosclerosis 5: 186-191.

214. Raines, E.W., Dower, S.K. and Ross, R. 1989. Interleukin-1 mitogenic activity for fibroblasts and smooth muscle cells is due to PDGF-AA. Science 243: 393-396.

215. Nakaki, T., Nakayama, M., Yamamoto, S. and Kato, R. 1989. Endothelin-mediated stimulation of DNA synthesis in vascular smooth muscle cells. Biochem. Biophys. Res. Commun. 158: 880-883.

216. Nemecek, G.M., Coughlin, S.R., Handley, D.A. and Moskowitz, M.A. 1986. Stimulation of aortic smooth muscle cell mitogenesis by serotonin. Proc. Natl. Acad. Sci. 83: 674-678.

217. Hultgardh-Nilsson, A., Nilsson, J., Jonson, B. and Dalsgaard, C.J. 1988. Coupling between inositol phosphate formation and DNA synthesis in smooth muscle cells stimulated with neurokinin A. J. Cell Physiol. 137: 141-145.

218. Nilsson, J., Sejersen, T., Nilsson, A.H. and Dalsgaard, C.J. 1986. DNA synthesis induced by the neuropeptide substance K correlates to the level of myc-gene transcripts. Biochem. Biophys. Res. Commun. 137: 167-174.

219. Nilsson, J., von Euler, A.M. and Dalsgaard, C.J. 1985. Stimulation of connective tissue cell growth by substance P and substance K. Nature 315: 61-63.

ADAPTATIONS OF MATURE AND DEVELOPING ARTERIES

TO LOCAL HEMODYNAMICS

B. Lowell Langille

Vascular Research Laboratory
The Toronto Hospital and Department of Pathology
The University of Toronto
Toronto, Canada

INTRODUCTION

Cardiovascular tissues are continuously exposed to large physical forces that have remarkable effects on their growth and function, and on the development of diseases of the circulation. Currently, there is much interest in the responses of vascular tissues to mechanical loads, how these loads are transduced by the cells of the vessel wall, and what types of interactions between these cells are responsible for integrating coordinated responses to changing hemodynamic conditions. These developing concepts are reshaping our view of long term regulation of circulatory function. It is probable that the next few years will see the development of highly innovative techniques as biologists seek to discriminate cellular responses to biomechanical and biochemical stimuli.

MECHANICAL LOADS ON ARTERIAL TISSUE

Analyzing the complex distribution of forces that result when pulsatile flows are driven through the three dimensional system of branching and curving tubes that make up the vascular system represents a daunting challenge to current bioengineers armed with the most sophisticated computer systems now available. Fortunately, however, the forces acting at any site can always be reduced to three basic types: pressure, tension (stretch), and shear (Figure 1). Pressure represents a force acting inward everywhere on the surface of a tissue element. Tension stretches the tissue along one axis, normally without constraining the tissue in other directions. Shear exerts tangential forces in opposite directions on opposite faces of tissue elements. A comprehension of how vascular tissues respond to these basic loading conditions is fundamental to our understanding of the control of vascular structure and function.

It is usual to express forces exerted on surfaces as forces per unit area, or "stresses" (tensile stress, shear stress and pressure). Similarly, larger tissues deform more when under the same stress as smaller tissues, so deformation is usually measured in relative terms, or "strains". Tensile strain (stretch) is change in length divided by initial length, compression is change in volume divided by initial volume, and shear strain is the relative displacement of opposite faces of a surface divided by the distance between the faces (Figure 1). When force and deformation are expressed as stress and strain, then the relationship between the two characterizes the deformability of the tissue per se; it is not affected by the geometry (length, diameter or curvature) of the vessel.

Atherosclerosis, Edited by A. I. Gotlieb *et al.*
Plenum Press, New York, 1991

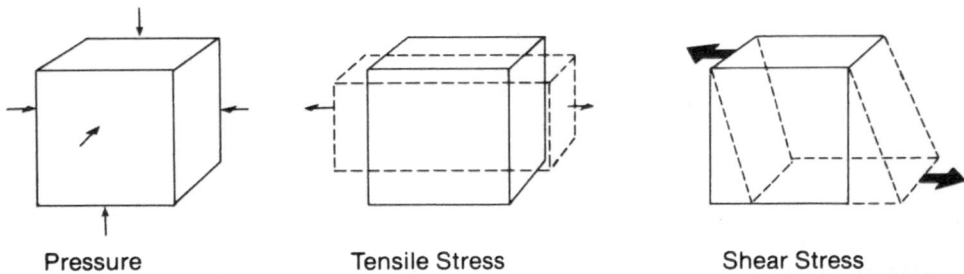

| Pressure | Tensile Stress | Shear Stress |

Figure 1. Pressure (left), tensile (centre) and shear (right) stresses (arrows) that can be imposed on tissue elements in the blood vessel wall.

Biological responses occur when physical forces cause strain. Soft tissues are highly incompressible so absolute pressures, at least at levels characteristic of the circulation, cause little strain and elicit negligible biological effects. On the other hand, differences in pressure across blood vessel walls induce tension, a major determinant of vascular tissue structure and function. This chapter focuses biological responses to shear stress and to the wall tension generated by transmural pressure gradients.

<u>Shear Stress</u>

Liquids adhere to, and remain stationary at, solid interfaces, so adjacent layers of blood must slide over each other in order to accommodate finite velocities at sites distant from the endothelial surface (Figure 2). Frictional forces are produced between layers and these shear stresses are transmitted to the vessel wall. The local shear stress in flowing blood is proportional to blood viscosity and to the speed at which adjacent layers of fluid slide past each. This "strain rate" is defined by the local rate of change of velocity with position. Thus,

$$\tau = \mu \cdot dv/dr$$

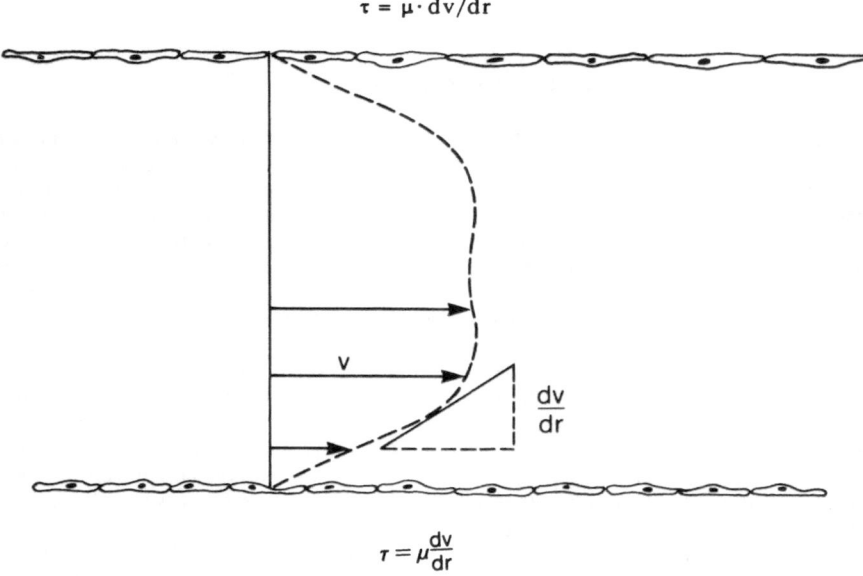

$$\tau = \mu \frac{dv}{dr}$$

Figure 2. Arrows indicate blood velocity at positions across a vessel, and their envelope (dashed line) which defines the "velocity profile". The slope of the velocity profile, dv/dr, is the "strain rate" which, when multiplied by blood viscosity (μ) gives the shear stress (τ).

where τ is shear stress, μ is viscosity, v is velocity and r is radial position. Shear stresses are small forces that cause very modest deformation of the full thickness of the artery wall. Thus, tensile strain of arteries can vary by more than 25% with physiological adjustments of arterial pressure[1], but physiological shear strains are below 1%[2]. Nonetheless, shear stress can elicit significant physiological and pathological responses from vascular tissues, primarily through their influence on endothelial cells, which are in direct contact with flowing blood.

Arterial Responses to Shear Forces

There is now much evidence that endothelial cells, which are in direct contact with blood flow, and are not embedded in matrix, are exquisitely sensitive to shear stress. Shear stress can alter endothelial cell structure and function and may significantly affect the pathogenesis of arterial diseases that originate in the intima, most notably atherosclerosis. Shear on endothelium can also elicit secondary responses from medial smooth muscle, including acute vasomotion and long term remodeling of medial structure.

Endothelial Responses. The sensitivity of endothelial cell structure to shear stress is readily apparent upon even a cursory inspection of these cells. In vivo, endothelial cells are long and thin and oriented in the direction of blood flow (Figure 3). Manipulations that alter blood flow cause the cells to change shape and to reorient in a manner consistent with the altered flow pattern, or to lose any preferred orientation if flow is eliminated[3-6]. Extremes of shear stress appear capable of causing injury to endothelium. While there is controversy over whether even very high shears cause gross damage or desquamation of the cell[7,8], sites of spontaneously high shear in vivo exhibit high endothelial cell replication rates, an indication of ongoing repair[9]. Furthermore, high cell replication can be induced by surgical manoeuvres that increase shear at sites where endothelium is normally quiescent[10]. Shears that are highly disorganized or turbulent also result in high endothelial cell turnover[10,11].

More recently, there is evidence that intracellular adaptations may limit shear-induced trauma to endothelium. F-actin microfilaments in endothelial cells undergo profound redistribution when in vivo shear stresses are altered[12,13]; that includes the formation of huge F-actin stress fibres in regions of high shear (see chapter by Gotlieb, this volume). These stress fibres complex with adhesion sites at the basal surface of the cells, probably as an adaptation that sustains endothelial integrity by enhancing cell-substrate adhesion[14-16]. This adaptation may limit pathogenesis of vascular disorders, such as atherosclerosis, if current theories implicating endothelial injury in this process are correct[17].

Acute Medial Responses. Shear stress mediates potent physiological responses of blood vessels to blood flow changes. Thus, arteries dilate when blood flow (shear stress) increases and constrict when flow decreases[18,19], such that shear stresses exerted on the vessel wall return toward initial levels. Physiologically, these responses enhance local control of blood flow. For example, when peripheral tissues are metabolically active or hypoperfused, local vasodilators increase tissue perfusion. The vasodilators act only on the microcirculation since they have no access to feed arteries extrinsic to the tissue; however, the initial increase in flow they produce stimulates dilation of upstream arteries through the effects of shear on endothelium. Thus vasodilation propagates upstream and the initial increase in blood flow is amplified. It is probable that such mechanisms are important in flow-regulatory responses of many vascular beds[20].

Flow-induced vasodilation is mediated through release of endothelium-derived relaxing factor (EDRF), a secondary extracellular messenger responsible for many types of agonist-induced vasodilatation[21]. Recently, Palmer et al.[22] produced evidence that EDRF is nitric oxide and that its action closely parallels those of the vasodilator nitrates, although it is possible that EDRF may be a family of agents that include other vasodilators as well[23]. Shear stress also induces endothelial cells to release the vasodilator, prostacyclin[24,25], but this agent appears to contribute little to flow-

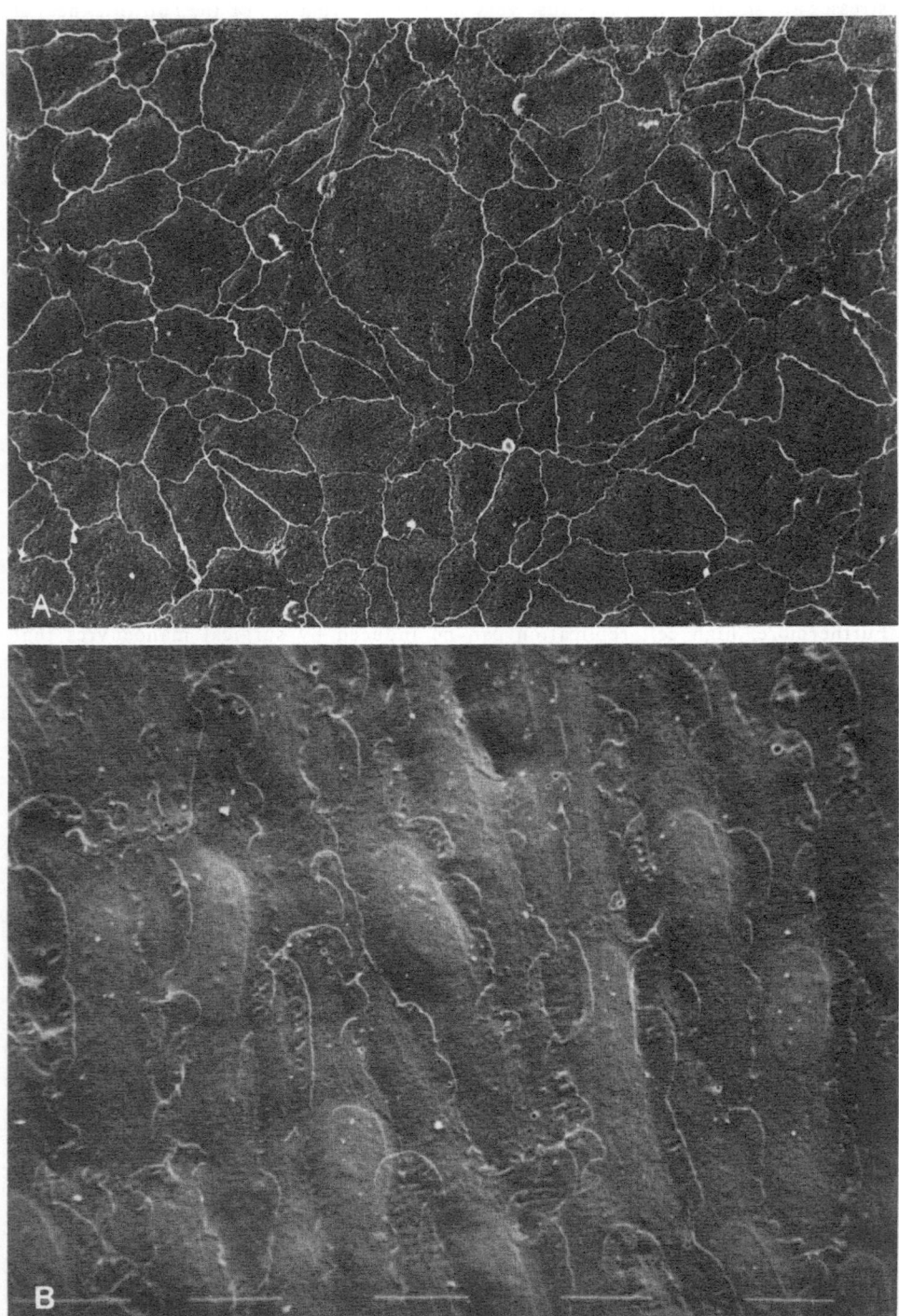

Figure 3. Scanning electron micrographs of endothelial cells grown in static culture (A) and *in vivo*, in the rabbit aorta (B).

induced vasomotion *in vivo*. Arteries also constrict in response to reduced blood flow, but it is not clear whether this is due to a reduction in tonic release of EDRF or to production of a vasoconstrictor.

Recent studies showed that atherosclerotic arteries lose their capacity to undergo EDRF-mediated vasomotion[26]. This impairment may limit vasodilator responses to ischemia since feed arteries will not be able to participate. It may also contribute to vasospasm, since many substances that cause EDRF-mediated vasodilatation cause vasoconstriction by acting directly on vascular smooth muscle when EDRF is inhibited. Thrombin and serotonin are important examples that could contribute to vasospasm at sites of vascular injury.

Shear at the blood-endothelium interface has other important short term effects in addition to eliciting vasomotor responses. Modulation of leukocyte-endothelium adhesion[27] in cell culture systems suggests important influences over inflammatory and immune responses. Similarly, effects on tumour cell-endothelium adhesion[28] may prove critical to metastatic processes. Furthermore, Diamond and coworkers[29] recently demonstrated that endothelial production of plasminogen activator is influenced by local shear stress, a phenomenon that may contribute to the induction of thrombus formation at sites of flow stasis. It is undoubtedly true that many more aspects of function of endothelial cells will prove sensitive to their hemodynamic environment.

Chronic medial responses

Chronic changes in blood flow rates through large arteries induce long term adjustments in arterial diameters that can have major effects on vascular function. For example, continuous changes in blood flow occur throughout development and it is believed that even the earliest growth of blood vessels is influenced by associated hemodynamic cues[30-33]. Long term changes in blood flow also accompany physiological responses of adults, including those associated with exercise training[34,35], reproductive cycles[36,37], pregnancy[38] and compensatory responses to nephrectomy[39]. Vascular structural adaptations have been linked to these blood flow changes[30,33,41,42]. Finally, numerous pathologic states produce clinical manifestations because they influence arterial blood flow and flow-induced diameter adjustments of arteries can significantly affect the progression of these disease states. In atherogenesis, for example, the disease process can be affected at several levels. Initially, the lesion narrows the vessel lumen and the resulting acceleration of blood flow through the lesion site elevates shear stress and Glagov *et al.*[43] presented evidence that the media subsequently expands to restore lumen diameter. Thus, flow-induced adaptations limit encroachment on the vessel lumen early in lesion development. Ultimately, however, growth of the lesion compromises blood flow and adjacent, healthy segments of the vessel wall experience reduced shear and will adapt by narrowing, a response that can exacerbate hypoperfusion. Consequently, this adaptive physiological response may be advantageous during early atherogenesis but disadvantageous in later stages. These chronic responses, like their acute counterparts are lost if endothelium is not intact (Figure 4)[2]. Furthermore, they are reinstituted by endothelial regrowth (personal observation). While the early phase of the response simply represents the flow-dependent vasomotor control described above, persistent flow changes induce medial remodeling that entrenches diameter changes[10]. In adults, this remodeling appears subtle, at least for adaptations to reduced flow, since it occurs without detectable changes in vessel wall DNA, elastin or collagen contents (Figure 5A).

Remodeling in the absence of altered amounts of wall constituents could involve cell turnover, with cell death matched by production of new cells that are at resting length at a reduced vessel circumference. Alternatively, matched degradation and resynthesis of extracellular protein could tether the vessel wall at a reduced diameter. An additional possibility is that intracellular turnover of contractile proteins in medial smooth muscle cells ultimately produces a contractile unit with a resting length shorter than that encountered in control vessels. Currently available data does not permit direct assessment of these possibilities.

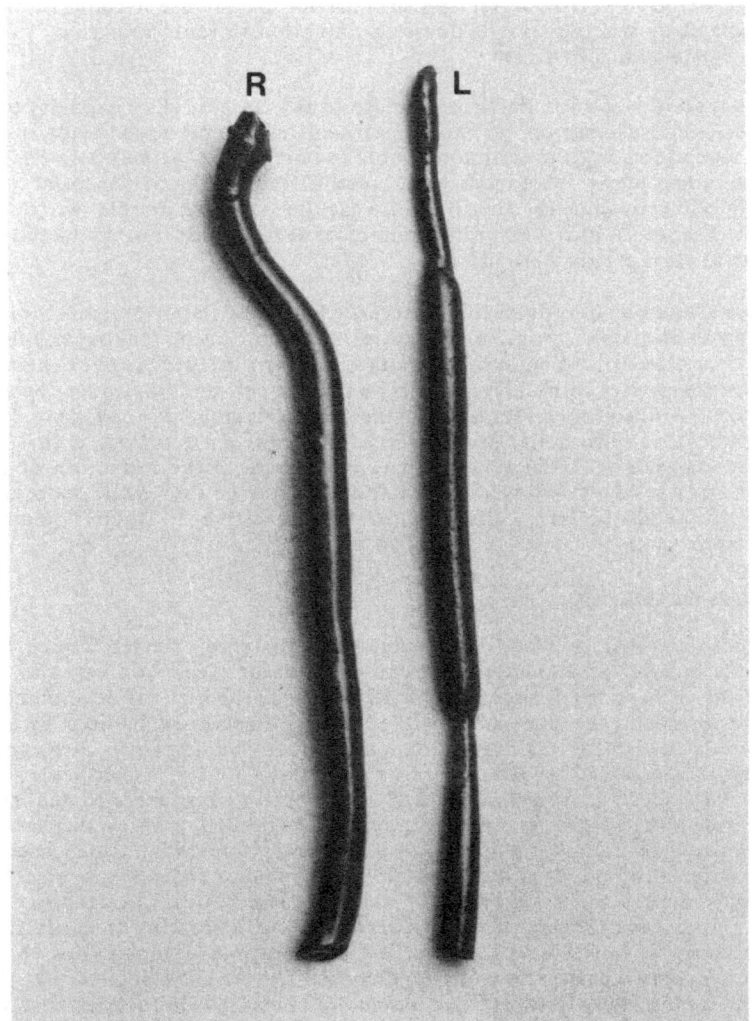

Figure 4. Methylmethacrylate casts of left (L) and right (R) common carotid arteries from a rabbit in which left carotid blood flow was reduced for 2 weeks by left external carotid ligation. Left carotid exhibits reduced diameter except where endothelium had been removed by a balloon catheter. Such vessel segments acquire a reduced diameter as endothelium regrows over denuded area.

In contrast to these findings in adult animals, arterial blood flow reductions during growth and development result in significant modulation of the rapid rates of synthesis of wall tissue constituents (Figure 5B) that occur at this time of life[44]. Elastin and DNA synthesis are significantly affected, although modulation of collagen contents was not detectable. The observation that experimental changes in blood flow modulate arterial growth strongly suggests that the continuous, ongoing changes in blood flow that occur throughout development are providing hemodynamic cues that influence subsequent arterial growth. This inference suggests that vascular growth is controlled, to a considerable extent, by feedback that reflects the changing perfusion requirements of developing peripheral tissues.

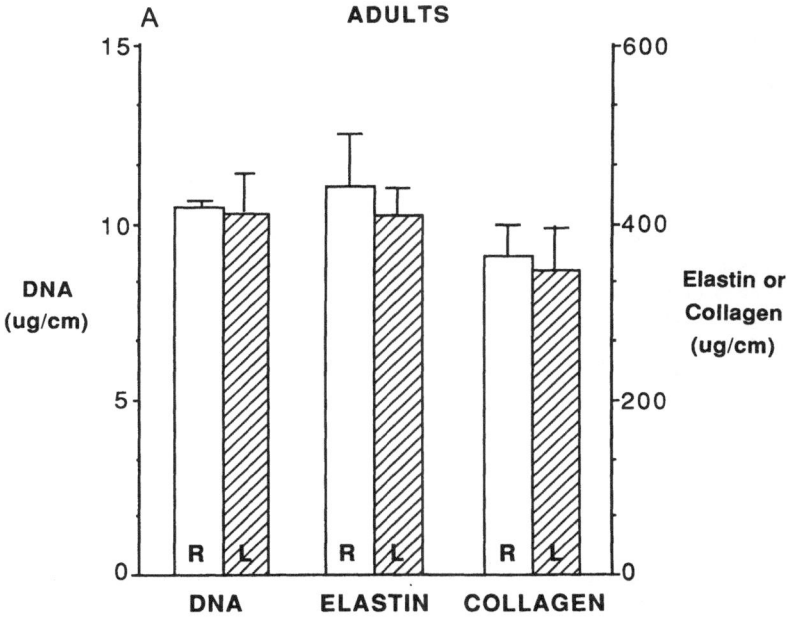

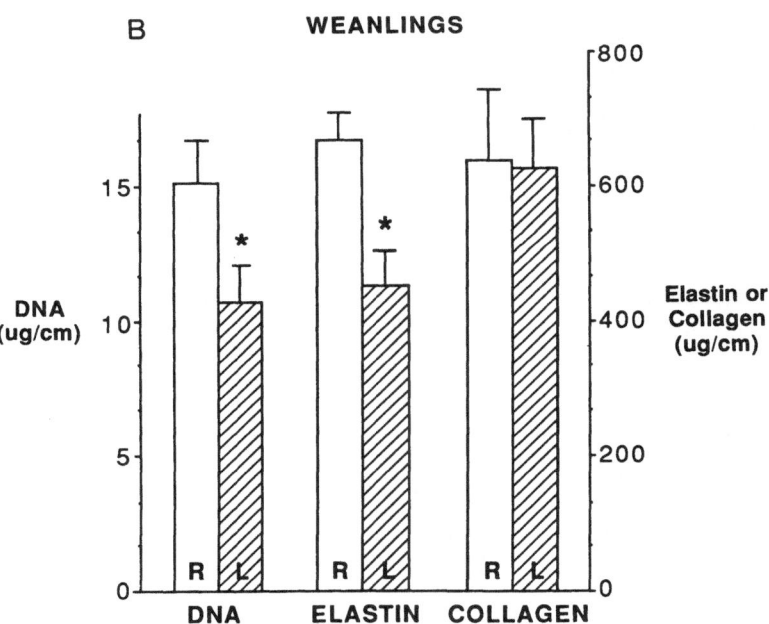

Figure 5. DNA, elastin and collagen contents per cm of carotid artery 1 month after left external carotid ligation in adult (A) or weanling (B) rabbits. *indicates a significant difference between left and right carotid arteries. N=4 per group for adult animals. N=8 to 12 per group for weanlings. Reference 44, by permission.

If developmental changes in blood flow influence developmental proliferation of arterial tissue, then the perinatal period ought to be characterized by intense modulation of vascular growth. This is because blood flow through almost all arteries changes suddenly and dramatically at parturition.

The most striking cardiovascular adjustments at birth relate to the cessation of placental blood flow and the initiation of pulmonary gas exchange. The consequent separation of lung and systemic vasculature, triggered by closure of the ductus arteriosus and foramen ovale, allows independent control of systemic and pulmonary arterial pressures. At the same time, nearly equal blood flows must enter the aorta and the pulmonary artery for the first time; a change that is related to profound increase in lung blood flow. Leung et al.[45] reported that the diameter of the two vessels, like the flows they carry, are tightly linked in young animals, whereas wall thickness follows intraluminal pressure changes (increasing systemic arterial pressure, decreasing pulmonary arterial pressure).

The perinatal pressure changes in the aorta and pulmonary artery are generalized throughout the systemic and pulmonary systems, but the blood flows carried by individual vessels show changes that are highly specific to the organ systems they supply. Thus, data collected post partum from lambs by Heymann et al.[46], demonstrated 50% increases in blood flow to the GI tract, 30-50% decreases in perfusion of the heart, kidneys, brain, adrenal glands, skin, skeletal muscle, and skeleton and a 10-fold increase in blood flow to the lungs.

An additional major change involves the loss of the fetal placental circulation at birth. Since the umbilical arteries are supplied by the lower abdominal aorta, birth causes a dramatic decrease in abdominal aortic blood flow. *In utero*, abdominal aortic blood flow amounts to 40% of combined ventricular output, but after birth it falls by more than 90% (unpublished observation). We have instrumented pre-term fetal lambs to allow monitoring of abdominal aortic diameter and flow, and arterial pressure through parturition[46a]. These experiments demonstrated that decreased abdominal aortic flow at parturition is accompanied by decreased diameter and inhibition of aortic tissue growth. These findings were particularly striking in that abdominal aortic diameter remained below *in utero* levels at 2 weeks of age despite 60% increases in the blood pressure distending the artery and a 60% increase in body weight in the post partum period. In contrast, flow through the thoracic aorta was only transiently decreased after birth and this segment of the vessel exhibited rapid tissue proliferation and significant increases in diameter in the first weeks post partum.

Following the transition to *ex utero* life, there is a generalized increase in perfusion that accompanies growth of most peripheral tissues. During this phase of vascular development there is evidence of a causal relation between blood flow and alterations in vessel size. Haworth and coworkers[47] showed that left lung vasculature failed to develop after ductal and left pulmonary artery ligation in the newborn piglet, whereas development of collateral vessels from the bronchial circulation was greatly enhanced. Coyle[48] has presented similar findings on expansion of collateral vessels following selective occlusion of cerebral vessels in rat pups. Guyton and Hartley[49] used pulsed Doppler flowmetry to monitor experimental flow reductions in rat pup carotid arteries. They established that an experimentally-induced change in blood flow could affect arterial development. These findings are consistent with our data on reduced wall tissue growth following flow reduction (Figure 5).

At present, the immediate mediators of flow-induced arterial growth modulation are unknown. Vascular endothelial and smooth muscle cells produce and/or respond to a variety of growth modulators including platelet-derived growth factor (PDGF)[50], epidermal growth factor (EGF)[51], fibroblast growth factor (FGF)[52] and transforming growth factor type ß (TGF-ß)[53]. Furthermore, interactions occur between different growth factors[54] and between these factors and other soluble agents [55] or matrix constituents[56], at least in tissue culture systems. While there is evidence for a developmental role for these growth factors in the blood vessel wall[57-59], very little is known about their effects on perinatal arterial growth. One of the most informative studies in this area was that of Seifert et al.[60]. These investigators showed that smooth muscle cells cultured from fetal and neonatal rat aorta produce 170 times the amount of PDGF, a potent smooth muscle mitogen, that was produced by similar cultures of adult rat aortic smooth muscle. This finding suggests autocrine stimulation of growth by arterial smooth muscle in the perinatal period. However, the *in vivo* importance of these factors and their sensitivity to hemodynamic changes during development remains in question.

Transduction of Shear Stress

Although many endothelial functions are affected by mechanical stress, the mechanism through which the cells transduce physical forces remains unknown. However, recent experiments from two laboratories have shed light on this question. Lansman et al.[61] showed that stretch, exerted by applying suction to a microelectrode at the cell surface, activated ion channels permeable to calcium. In addition, Olesen et al.[62] showed that shear stress activates endothelial cell potassium channels.

Activation of ion channels may represent the direct mechanism by which endothelial cells respond to shear, but it is also possible that ion channel activation is downstream of a more fundamental transduction process. Ingber and Folkman[63] hypothesize that transmission of physical forces to cytoskeletal elements represents the primary means of mechano-transduction. It is also possible that membrane-associated structures are redistributed by shear in a phenomenon similar to the "capping" of receptors that is driven by ligand binding in some cell types, and that this process elicits intracellular responses. A more basic process might involve shear-induced conformational changes of transmembrane proteins that drive intracellular responses in a fashion analogous to many receptor systems. However, these proposals are totally speculative, and evidence to date relates only to activation of ion channels.

WALL TENSION

Concepts of tension are normally applied to strips of material rather than cylinders; however, any cylinder can be considered a composite of longitudinal strips each bearing the same circumferential tension (Figure 6). Tension, by definition, is the force per unit length (F/L) imposed on each such strip. Tension can be related to wall stretch to assess how stiff the vessel is; however, this relationship cannot discriminate stiffness attributable to the thickness of the strip from that attributable to the tissue itself. To characterize the latter, tensile stress, not tension must be used. Tensile stress in this case is simply tension per unit wall thickness, which is equal to the force per unit cross-sectional area over which the force is exerted (stippled area in Figure 6).

With a cylindrical geometry, a conceptual difficulty arises when considering how pressures, which act outwardly at each site on the luminal surface, can give rise to circumferential forces in the vessel wall. This situation is possible because wall tension at any one site is not produced by a locally acting pressure; instead, it is an integrated effect of the pressures exerted at all other points around the vessel circumference.

To see how this arises, consider the two diametrically opposite strips of tissue seen in the end-view of the arterial segment in Figure 6, which divide the artery into two half cylinders. Pressure stresses imposed at each site on the upper half cylinder can be resolved into horizontal and vertical components. The horizontal components make no contribution to the vertically-directed circumferential stresses in the two strips under consideration (instead, they contribute to tension in tissue at the top and bottom of the vessel). However, the vertical components provide a net upward force on the upper half of the artery. Similarly, a downward force acts on the lower half, so that circumferential tension is imparted to the strips of tissue. Cylindrical symmetry implies that a similar argument can explain tension generation at other sites around the vessel circumference.

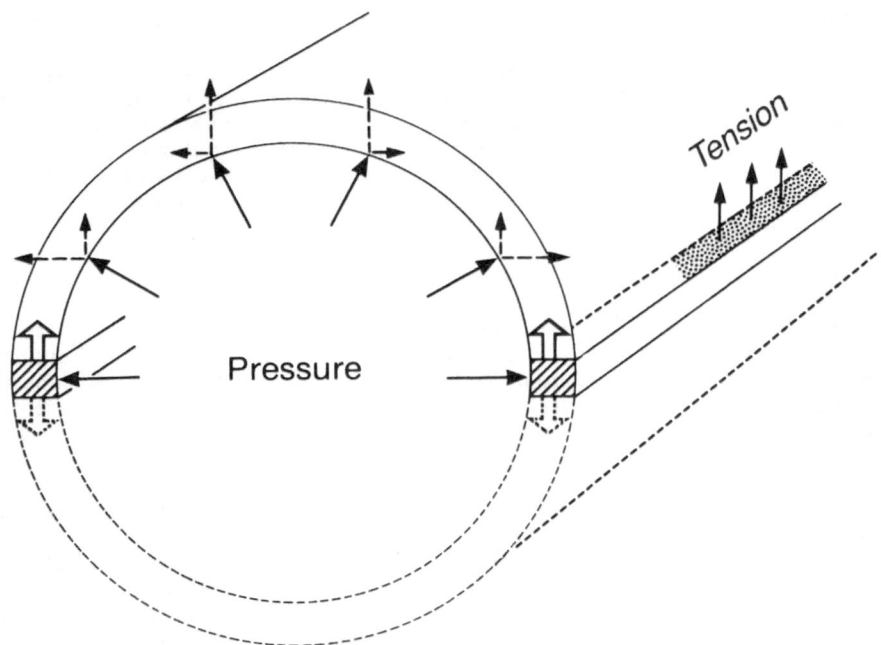

Figure 6. Tension and tensile stress are defined for each longitudinal strip of tissue that makes up a blood vessel wall. Tension is force per unit length of vessel and tensile stress is tension divided by wall thickness. The end view of the arterial segment illustrates tension generation in two diametrically opposed longitudinal segments of the vessel, which divide the vessel into half cylinders. Pressures acting at each point on the upper half cylinder produce vertical and horizontal force components on the vessel wall. The vertical components contribute to a net upward force on the half cylinder that is transmitted to the two longitudinal strips of tissue (open arrows). Downward components of local pressures exerted on the lower half of the vessel contribute a similar but opposite force to the lower side of the strips, so a net wall tension is produced. Horizontal components of these forces contribute to identical tensions exerted on tissue at the top and bottom of the vessel segments.

The pressures depicted in Figure 6 represent forces per unit area (they are stresses). It follows that the total force they generate, and the tension transmitted to the wall, is proportional to the area over which they are exerted, i.e., the inner surface area of the two halves of the vessel. Since this area is simply $\pi \cdot R \cdot L$ (R = radius, L = segment length), it follows that wall tension (forces per unit length) is proportional to vessel radius as well as blood pressure. For cylinders, the constant of proportionality is one and the Law of Laplace results.

$$T = P \cdot R \qquad (1)$$

The physical load imposed on wall tissue, the tensile wall stress, is

$$s = P \cdot R/W \qquad (2)$$

where W is wall thickness. Equation (2) indicates that if vessels of different radii have the same wall thickness then a given pressure will impose greater stresses on the tissues of the larger vessel. If wall thicknesses are in proportion to radii, then the wall tissues will be equally stressed.

Tensile stresses are borne primarily by the media in normal healthy arteries. The tensile stresses defined by equation (2) represent an average stress, but in some cases different layers of the media bear significantly different loads. For example, in medium-sized and small muscular arteries, a relatively thick wall and a large range of vasomotor tone imply that contraction can narrow the vessel until the innermost smooth muscle cells shorten to their maximally contracted lengths. Inner layers are then unloaded and further contraction of outer layers can even drive these tissues into compression. In this instance, they are used as a passive plug to close the vessel further[64]. When such vessels are constricted, wall tension varies greatly across the media.

Diseased large arteries also must be given special attention. In early atherosclerosis, for example, lesion tissue accumulates inside a stressed media, but it is not certain that this tissue assumes part of the tensile load on the vessel, at least in the short term. However, if the media expands to preserve luminal diameter during early atherogenesis (previous section), lesion tissues in the intima may assume a tensile load as the vessel expands.

Longitudinal Tension

Arteries are under longitudinal, as well as circumferential, stress and it is for this reason that they retract by 20-40% when excised[65]. The forces that generate this lengthwise stretch are associated with arterial tethering to contiguous structures via connective tissues and daughter vessels. The origin of *in situ* longitudinal tension is not well understood. One possibility is that skeletal growth rates exceed those of the vasculature during development, and a strong age-dependence of longitudinal arterial strain[66] implicates some role for development. To date, no evidence has revealed important functional implications of longitudinal arterial tension.

Arterial Tissue Responses to Wall Stretch

Acute Responses. Acute hypertension causes increased distension of large and small arteries alike. This stretch causes immediate vasoconstriction in many vessels. Conversely, decreased blood pressure induces vasodilatation. In many beds, baroreceptor-mediated changes in sympathetic drive override this direct "myogenic" response, but it is partly responsible for the capacity of some other vascular beds to autoregulate, i.e., to maintain constant blood flow when driving pressure changes. The coronary and cerebral circulation are important examples. In addition, several acute pathological events including vasospasm, thrombosis and thromboembolism depressurize downstream vasculature. Myogenic relaxation contributes a vasodilator response that limits hypoperfusion following such events.

One recent study argues that myogenic responses are mediated by endothelium[67]. If this finding is confirmed, then impairment of endothelium-mediated dilatation with atherosclerosis[26] or hypertension[68] may compromise the capacity of vascular beds to dilate in response to acute hypoperfusion secondary to the complications of these diseases.

Chronic Responses. When blood pressure elevation is sustained chronically, then remodeling of the blood vessel wall occurs. This includes synthesis of wall matrix materials and increased smooth muscle cell content. An increased replication of endothelial cells is transient and probably represents turnover (repair) resulting from acute endothelial injury. Smooth muscle cell responses, on the other hand, contribute to increased wall mass. Under normal conditions these responses play an important role in the developmental proliferation of vascular tissue. Systemic arterial wall tension increases dramatically throughout development due to increases in both arterial pressure and arterial diameters. A thickening of arterial walls lessens the increases in loads these changes impose on wall tissues.

It is very likely that these growth responses reflect, at least in part, a direct sensitivity to tensile forces. Leung et al.[45] demonstrated that early post partum wall thickening of both the aorta and the main pulmonary artery of rabbits closely matched changes in respective wall tensions, even though the changes were very different for the two vessels. Aortic wall tension rose very quickly after birth, but increases in pulmonary artery wall tension were very limited because of post partum decreases in pulmonary arterial pressure (they still occurred due to increases in vessel radius, according to the Law of Laplace). It is noteworthy, however, that tensile wall stresses, do not remain constant through later stages of development. Thus, increasing blood pressure and decreasing wall thickness to diameter ratios combine to create wall stresses in adults that are several fold above neonatal levels. Increased blood pressure in adults also elicits an arterial proliferative response. Thus hypertension, whether spontaneous or experimentally-induced, elicits arterial smooth muscle proliferation and synthesis of extracellular wall constituents (see below).

An intriguing possibility is that acute vasomotor responses and chronic proliferative responses to elevated blood pressure are controlled, at least in part, by the same mediators. Platelet-derived growth factor, a critically important mitogen for smooth muscle[69], appears to be preferentially synthesized by vascular tissues during development[60]. Recent studies indicate that this substance and other growth factors are also vasoconstrictors[70,71]. Furthermore, some vasoconstrictors exhibit potent growth promoting capabilities[72].

Non-cylindrical Vessels

Tensile wall stresses become far more complex when vessel geometry is non-cylindrical and sophisticated computer modeling is required to even approximate tissue loads. As a general rule, branch ostia, sites of vascular surgery or prosthetic implantation, or other non-uniformities will generate neighbouring regions that experience highly divergent tensile stresses. It is probable that these local stress loads contribute to proliferative responses at implant sites, and to the natural anatomy of non-uniform vessel segments.

Wall Tension and the Pathogenesis of Hypertension

Artery wall tissue responses to hypertension. Theories on the etiology of essential hypertension abound, but its pathogenesis remains obscure. In contrast, vascular tissue responses to hypertension, or more precisely the elevation of wall tension that accompanies hypertension, are well characterized. These vascular tissue responses take on greater importance because of the abnormal status of the peripheral vasculature in hypertension: arterial pressure is usually high because normal cardiac outputs are driven through an elevated peripheral vasculature resistance. It follows that vascular tissue reactions to early increases in blood pressure will affect progression of the disease.

Arterial and arteriolar wall thickening occurs in hypertension and is accomplished through a synthesis of elastic and connective tissues followed by increases in smooth muscle cell content[73-75]. Some controversy persists whether hypertrophy or hyperplasia accounts for smooth muscle accumulation, since increases in both vessel wall DNA and in smooth muscle cell size have been reported. Owens and Schwartz[75] partially resolved this question by demonstrating polyploidy of hypertrophied smooth muscle cells in aortas of hypertensive animals; however, polyploidy appears to depend on both the mechanism of induction of experimental hypertension and the level of the arterial system that is examined. Furthermore, the applicability of these animal studies to essential hypertension in humans has not been established.

Arterial wall thickening in hypertension is "adaptive" since it reduces arterial wall stresses toward normal levels; however, artery wall thickening may also contribute to the progression of hypertension by increasing vascular flow resistance. Folkow[76] proposed that wall thickening encroaches on the lumen of resistance vessels and he also described how wall thickening provides a mechanical advantage for constriction when smooth muscle is stimulated. Part of this mechanical advantage is due to stimulation of a greater mass of muscle by given agonist concentrations. In addition, however, thickened arteries or arterioles narrow their lumen more for a given level of constriction than do normal vessels. Thus, for example, if a normal and hypertrophied artery with the same diameters constrict by 50%, then the thickness of both vessel walls will approximately double as diameter decreases. The hypertrophied wall will then encroach more on the vessel lumen than the normal wall and flow resistance will be more affected. A dependence of flow resistance on the inverse fourth power of internal radius means that very modest hypertrophy, at a level difficult to detect morphometrically, can have a significant impact on flow resistance and hence arterial pressure.

HEMODYNAMICS AND ATHEROGENESIS

Early atherogenesis involves many blood and artery wall constituents; however, a compromised endothelium appears to be critical. For example, lipid insudation and monocyte/macrophage uptake and accumulation can be caused by a loss of endothelial integrity[77,78]. In addition, particular attention has focused on the role of endothelial integrity in intimal proliferation of smooth muscle cells and their synthesis of matrix materials. Loss of endothelium causes platelet attachment and release of platelet-derived growth factor (PDGF), a chemotactic mitogen for vascular smooth muscle. Denudation of endothelium is not characteristic of very early disease, but it may occur later in lesion progression[79]. Furthermore, chronic non-denuding injury may permit sufficient platelet-vessel wall interaction during cell turnover to influence smooth muscle growth in the long term.

As the lesion develops other cells appear to produce mitogens. Thus Wilcox *et al.*[80] demonstrated that both intimal smooth muscle cells in human lesions, and endothelial cells that cover these lesions, expressed the gene for PDGF. Macrophages, the other cells found in the lesions, did not. There is evidence that injured endothelium produces PDGF[81] and that PDGF production is particularly characteristic of intimal smooth muscle cells[82]. At present, the role for other vascular cell mitogens in atherogenesis is more speculative.

If endothelial dysfunction precedes lesion development, then the cause of this dysfunction is of paramount importance. The localization of atherosclerotic lesions around arterial bends and branch sites early in disease development led to the widespread belief that hemodynamic factors are involved in its pathogenesis. Lesion redistribution with experimental manipulation of local hemodynamics in experimental animals confirm this hypothesis[83-85]. Nonetheless, the relative importance of mechanical factors in the localization of the disease remains unclear, since developmental, anatomic and other potential contributors to the focal nature of the disease have received little or no study.

Shear Stress

An historical focus on shear in atherogenesis has been justified by the capacity of this stress to influence intimal tissues, where lesions originate. For example, both high and rapidly fluctuating shear stresses are associated with altered endothelial structure and function (see above) and with high endothelial cell turnover rates, an indicator of endothelial injury. These shear conditions also can inhibit the capacity of endothelium to repair[10].

Effects of shear stress on endothelium, coupled with the capacity of shear stress to influence interactions of platelets[86] and leukocytes[87] with the vessel wall, suggest multiple mechanisms through which this stress could influence atherogenesis. Nonetheless, progress has been slow in this area, and controversy continues between advocates of specific features of shear as causative factors (e.g., high versus low versus rapidly fluctuating shear). Reasons for this lack of consensus include inadequate information on early lesion distribution, variability in lesion distribution from site to site in human arteries and between sites in different animal models, and the extreme complexity of shear distribution in large arteries.

Shear distribution is complex in large vessels because of the pulsatile nature of flow and because "secondary flow" effects are large near major systemic branch sites. Secondary flow refers to flow disturbances that result when the momentum of blood entering branch ostia causes flow trajectories (streamlines) to deviate from the direction of the vessel axis. These complex, but organized, flow patterns are not turbulence, which refers to situations in which fluid momentum has rendered flow so unstable that lateral fluid motion has become randomized. Turbulence in healthy arteries is rare.

At a qualitative level, shears are high near flow dividers of branch sites where high velocity flows are delivered to the vicinity of the wall, and they fluctuate in direction lateral to branches where flow trajectories depend on blood velocity (Figure 4.13). Intuitively, low shears may be anticipated on the upstream side of branch ostia, but flows here may be especially complex and some model studies predict localized elevation of shear[88].

In recent years there has been a shift in bias toward the hypothesis that low rather than high shear stress is atherogenic. Low shears are unlikely to directly traumatize endothelium, but they may favour vessel wall interactions with blood cells or platelets by increasing the "residence time" of cells near the vessel wall[89]. Emphasis on low shear stress has been based largely on careful studies correlating human lesion distribution in the carotid bifurcation region with flow conditions prevailing in hydraulic models of this site[89]. These studies demonstrated that the region of low shear on the lateral side of the proximal internal carotid artery develops lesions whereas the flow divider region separating the internal and external carotids, a region of high shear stress, is spared. Some caution is needed when interpreting these findings. First, as the investigators point out, this low shear site is also characterized by rapid fluctuations in shear and these two flow conditions can have different effects. Second, many studies on these vessels demonstrate correlations between shear and lesion distribution, but do not establish cause and effect relationships. Finally, lesions at other vascular sites do not exhibit obvious correlations with specific features of shear distribution. Specifically, early human lesions can occur in regions of high shear, low shear or shear that rapidly fluctuates in direction[90]. It appears probable that low shear/long residence times promotes atherosclerosis but, at present, it is probably safest to recognize the multifaceted effects of local blood flow conditions on the many processes involved in atherogenesis, and to accept that variant extremes of shear stress may promote the disease.

Local Wall Tension and Atherogenesis

Recently, Thubrikar[91] has revived an hypothesis[92] that local elevations of wall tension at bends and branch sites favour lesion formation. His calculations indicate that stresses around branch ostia are affected little by the properties of the daughter vessel. Under these conditions, tensile stresses in tissue surrounding the ostia are subject to local stress concentration. Presumably, increases in stress at these sites cause tissue injury or growth stimulation and thereby contribute to atherogenesis. Thubrikar found that computer maps of tension distribution in experimental animals correlate with lesion sites. Furthermore, some experiments suggest that reducing wall tension inhibits lesion development[91], although these studies suffer from an inability to measure tensile stresses at the sites of interest. Certainly further work on this intriguing hypothesis is merited.

REFERENCES

1.   Arndt, J.O., Stegall, H.F. and Wicke, H.J. 1971. Mechanics of the aorta *in vivo*. Circ. Res. 28: 693.

2.   Langille, B.L. and O'Donnell, F. 1986. Reductions in arterial diameter produced by chronic decreases in blood flow are endothelium-dependent. Science 231: 405.

3.   Flaherty, J.T., Pierce, J.E., Ferrans, V.J., Patel, D.J., Tucker, W.K. and Fry, D.L. 1972. Endothelial nuclear patterns in the canine arterial tree with particular reference to hemodynamic events. Circ. Res. 30: 23.

4.   Langille, B.L. and Adamson, S.L.   1981.   Relationship between blood flow direction and endothelial cell orientation in arterial branch sites in rats and mice. Circ. Res. 48: 481.

5.   Reidy, M.A. and Langille, B.L. 1980. The effect of local blood flow patte   on endothelial cell morphology. Exp. Mol. Pathol. 32: 276.

6.   Silkworth, J.B. and Stehbens, W.E. 1975. The shape of endothelial cells in en face preparations of rabbit blood vessels. Angiology 26: 474.

7.   Fry, D.L. 1968. Acute vascular endothelial changes associated with increased blood velocity gradients. Circ. Res. 22: 165.

8.   Langille, B.L. 1984. Integrity of arterial endothelium following acute exposure to high shear stress. Biorheology 21: 333.

9.   Kunz, J., Schreiter, B., Schubert, B., Voss, K. and Krieg, K. 1978. Experimentelle Untersuchungen uber die Regeneration der Aortenendothelzellen. Automatische und visuelle Auswertung von Autoradiogrammen. Acta Histochem 61: 53.

10.   Langille, B.L., Reidy, M.A. and Kline, R.L.   1986.   Injury and repair of endothelium at sites of flow disturbances near abdominal aortic coarctations in rabbit. Arteriosclerosis 6: 146.

11.   Davies, P.F., Remuzzi, A., Gordon, E.J., Dewey, E.F. and Gimbrone Jr., M.A. 1986. Turbulent fluid shear stress induces vascular endothelial cell turnover in vitro. Proc. Nat. Acad. Sci. USA 83: 2114.

12.   Kim, D.W., Gotlieb, A.I. and Langille, B.L.   1989.   In vivo modulation of endothelial F-actin microfilaments by experimental alterations in shear stress. Arteriosclerosis 9: 439.

13.   Kim, D.W., Langille, B.L., Wong, M.K K. and Gotlieb, A.I.   1988.   Patterns of endothelial microfilament distribution in the rabbit aorta in situ. Circ. Res. 64: 21.

14.   Lloyd, C.W., Smith, C.G., Woods, A. and Rees, D.A. 1977. Mechanisms of cellular adhesion.   II. The interplay between adhesion, the cytoskeleton and morphology in substrate-attached cells. Exp. Cell Res. 110: 427.

15.   Mangeat, P.H. and Burridge, K. 1984. Actin-membrane interaction in fibroblasts: What proteins are involved in this association? J. Cell Biol. 99: 95s.

16.   Wehland, J., Osborn, M. and Weber, K. 1979. Cell-to-substratum contacts in living cells: A   direct   correlation   between   interference-reflexion   and   indirect-immunofluorescence microscopy using antibodies against actin and alpha-actinin. J. Cell Sci. 37: 257.

17.   Ross, R. 1986. The pathogenesis of atherosclerosis - an update. New Engl. J. Med. 314: 488.

18.   Pohl, U., Holtz, J., Busse, R. and Bassenge, E. 1986. Crucial role of endothelium in the vasodilator response to increased flow in vivo. Hypertension 8: 37.

19.   Smiesko, V., Kozik, J. and Dolezel, S. 1985. Role of endothelium in control of arterial diameter by blood flow. Blood Vessels 22: 247.

20. Griffith, T.M. 1989. Endothelium-influenced vasomotion: models and measurements. In: Vascular Dynamics Physiological Perspectives. Westerhof, N. and Gross, D.R. (eds), p. 177. Plenum, New York.

21. Furchgott, R.F. 1983. Role of endothelium in responses of vascular smooth muscle. Circ. Res. 53: 557.

22. Palmer, R.M.J., Ferrige, A.G. and Moncada, S. 1987. Nitric oxide accounts for the biological activity of endothelium-derived relaxing factor. Nature 327: 524.

23. Vanhoutte, P.M. 1987. The end of the quest? Nature 327: 459.

24. Van Grondelle, A., Worthen, G.S., Ellis, D., Mathias, M.M., Murphy, R.C., Strife, R.J., Reeves J.T. and Voelkel, N.F. 1984. Altering hydrodynamic variables influences PGI_2 production by isolated lungs and endothelial cells. J. Appl. Physiol. 57: 388.

25. Frangos, J.A., Eskin, S.G., McIntire, L.V. and Ives, C.L. 1985. Flow effects on prostacyclin production by cultured human endothelial cells. Science 227: 1477.

26. Freiman, P.C., Mitchell, G.G., Heistad, D.D., Armstrong, M.L. and Harrison, D.G. 1986. Atherosclerosis impairs endothelium-dependent vascular relaxation to acetylcholine and thrombin in primates. Circ. Res. 58: 737.

27. Lawrence, M.B., Smith, C.W., Eskin, S.G. and McIntire, L.V. 1990. Effect of venous shear stress on CD18-mediated neutrophil adhesion to cultured endothelium. Blood 75: 227.

28. Bastida, E., Almirall, L., Bertomeu, M.C. and Ordinas, A. 1989. Influence of shear stress on tumor-cell adhesion to endothelial-cell extracellular matrix and its modulation by fibronectin. Int. J. Cancer 43: 1174.

29. Diamond, S.L., Eskin, S.G. and McIntire, L.V. 1989. Fluid flow stimulates plasminogen activator secretion by cultured human endothelial cells. Science 243: 1483.

30. Liebow, A.A. 1963. Situations which lead to changes in vascular patterns. In: Handbook of Physiology, Circulation II Sec 2.

31. Meyer, W.W., Walsh, S.Z. and Lend, J. 1980. Functional morphology of human arteries during fetal and postnatal development. In: Structure and Function of the Circulation, C.J. Schwartz, N.T. Werthessen and S. Wolf (eds), Plenum, New York.

32. Pallie, W. 1980. Embryology of the human arterial system (arteriogenesis). In: Structure and Function of the Circulation, C.J. Schwartz, N.T. Werthessen and S. Wolf (eds), Plenum, New York.

33. Thoma, R. 1983. Untersuchagen iiber die Histogenese und Histomechanik des Gefassystems, Stuttgart, Enke.

34. Longhurst, J.C., Kelly, A.R., Gonyea, W.J. and Mitchell, J.H. 1981. Chronic training with static and dynamic exercises: cardiovascular adaptation, and response to exercise. Circ. Res. 48 (Supp I) 1-171.

35. Rowell, L.B. 1974. Human cardiovascular adjustments to exercise and thermal stress. Physiol Rev 54: 75.

36. Bruce, N.W. and Moor, R.M. 1976. Capillary blood flow to ovarian follicles, stroma and corpora lutea of anaesthetized sheep. J. Reprod. Fertil. 46: 299.

37. Hossain, M.J., Lee, C.S., Clack, I.J. and O'Shea, J.D. 1979. Ovarian and luteal blood flow, and peripheral plasma progesterone levels, in cyclic guinea pigs. J. Reprod. Fertil. 57: 167.

38.   Rosenfeld, C.R. 1975. Distribution of cardiac output in ovine pregnancy. Am. J. Physiol. 232: H231.

39.   Kaufman, J.M., Siegel, N.J. and Hayslett, J.P. 1975. Functional and hemodynamic adaptation to progressive renal ablation. Circ. Res. 36: 286.

40.   McNay, J.L. and Miyazaki, M. 1973. Regional increases in mass and flow during compensatory renal hypertrophy. Am. J. Physiol. 224: 219.

41.   Azmi, T.I. and O'Shea, J.D. 1984. Mechanisms of deletion of endothelial cells during regression of the corpus luteum. Lab. Invest. 51: 206.

42.   Duling, B.R., Hogan, R.D., Langille, B.L., Lelkes, P., Segal, S.S., Vatner, S.F., Weigelt, H. and Young, M.A. 1987. Vasomotor Control: functional hyperemia and beyond. Fed. Proc. 46: 251.

43.   Glagov, S., Weisenberg, E., Zarins, C.K., Stankunavicius, R. and Kolettis, G.J. 1987. Compensatory enlargement of human atherosclerotic coronary arteries. New Engl. J. Med. 316: 1371.

44.   Langille, B.L., Bendeck, M.P. and Keeley F.W. 1989. Adaptations of carotid arteries of young and mature rabbits to reduced carotid blood flow. Am. J. Physiol. 256:H931.

45.   Leung, D.Y.M., Glagov, S. and Mathews, M.B. 1977. Elastin and collagen accumulation in rabbit ascending aorta and pulmonary trunk during postnatal growth. Circ. Res. 41: 316.

46.   Heymann, M.A., Iwamoto, H.S. and Rudolph, A.M. 1981. Factors affecting changes in the neonatal systemic circulation. Ann. Rev. Physiol. 43: 371.

46a.   Langille, B.L., Brownlee, R.D. and Adamson, S.L. 1990. Perinatal aortic growth in lambs: Relation to blood flow changes at birth. Am. J. Physiol. 259: H1247-1253.

47.   Haworth, S.G., de Leval, M. and Macartney, F.J. 1981. How the left lung is perfused after ligating the left pulmonary artery in the pig at birth: clinical implications for the hypoperfused lung. Cardiovas. Res. 15: 214.

48.   Coyle, P. 1985. Interruption of the mid cerebral artery in 10-day-old rat alters normal development of distal collaterals. Anat. Rec. 212: 179.

49.   Guyton, J. R. and Hartley, C. J. 1985. Flow restriction of one carotid artery in juvenile rats inhibits growth of arterial diameter. Am. J. Physiol. 248: H540.

50.   Ross, R., Raines, E.W. and Bowen-Pope, D.F. 1986. The biology of platelet-derived growth factor. Cell 46: 155.

51.   Gospodarowicz, D., Hirabayashi, K., Giguere, L., and Tauber, J.P. 1981. Factors controlling the proliferative rate, final cell density, and life span of bovine smooth muscle cells in culture. J. Cell Biol. 89: 568.

52.   Gospodarowicz, D., Moran, J., Braun, D. and Birdwell, C. 1976. Clonal growth of bovine vascular endothelial cells; FGF as a survival agent. Proc. Natl. Acad. Sci. U.S.A. 73: 4120.

53.   Heimark, R.L., Twardzik, D.R., and Schwartz, S.M. 1986. Inhibition of endothelial cell regeneration by type-beta transforming growth factor from platelets. Science 233: 1078.

54.   Leof, E.B., Proper, J.A., Goustin, A.S., Shipley, G.D., DiCorleto, P.E. and Moses, H.L. 1986. Induction of c-sis mRNA and activity similar to platelet-derived growth

factor by transforming growth factor b: a proposed model for indirect mitogenesis involving autocrine activity. Proc. Nat. Acad. Sci. U.S.A. 83: 2453.

55.   Pash, J.M. and Bailey, J.M. 1988. Inhibition by corticosteroids of epidermal growth factor-induced recovery of cyclooxygenase after aspirin treatment. FASEB J. 2: 2613.

56.   Vlodavsky, I., Folkman, J., Sullivan, R., et al. 1987. Endothelial cell-derived basis fibroblast growth factor; synthesis and deposition into subendothelial extracellular atrix. Proc. Nat Acad. Sci. U.S.A. 84: 2292.

57.   Moscatelli, D., Presta, M. and Rifkin, D.B. 1986. Purification of a factor from human placenta that stimulates capillary endothelial cell protease production, DNA synthesis, and migration. Proc. Nat. Acad. Sci. U.S.A. 83: 2091.

58.   Burgos, H. 1986. Angiogenic factor from human term placenta. Purification and partial characterization. J. Cell Physiol. 138: 115.

59.   Goustin, A.S., Betsholtz, C., Pfeifer-Ohlsson, S., Persson, H., Rydenert, J., Bywater, M., Holmgren, G., Heldin, C.-H., Westermark, B. and Ohlsson, R. 1985. Coexpression of the sis and myc proto-oncogenes in developing human placenta suggests utocrine control of trophoblast growth. Cell 41: 301.

60.   Seifert, R.A., Schwartz, S.M. and Bowen-Pope, D.F. 1984. Developmentally-regulated production of platelet-derived growth factor-like molecules. Nature 311: 669.

61.   Lansman, J.B., Hallam, T.J. and Rink, T.J. 1987. Single stretch-activated ion channels in vascular endothelial cells as mechanotransducers. Nature Lond. 325: 811.

62.   Olesen, S.P., Clapham, D.E. and Davies, P.F. 1988. Haemodynamic shear stress activates a K^+ current in vascular endothelial cells. Nature Lond. 331: 168.

63.   Ingber, D.E. and Folkman, J. 1989. Tension and compression as basic determinants of cell form and function: utilization of cellular tensegrity mechanism. In: Cell Shape: Determinants Regulation and Regulatory Role. W.D. Stein and F. Bonner (eds), Academic Press, Inc., San Diego.

64.   McIntyre, T. W. 1969. An analysis of critical closure in the isolated ductus arteriosus. Biophys. J. 9: 685.

65.   Learoyd, B.M. and Taylor, M.G. 1966. Alterations with age in the viscoelastic properties of human arterial walls. Circ. Res. 18: 278.

66.   Patel, D.J. and Vaishnav, R.N. 1980. Basic hemodynamics and its role in disease processes. University Park, Baltimore.

67.   Katusic, Z.S., Shepherd, J.T. and Vanhoutte, P.M. 1987. Endothelium-dependent contraction to stretch in canine basilar arteries. Am. J. Physiol. 252: H671.

68.   Van de Voorde, J., Vanheel, B. and Leusen, I. 1988. Depressed endothelium-dependent relaxation in hypertension: relation to increased blood pressure and reversibility. Pflugers Arch. 411: 500.

69.   Ross, R., Raines, E.W. and Bowen-Pope, D.F. 1986. The biology of platelet-derived growth factor. Cell 46: 155.

70.   Berk, B.C., Alexander, R.W., Brock, T.A., Gimbrone Jr., M.A. and Webb, R.C. 1986. Vasoconstriction: a new activity for platelet-derived growth factor. Science 232: 87.

71. Berk, B.C., Brock, T.A., Webb, R.C., Taubman, M.B., Atkinson, W.J., Gimbrone Jr., M.A. and Alexander, R.W. 1985. Epidermal growth factor, a vascular smooth muscle mitogen, induces rat aortic contraction. J. Clin. Invest. 75: 1083.

72. Geisterfer, A.A.T., Peach, M.J. and Owens, G.K. 1988. Angiotensin II induces hypertrophy, not hyperplasia, of cultured rat aortic smooth muscle. Circ. Res. 62: 749.

73. Bevan, R.D., Eggena, P., Hume, W.R., Marthens, E.V. and Bevan, J.A. 1980. Transient and persistent changes in rabbit blood vessels associated with maintained elevation of arterial pressure. Hypertension 2: 63.

74. Keeley, F.W. and Johnson, D.J. 1986. The effects of developing hypertension on the synthesis and accumulation of elastin in the aorta of the rat. Biochem. Cell Biol. 64: 38.

75. Owens, G.K. and Schwartz, S.M. 1982. Alterations in vascular smooth muscle mass in the spontaneously hypertensive rat. Circ. Res. 51: 280.

76. Folkow, B. 1982. Physiological aspects of primary hypertension. Physiol. Rev. 62: 347.

77. Carew, T.E., Pittman, R.C., Marchand, E.R. and Steinberg, D. 1984. Measurement in vivo of irreversible degradation of low density lipoprotein in the rabbit aorta. Arteriosclerosis 4: 214.

78. DiCorleto, P.E. and de la Motte, C.A. 1985. Characterization of the adhesion of the human monocytic cell line U937 to cultured endothelial cells. J. Clin. Invest. 75: 1153.

79. Faggiotto, A. and Ross, R. 1984. Studies of hypercholesterolemia in the non-human primate. II. Fatty streak conversion to fibrous plaque. Arteriosclerosis 4: 341-356.

80. Wilcox, J.N., Smith, K.M., Williams, L.T., Schwartz, S.M. and Gordon, D. 1988. Platelet-derived growth factor mRNA detection in human atherosclerotic plaques by in situ hybridization. J. Clin. Invest. 82: 1134.

81. Fox, P.L. and DiCorleto, P.E. 1984. Regulation of production of a platelet-derived growth factor like protein by cultured bovine aortic endothelial cells. J. Cell Physiol. 121: 298.

82. Walker. L.N., Bowen-Pope, D.F., Ross, R. and Reidy, M.A. 1986. Production of platelet-derived growth factor-like molecules by cultured arterial smooth muscle cells accompanies proliferation after arterial injury. Proc. Nat. Acad. Sci. U.S.A. 83: 7311.

83. Bomberger, R.A., Zarins, C.K. and Glagov, S. 1981. Subcritical stenosis enhances distal atherosclerosis. J. Surg. Res. 30: 205.

84. Roach, M.R. and Fletcher, J. 1977. Altered renal flow in the localization of sudanophilic lesions in rabbit aortas. In: Proc. Int. Workshop Conf. Atherosclerosis. Manning, G.W. and Haust, D. (eds), Plenum, New York.

85. Zarins, C.K., Bomberger, R.A. and Glagov, S. 1981. Local effects of stenosis: increased flow velocity inhibits atherogenesis. Circulation 64(suppl II): II-221.

86. Baumgartner, H.R. and Haudenschild, C. 1972. Adhesion of platelets to subendothelium. Ann. N.Y. Acad. Sci. 201: 22.

87. Worthen, G.S., Smedley, L.A., Tonnesen, M.G., Ellis, D., Voelkel, N.F., Reeves, J.T. and Henson, P.M. 1987. Effects of shear stress on adhesive interaction between neutrophils and cultured endothelial cells. J. Appl. Physiol. 63: 2031.

88.   Lutz, R.J., Cannon, J.N., Bischoff, K.B., Dedrick, R.L., Stiles, R.K. and Fry, D.L. 1977. Wall shear stress distribution in a model canine artery during steady flow. Circ. Res. 41: 391.

89.   Ku, D.N., Giddens, D.P., Zarins, C.K. and Glagov, S. 1985. Pulsatile flow and atherosclerosis in the human carotid bifurcation: Positive correlation between plaque location and low and oscillating shear stress. Arteriosclerosis 5: 293.

90.   Flaherty, J.T., Ferrans, V.J., Pierce, J.E., Carew, T.E. and Fry, D.L. 1972a. Localizing factors in experimental atherosclerosis. In: Atherosclerosis and Coronary Heart Disease. Likoff, W., Segal, B.L., Insull, W. and Moyer, J.H. (eds), p. 40, Grune and Stratton, New York.

91.   Thubrikar, M.J., Baker, J.W. and Nolan, S.P. 1988. Inhibition of atherosclerosis associated with reduction of arterial intraluminal stress in rabbits. Arteriosclerosis 8: 410-420.

92.   Niimi, H. 1979. Role of stress concentration in arterial walls in atherogenesis. Biorheology 16: 223.

THE CYTOSKELETON OF ARTERIAL SMOOTH MUSCLE CELLS

DURING DEVELOPMENT, ATHEROMATOSIS AND TISSUE CULTURE

Giulio Gabbiani and Laura Rubbia

Department of Pathology
University of Geneva, CMU
1 rue Michel-Servet
1211 Geneva 4, Switzerland

ABSTRACT

Arterial smooth muscle cells (SMC) assume cytoskeletal features of embryonic cells during human and experimental atheromatosis, as well as when placed in culture. The expression of α-smooth muscle (SM) actin can be used to monitor these changes. The synthesis of α-SM actin is decreased early after plating in cultured cells and early after endothelial lesion in animals. The amount of mRNA for α-SM actin is not affected in cultured cells, while it is drastically decreased in the carotid artery after endothelial injury. Heparin does not modify *in vivo* the early decrease of α-SM actin mRNA and synthesis, but, by inhibiting the entry of SMC into the S phase of the cell cycle, it produces an early reinduction of α-SM actin. When heparin is incubated with cultured SMC, it reduces SMC proliferation and increases α-SM actin expression. This action is not always dependent on SMC proliferation, suggesting that heparin may act directly on SMC differentiation.

Medial smooth muscle cells (SMC) proliferate and migrate towards the intima to form the bulk of the atheromatous plaque. Evaluation of cytoskeletal features of SMC during human early atheromatous plaques and rat experimental intimal thickenings has furnished new information concerning the changes of SMC phenotype during atheromatosis. Data were mainly obtained by means of immunofluorescent staining with affinity purified polyclonal antibodies against vimentin or desmin (1) and with a monoclonal antibody against α-smooth muscle actin (2). Actin isoforms as percentage of total actin were estimated by laser beam densitometric analysis of two-dimensional gel electrophoresis (3). To localize the levels of synthesis regulation, Northern blot hybridizations of total RNA were performed with rat total actin and specific α-SM actin mRNA probes (4).

Normal adult rat aortic medial SMC, despite an homogeneous morphology, are heterogeneous as far as intermediate filament pattern expression is concerned: 51% are vimentin positive, 48% are vimentin plus desmin positive and 1% are desmin positive (1). Another main feature is a pattern of actin isoforms with α-SM actin predominance (1). One of the most used experimental models for the atheromatous plaque is the removal of rat aortic endothelium with an inflated balloon catheter. In 15-day-old experimental rat aortic intimal thickening, before endothelial regeneration and while SMC are actively replicating, the proportion of SMC in the intima containing only vimentin increases (79%); moreover, a switch to a ß-actin predominance is observed, accompanied by an increase in ɣ-actin (1). Similar changes are seen in human fibrous atheromatous plaques (5) as well as in poorly differentiated fetal and newborn aortic SMC (3). Thus, pathological SMC assume a "dedifferentiated" phenotype, as far as their cytoskeleton is concerned; this cytoskeletal remodelling is related, at least in part, to cell replicative activity. Sixty days after experimental injury, when endothelium continuity is completely re-established and SMC have again stopped replicating, SMC switch back to α-SM actin predominance and 50% of them are positive to both vimentin and desmin (6). Thus, *in vivo*, rat SMC can redifferentiate despite their ectopical intimal location. In human atheromatosis, desmin positive SMC reappear in complicated plaques, however a ß-actin predominance is maintained (5); this is probably due to blood born cells infiltration which contain only ß and γ actins. *In vitro*, proliferating SMC develop cytoskeletal features similar to those observed in normal fetal or pathological SMC (7); this model being useful for understanding mechanisms leading to atheromatous lesions and SMC differentiation. A slight increase in α-SM actin can be seen in adult SMC kept in 10% FCS culture 7 days after confluence or kept in a supplemented serum-free medium. During these situations, SMC rarely proliferate (see below). However, the expression of α-SM actin isoform is never as great as in normal medial SMC sixty days after endothelial removal (4,7). During culture in non-proliferating conditions, the index of thymidine cellular labeling is greater (1-5%) than in normal media *in vivo* (0.06%) (8); thus, true quiescence is never obtained *in vitro*. This could in part explain why, *in vitro*, SMC do not reacquire an α-SM actin predominance. To better understand the mechanisms underlying the changes in actin isoform expression, expression of actin mRNAs was studied *in vivo* and *in vitro* with the help of the two specific probes previously described (4). α-SM actin mRNA expression increases during the development of the aorta, reaching about 90% of total actin mRNA in adult tissue (4,6). In experimental aortic thickening, α-SM actin mRNA expression decreases 15 days after injury (32%) and is reacquired 60 days after injury (87%) (4). This is similar to what is observed above for the protein. Thus, the expression of α-SM actin isoform *in vivo* follows the level of its mRNA, situating the regulation of protein synthesis at the transcriptional or post-transcriptional level. Aortic SMC in primary culture exhibit a decrease in α-SM actin expression (7) and synthesis (9) by 48 hours after plating (even in a serum-free medium); these features progressively decline up to confluence (4,7). Surprisingly, α-SM actin mRNA expression is still predominant after 6 days of primary culture, accounting for 83% of total actin mRNA. Thus, in primary culture, regulation of α-SM actin synthesis seems to take place, to a certain extent, at the level of mRNA translation. SMC grown to passage 5, however, express cytoplasmic actin mRNAs' predominance; α-SM actin mRNA representing only 20% of total actin message. At this stage, the proportion of α-SM actin mRNA becomes comparable, as *in vivo*, to the level of protein expression.

We have seen above that quiescent SMC in normal artery express predominantly the α-SM actin isoform. This pattern is altered to ß-actin predominance during development, pathological conditions as well as culture, when SMC proliferate. From these data, however, the relationships between entry into the cell cycle and the changes in the expression of actin isoforms in SMC are not clear. To study these relationships, the model of the rat carotid artery injury was chosen (10). In this model, after the injury, 30% of SMC enter the cell cycle as a synchronous wave. Protein synthesis experiments were performed by 35S-methionine infusion and *in vitro* translation of total mRNA experiments were performed by using rabbit reticulocyte lysate. In the first 8-24 hours after carotid injury, before cells had left the G0/G1 phase of cell cycle, a decrease of α-SM actin mRNA expression and an increase of ß- and γ-actin mRNAs were seen (11). α-SM actin *in vitro* translation and *in vivo* synthesis also declined in the first 24 hours, indicating a decrease in the level of functional α-SM actin mRNA, its amount being proportional to the amount of mRNA present. Synthesis of ß- and γ-actin isoforms increased. DNA synthesis starts around 27 hours after injury (12) and, at 36 hours, only the S/G2 SMC population showed a small decrease in α-SM actin isoform, accompanied by a major decrease in α-SM actin mRNA (11). α-SM actin isoform in the whole SMC population declined only at 5 days after injury (11). During this period, total actin content remained the same. To determine whether the changes in actin isoforms are related to entry into the S phase of the cell cycle, heparin, which blocks SMC in late G1 (13), was administered with an osmotic pump. Heparin did not prevent the early (8-24 hours) changes in actin mRNAs and actin isoform synthesis. However, α-SM mRNA and α-SM actin expression were reinduced at 5 days in heparin treated animals. These findings suggest that actin isoform changes after injury follow variation in the mRNA levels. These variations in mRNAs take place during G0/G1 phase, whether or not the cell will subsequently enter the S phase of the cell cycle. They become greater in the S/G2 population compared to those remaining in G0/G1. These early changes are not prevented by heparin; however, heparin blocks SMC proliferation and reinduces a quiescent phenotype with α-SM actin predominance earlier after injury than in heparin untreated animals. Heparin or related proteoglycans could have a role in maintaining *in vivo* SMC in a quiescent state, by controlling α-SM actin mRNA and α-SM actin expression.

In order to better understand the possible mechanisms of heparin action on SMC, we have investigated the effect of heparin on proliferation and actin isoform expression in cultured rat SMC (14). Heparin treated primary and passage 5 SMC showed, in the presence of 10% fetal calf serum, a decrease of proliferation and an increase of α-SM actin (measured by Western blots or two-dimensional gel electrophoresis) compared to untreated SMC. When SMC were cultured in the presence of 10% plasma derived serum, no proliferation occurred and heparin did not modify α-SM actin expression. This suggests that the action of heparin is related to its anti-proliferative activity. SMC cultured in the presence of 10% FCS plus heparin had the same level of proliferation as SMC cultured in 5% FCS, but a higher content of α-SM actin. SMC cultured in 20% rat whole blood serum had a similar proliferation rate to that observed in SMC cultured in 10% FCS, but a higher content of α-SM actin. Moreover, in SMC cultured in 20% whole blood serum, heparin inhibited SMC proliferation but did not modify α-SM actin expression. Thus, the action of heparin on α-SM actin expression appears to be partially independent of proliferation and related to culture conditions. The proportion of α-SM actin mRNA, as measured by Northern blots with an α-SM actin mRNA specific probe, was increased by heparin compared to cells cultured in 10% FCS; this suggests that heparin acts at the transcriptional or post-transcriptional levels. Our results show that heparin acts not only on SMC proliferation but also on SMC differentiation; work on these lines may help in the understanding of the mechanisms of SMC adaptation during the atheromatous process.

REFERENCES

1.   Kocher, O., Skalli, O., Bloom, W.S. and Gabbiani, G. 1984. Cytoskeleton of rat aortic smooth muscle cells: normal conditions and experimental intimal thickening. Lab. Invest. 50: 645-652.

2.   Skalli, O., Ropraz, P., Trezciak, A., Benzonana, G., Gillessen, D. and Gabbiani G. 1986. A monoclonal antibody against α-smooth muscle actin: a new probe for smooth muscle differentiation. J. Cell. Biol. 103: 2787-2796.

3.   Kocher, O., Skalli, O., Cerutti, D., Gabbiani, F., Gabbiani, G. 1985. Cytoskeletal features of rat aortic cells during development. Circ. Res. 56: 829-838.

4.   Kocher, O. and Gabbiani, G. 1987. Analysis of α-smooth muscle actin mRNA expression in rat aortic smooth muscle cells using a specific cDNA probe. Differentiation 34: 201-209.

5.   Kocher, O. and Gabbiani, G. 1986. Cytoskeletal features of normal and atheromatous human arterial smooth muscle cells. Hum. Pathol. 17: 875-880.

6.   Kocher, O. and Gabbiani, G. 1986. Expression of actin mRNA in rat aortic smooth muscle cells during development, experimental intimal thickening, and culture. Differentiation 32: 245-251.

7.   Skalli, O., Bloom, W.S., Ropraz, P., Azzarone, B., Gabbiani, G. 1986. Cytoskeletal remodeling of rat aortic smooth muscle cells *in vitro*: relationships to culture conditions and analogies to *in vitro* situations. J. Submicrosc. Cytol. 18: 481-493.

8.   Clowes, A.W., Reidy, M.A. and Clowes, M.M.   1983.   Kinetics of cellular proliferation after arterial injury. I. Smooth muscle cell growth in the absence of endothelium. Lab. Invest. 49: 327-333.

9.   Barja, F., Coughlin, C., Belin, D. and Gabbiani, G. 1986. Actin isoform synthesis and mRNA levels in quiescent and proliferating rat aortic smooth muscle cells *in vivo* and *in vitro*. Lab. Invest. 55: 226-233.

10.   Clowes, A.W. and Clowes, M.M. 1986. Kinetics of cellular proliferation. IV. Heparin inhibits rat smooth muscle mitogenesis and migration. Circ. Res. 58: 839-845.

11.   Clowes, A.W., Clowes, M.M., Kocher, O., Ropraz, P., Chaponnier, C. and Gabbiani, G. 1988. Arterial smooth muscle cells *in vivo*: Relationship between actin isoform expression and mitogenesis of arterial smooth muscle cells *in vivo*. Modulation by heparin. J. Cell Biol. 107: 1939-1945.

12.   Majesky, M.W., Schwartz, S.M., Clowes, M.M. and Clowes, A.W. 1987. Heparin regulates smooth muscle S phase entry in the injured rat carotid artery. Circ. Res. 61: 296-300.

13.   Clowes, A.W. and Clowes, M.M. 1985. Kinetics of cellular proliferation after arterial injury. II. Inhibition of smooth muscle growth by heparin. Lab. Invest. 52: 611-616.

14.   Desmoulière A., Rubbia-Brandt, L. and Gabbiani, G. Modulation of actin isoform expression in cultured arterial smooth muscle cells by heparin and culture conditions. Arteriosclerosis, in press.

SECTION 2

VASCULAR WALL INJURY AND REPAIR

REPAIR OF ARTERIAL ENDOTHELIUM

Michael A. Reidy and Volkhard Lindner

Department of Pathology
University of Washington
Seattle, Washington

In recent years there has been an increase in the number of studies which have examined the growth of vascular endothelial cells. A major reason for this interest stems from the clinical data which shows that vascular prostheses are rarely endothelialized (1-5), and perhaps this fact alone is responsible for the occlusion of many grafts. There is also concern as to whether arteries subjected to angioplasty ever achieve total re-endothelialization. This chapter will deal with aspects controlling endothelial cell growth with a particular emphasis on growth of these cells *in vivo*.

ENDOTHELIAL GROWTH IN ARTERIES

There is surprisingly little information relating to the ability of endothelial cells to replicate, which in part reflects the lack of animal models to study this problem. In this laboratory, we have developed at least two *in vivo* models where endothelial regrowth is severely limited (4,5) and have used these approaches to understand what factors are important in controlling endothelial cell growth.

Many investigators, including ourselves, have found that loss of endothelium *in vivo* is immediately followed by a significant increase in the replication rate of those endothelial cells adjoining the wound (6) and, provided the area of denudation is small, cell replication continues until re-endothelialization is complete (7). This response seems to vary as a function of time since the replication rate of these cells significantly decreases after a few days, and endothelial cells have ceased proliferating after several weeks (5,8). This cessation of endothelial growth is not because endothelialization is complete since large areas of the arteries are still denuded at this time. This response of endothelial cells has been seen in denuded arteries of rats and rabbits (8,9) and in synthetic vascular grafts in primates (10). The key feature in all these studies is that a large area of the artery is denuded of endothelium and that regrowth does not occur from many sites. This last point is important since when multiple sites of endothelium are available, such as the intercostal arteries of the thoracic aorta, then regrowth is faster and often complete (11). This is in contrast to a denuded carotid artery of approximately equal length where there are only two sites for regrowth, and in these circumstances endothelial regrowth is slow and incomplete. These data in no way imply that endothelial cell growth from different arteries have different growth potentials, but simply that *in vivo* this retardation of endothelial cell growth can only be observed several weeks post-injury. Therefore, if the denuded artery can be totally healed within a relatively short period of time, then no retardation of endothelial cell growth can be detected.

Atherosclerosis, Edited by A. I. Gotlieb *et al.*
Plenum Press, New York, 1991

Table 1. Thymidine Index of Rabbit Carotid Endothelium 72 Hours after Narrow Scratch Injury

Site of injury	23.52 ± 5.45 %
Zone away from injury	0.02 ± 0.005 %

Values are means ±SE; p<0.001

One early explanation for this lack of growth was that rapid and prolonged replication initiated after injury could ultimately lead to endothelial cell senescence. If this were true, than it might explain why endothelial cell growth slowed only after a prolonged period of high replication. To test this hypothesis, carotid arteries were totally denuded of endothelium and after approximately 6 weeks endothelial regrowth had stopped. These partially endothelialized arteries were then mechanically reinjured and endothelial replication was assessed by ^{3}H-thymidine uptake and *en face* autoradiography as compared to control cells (Fig. 1) (4). These endothelial cells showed a marked increase in their replication after reinjury (Table 1), and the speed of healing in this wound was identical to that observed in unmanipulated arteries. Interestingly, despite this increase in cell growth, the endothelial monolayer as a whole was not restimulated to grow into the previously denuded zones of the artery. An important conclusion from this study was that "inhibited" endothelial cells are indeed capable of replicating but that injury alone was not a sufficient stimulus to induce regrowth onto the previous denuded segments of the artery.

Since we and others had only observed inhibition of endothelial cell growth on an injured artery where smooth muscle formed the luminal surface (pseudo-endothelium) (6,8,9), one possibility was that these cells somehow interfered with the regrowth of endothelium. In fact, one obvious possibility was that endothelial-smooth muscle cell contact was inhibiting endothelial growth. Therefore, an experiment was undertaken in which smooth muscle cell growth onto the denuded surface of the artery was inhibited and the extent of endothelial cell regrowth examined. In this experiment, totally inert vascular grafts were transplanted into the carotid arteries of rabbit (1) (Fig. 2). Thus, endothelial cells which grow from the cut end of the artery onto a graft surface would not encounter any smooth muscle cells on the luminal surface. The rate of endothelial cell replication and the total endothelial outgrowth onto these grafts were found to be similar to those observed in an artery subjected to balloon injury (Table 2), with the exception that there was a slight increase in replication of the endothelial cells near the anastomosis. Examination of these grafts by electron microscopy showed that no other cells except platelets were present on the luminal surface and that the ingrowing endothelium was not in contact with any cells on the luminal surface. This

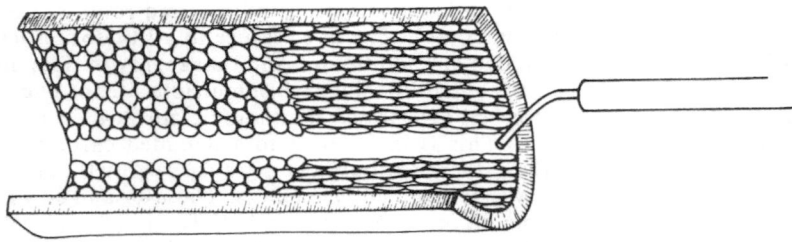

Figure 1. Schematic illustration of reinjury of rabbit carotid artery. The vessel was denuded of endothelium with a balloon catheter and reinjured after 12 weeks with a nylon filament which removes a zone approximately 10 cells wide.

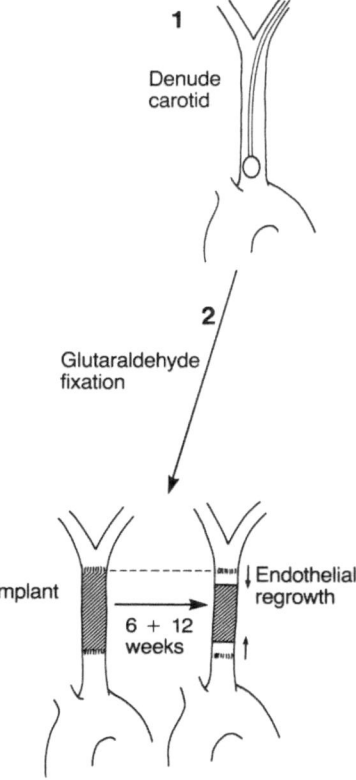

Figure 2.   Schematic illustration of experimental design.   After balloon catheter denudation of their left carotid artery (1), rabbits were perfusion fixed with 2% glutaraldehyde (2). The carotids were washed in glycine buffer (0.1M) for 3 days and in sterile PBS for a further 2 days and then implanted into the left carotid artery of recipient rabbits. After 6 and 12 weeks, these animals were given ^{3}H-thymidine and perfusion fixed.

result led us to conclude that the progress of re-endothelialization was not retarded by the presence of smooth muscle cells on the luminal surface. These data appear to be at variance with the recent data from D'Amore and coworkers (12) who have shown that pericytes and smooth muscle cells, when in contact with endothelial cells *in vitro*, inhibit their growth and do so by activating TGF-ß (13). It should be remembered, however, that the aim of our studies *in vivo* was to determine whether any other cells on the luminal surface would interfere with endothelial regrowth. Endothelial cells, however,

Table 2. Endothelial Cell Outgrowth onto Rabbit Carotid Vessels

Duration of outgrowth	Denuded Carotid	Fixed Carotid Graft
6 weeks	4.1 ± 0.6 mm (N = 10)	3.0 ± 0.6 mm (N = 2)
12 weeks	5.3 ± 1.9 mm (N = 10)	2.9 ± 0.8 mm (N = 10)

form contacts with smooth muscle cells via their basal surface and we did observe that as the endothelial cells grew onto the graft, they were accompanied by underlying smooth muscle cells (1). Thus, even though there were no other cells present on the denuded surface, the regrowing endothelium was always in close approximation with underlying smooth muscle cells. The possibility, therefore, still exists that a physical connection between these two cell types can play a role in inhibiting endothelial cell growth.

IN VIVO CONTROL OF ENDOTHELIAL CELL GROWTH

Until recently, we and others have had no success in stimulating complete endothelial regrowth over a large area, e.g., a denuded carotid artery, and so we have suggested that endothelial cells might have a finite ability to replicate *in vivo*. We have now found this conclusion to be incorrect and that endothelial cells have the ability to spontaneously repopulate large denuded areas. In a recent series of experiments, rat carotid arteries were denuded of endothelium using a less traumatic device which does not disrupt the underlying smooth muscle cells (5). As in carotids denuded with a balloon catheter, endothelial regrowth occurred from either end of the vessel but these cells maintained a high proliferative rate (Fig. 3) until the vessel was totally repopulated with new endothelium (Fig. 4). At present, we do not fully understand why the growth rates of endothelium are different, but this finding clearly illustrates that endothelial cells do have the potential for continuous growth and that total regrowth of large denuded areas can be achieved under certain conditions.

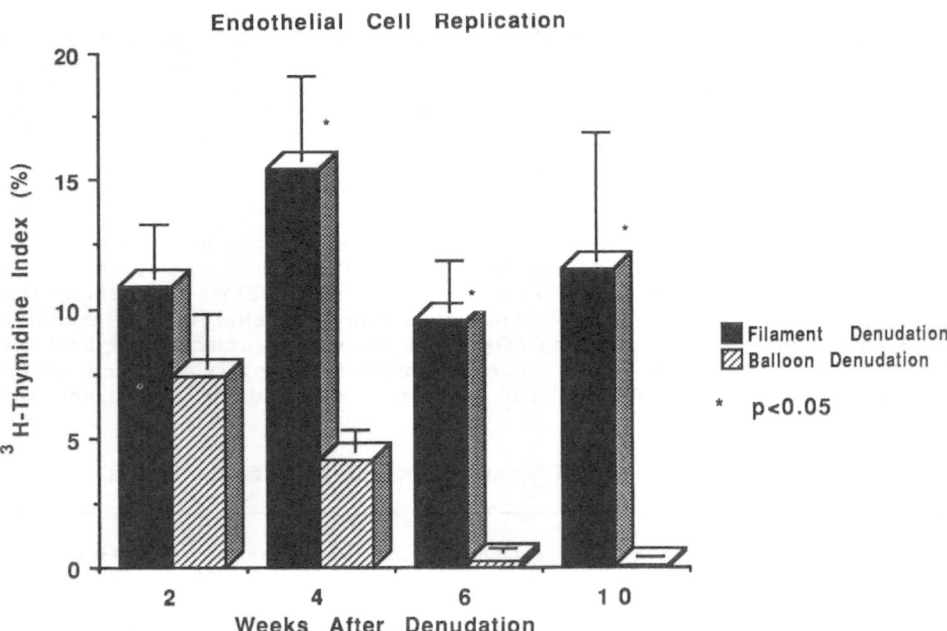

Figure 3. Endothelial replication in the rat carotid artery after denuding injury. Replication was measured by ^{3}H-thymidine index which was calculated for endothelial cells that were within 1 mm of the distal endothelium-smooth muscle cell interface. After 2 weeks, significantly higher endothelial replication rates were found in vessels denuded with the filament loop (means ± SEM).

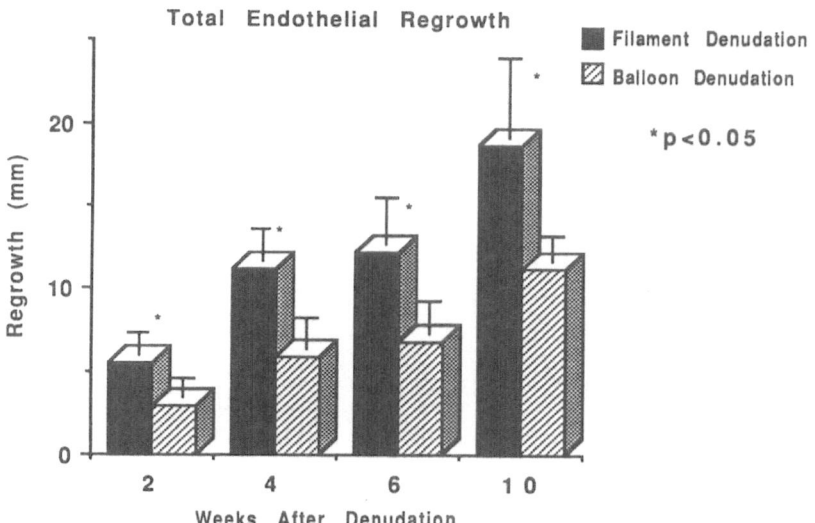

Figure 4. After denuding injury of rat carotid artery, endothelial regrowth was measured from the aortic arch and the bifurcation of common carotid artery. Regrowth was significantly higher all times in filament denuded vessels (means ± SD).

What factors are involved in the process are still unclear but one indication came from an experiment where we examined the endothelial cells for the presence of basic fibroblast growth factor (FGF). In rat carotid arteries, quiescent endothelium was found to stain only faintly with an FGF antibody but regenerating endothelial cells stained strongly (5). Of particular interest was the fact that 6 weeks after balloon catheter injury, when endothelial cell growth had stopped, no cells stained with the basic FGF antibody, whereas the endothelium in arteries denuded by the new "gentle technique", which was still replicating, stained strongly for this mitogen. This lead us to consider that basic FGF, possibly synthesized by the endothelial cells, was important for their growth.

To establish that basic FGF was indeed effective on rat arterial endothelial cells *in vivo*, the mitogen was infused into animals whose carotid arteries had been previously denuded with a balloon catheter. A single bolus injection of FGF caused a highly significant increase in endothelial cell replication (Fig. 5) (14). To our knowledge, this is the first time that endothelial cell replication in large arteries has been restimulated after the cells had spontaneously stopped growing. One interesting fact was that the response of the endothelial cells in these animals was not uniform and only those endothelial cells in the still denuded artery showed this dramatic increase in their replication rates (see Fig. 5). The endothelium of adjacent untraumatized arteries, i.e., aorta or right carotid, showed an increased replication when compared with control animals, but the increase was approximately 30-fold less than in the injured arteries. No data are currently available to explain why only certain cells respond so well to basic FGF. In order to answer this problem, future studies will need to address the availability of receptors on these cells and the presence of inhibitors known to occur on confluent but not sparse endothelium.

An increase in cell replication might imply new endothelial growth onto a denuded surface but in previous work we have shown that replication can occur without any detectable outgrowth (15-16). An experiment was therefore designed to test whether bFGF would stimulate the growth of endothelial cells onto a denuded artery. Rat carotid arteries were denuded of endothelium with a balloon catheter and basic FGF (12 µg) was injected daily for a period of 2 weeks (14). The extent of endothelial growth onto this denuded surface was found to be significantly increased after FGF

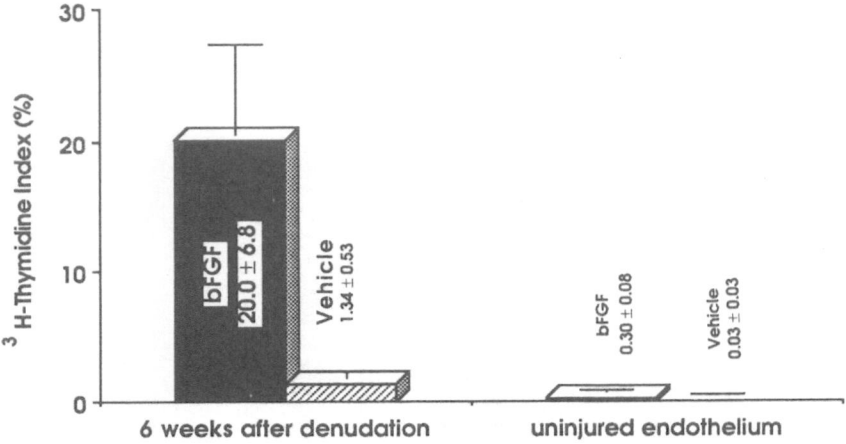

Figure 5. Six weeks prior to the intravenous injection of bFGF (12 µg), the left carotid artery of rats had been denuded of endothelium with a balloon catheter. After labelling with ^{3}H-thymidine, the animals were perfusion fixed 40 hours after injection of bFGF and endothelial replication was determined as described above. Replication was significantly higher after bFGF injection (p<0.03, means ± SEM).

administration when compared with animals receiving the vehicle above (Fig. 6). It is apparent from these data that bFGF is effective in stimulating endothelial cell growth in denuded arteries and perhaps this mitogen may have a role in promoting endothelial cell growth in vascular grafts.

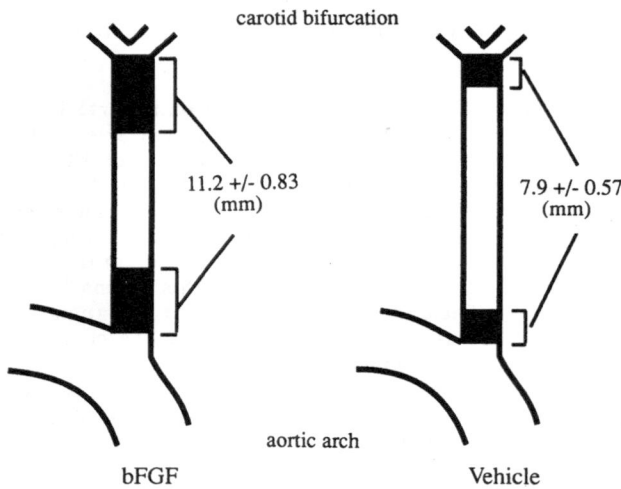

Figure 6. Endothelial regrowth in the common carotid artery of the rat. Six weeks after balloon catheter denudation of the left carotid artery, 12 µg of bFGF in 0.2 ml saline or an equal amount of vehicle in saline was injected intravenously via the tail vein once daily over a period of 12 days. Regrowth occurring from the aortic arch and the bifurcation was measured after the treatment period. Significantly more endothelial regrowth occurred in the bFGF treated animals (p≤0.013, n=6 animals in each group).

Smooth muscle cells respond to FGF *in vitro* (17) and we found that the systemic administration of FGF to these animals caused a dramatic increase in the growth of smooth muscle cell in balloon catheter denuded arteries (18,19). In the experiment described above, when bFGF was administered for several weeks, this increase in proliferation results in an intima which was approximately twice the size of that in the control animals. Thus, while bFGF might be advantageous in promoting the growth of endothelium, the regrowth of smooth muscle cells could be a serious complication in terms of vessel patency.

In summary, we have shown that the response of arterial endothelial cells to widespread denudation is limited, and large areas of the denuded vessel may be permanently devoid of a continuous endothelium. The reasons for the inability of endothelial cells to grow in these circumstances are not yet clear but our studies have eliminated cell senescence and the presence of smooth muscle cells on the surface of these arteries as possible candidates. Our data, however, does show that these endothelial cells can be restimulated to replicate by the administration of FGF and that multiple doses can lead to complete endothelial regrowth. Any possible use of FGF to initiate total endothelial cell regrowth must, however, consider that this mitogen can act on many cells. Studies are now ongoing in this laboratory to see if other more specific mitogens can stimulate endothelial cells in a more specific and selective manner.

REFERENCES

1.   Reidy, M.A. 1988. Endothelial regeneration. VIII. Interaction of smooth muscle cells with endothelial regrowth. Lab. Invest. 59: 36-43.

2.   Berger, K., Sauvage, L.R., Rao, A.M., et al. 1972. Healing of arterial prostheses in man: its incompleteness. Ann. Surg. 175: 118-127.

3.   Sauvage, L.R., Berger, K.E., Wood, S.J., et al. 1974. Interspecies healing of porous arterial prostheses. Arch. Surg. 109: 698-705.

4.   Reidy, M.A., Clowes, A.W. and Schwartz, S.M. 1983. Endothelialization regeneration. V. Inhibition of endothelial regrowth in arteries of rat and rabbit. Lab. Invest. 49: 569-575.

5.   Lindner, V., Reidy, M.A. and Fingerle, J. 1989. Regrowth of arterial endothelium. Denudation with minimal trauma leads to complete endothelial cell regrowth. Lab. Invest. 61: 556-563.

6.   Schwartz, S.M., Stemerman, M.B. and Benditt, E.P. 1975. The aortic intima. II. Repair of the aortic lining after mechanical denudation. Am. J. Pathol. 81: 15-42.

7.   Reidy, M.A. and Schwartz, S.M. 1981. Endothelial regeneration. III. Time course of intimal changes after small defined injury to rat aortic endothelium. Lab. Invest. 44: 301-308.

8.   Reidy, M.A., Standaert, D. and Schwartz, S.M. 1982. Inhibition of endothelial cell regrowth. Cessation of aortic endothelial cell replication after balloon catheter denudation. Arteriosclerosis 2: 216-220.

9.   Goff, S.G., Wu, H.D., Sauvage L.R., et al. 1988. Differences in reendothelialization after balloon catheter removal of endothelial cells, superficial endarterectomy, and deep endarterectomy. J. Vasc. Surg. 7: 119-129.

10.   Reidy, A.M., Chao, S.S., Kirkman, T.R. and Clowes, A.W. 1986. Endothelial regeneration. VI. Chronic nondenuding injury in baboon vascular grafts. Am. J. Pathol. 123: 432-439.

11. Haudenschild, C.C. and Schwartz, S.M. 1979. Endothelial regeneration. II. Restitution of endothelial continuity. Lab. Invest. 41: 407-418.

12. Orlidge, A. and D'Amore, P.A. 1987. Inhibition of capillary endothelial cell growth by pericytes and smooth muscle cells. J. Cell Biol. 105: 1455-1462.

13. Antonelli-Orlidge, A., Saunders, K.B., Smith, S.R., et al. 1989. An activated form of transforming growth factor beta is produced by co-culture of endothelial cells and pericytes. Proc. Natl. Acad. Sci. USA 986: 4544-4548.

14. Lindner, V., Majack, R.R. and Reidy, M.A. 1990. Basic FGF stimulates endothelial regrowth and proliferation in denuded arteries. J. Clin. Invest., in press.

15. Reidy, M.A. and Schwartz, S.M. 1983. Endothelial injury and regeneration. IV. Endotoxin: a nondenuding injury to aortic endothelium. Lab. Invest. 48: 25-34.

16. Reidy, M.A., Chopek, M., Chao, S., et al. 1989. Injury induces increase in von Willebrand Factor in rat endothelial cells. Am. J. Pathol. 134: 857-864.

17. Gospodarowicz, D., Ferrara, N., Haaparanta, T. and Neufeld, G. 1988. Basic fibroblast growth factor: expression in cultured bovine vascular smooth cells. Eur. J. Cell Biol. 46: 144-151.

18. Lindner, V., Majack, R.A. and Reidy MA. 1989. Basic FGF stimulates endothelial regrowth in denuded arteries *in vivo*. J. Cell Biol. 109: 98a.

19. Lindner, V., Majack, R.A. and Reidy, M.A. 1990. Basic FGF induces the proliferation of vascular cells in injured arteries. FASEB J. 4: A623.

THE ROLE OF THE CYTOSKELETON IN ENDOTHELIAL REPAIR

Avrum I. Gotlieb

Vascular Research Laboratory, Department of Pathology
and Banting and Best Diabetes Centre
Faculty of Medicine, University of Toronto
Toronto Hospital Research Centre - General Division
Toronto, Ontario, Canada

INTRODUCTION

Based on a solid foundation of traditional morphology, modern cell and molecular biology is providing information on structure-function relationships of endothelial cells. It has become apparent that, in addition to acting as a thromboresistant surface and macromolecular barrier, endothelial cells are very active metabolically during the initiation and subsequent growth of fibrofatty atherosclerotic plaques[1]. In addition, many endothelial functions are inducible[2] and important interactions occur at the vessel wall-blood interface that regulate endothelial cell function. The subendothelium[2], hemodynamic forces[3,4,5], and many constituents of the blood, including platelets[6], lipoproteins[7], coagulation factors [8,9,10], fibrinolytic substances[11,12] and leukocytes[13,14,15] are involved in these processes. In addition, the endothelial cells are active participants since they synthesize and secrete coagulation[16,17] and anticoagulation factors[18], profibrinolytic factors[19], platelet antiaggregation factors[20], cytokines[21,22], growth factors[23,24] and many other important agents.

ENDOTHELIAL INTEGRITY

Although the sequence of events that lead to the development and growth of the fibrofatty atherosclerotic plaque are not fully known, the loss of endothelial integrity is important in the initiation of the lesion[1,25]. Loss of integrity may occur due to structural and/or functional changes. Exposure of subendothelium may occur due to frank loss of endothelial cells, due to retraction of adjacent cells leaving endothelial gaps, or due to subtle changes in intercellular adhesion. Little is known about the last category. The ability of the endothelial cells to adapt to these changes and to repair them once they occur is very important in the maintenance of endothelial integrity. The endothelial cytoskeleton consists of intracellular systems which are likely to regulate both the structural integrity of the endothelium and important repair processes.

ENDOTHELIAL DENUDATION

Denudation of endothelium, even only several cells wide, is not found in the normal artery[26]. Previous findings of denudation are considered artefacts due to fixation without perfusion at physiological pressures and due to tissue processing. There is a need to control pressure, temperature, and osmolality during fixation; to avoid heparin; and to handle the specimens carefully during post fixation in order to reduce artefacts[27]. Studies of the endothelium which covers the surface of early

Atherosclerosis, Edited by A. I. Gotlieb *et al.*
Plenum Press, New York, 1991

atherosclerotic lesions in hyperlipidemic cynomologous monkeys did not reveal degeneration, disruption, or sloughing of endothelial cells. Instead, structural adaptations consisting of increased cellular organelles, attenuation and reshaping of cells and a decrease in the extent and complexity of lateral contact regions between adjacent endothelial cells were observed[27]. In other studies, the surface of prominent fatty streak lesions showed gaps between endothelial cells covering the lesion[14]. Denudation also occurs on the surface of well developed fibrofatty plaques, especially when they are complicated by surface erosion and ulceration. When focal denudation does occur[28], it is most likely due to severe cell injury or cell death[29]. Hemodynamic forces are unlikely, by themselves, to cause frank denudation under acute conditions[30]; however, chronic hemodynamic abnormalities will result in denudation and slow repair[31]. Based on a study of endotoxic endothelial injury in rats, it has been postulated that endothelial cells which die are lifted off from the monolayer by lamellipodia from adjacent cells which undermined the dead cells[32]. This method of endothelial repair thus limits the area of frank denudation and minimizes the exposure of the subendothelium to the blood stream, and is thus a very important method of rapidly restoring structural endothelial integrity.

ENDOTHELIAL REGENERATION

In Vivo Models

The endothelium is a single layer of very slowly replicating cells[33] that show a high degree of density-dependent inhibition of growth. Cell shape and orientation are strongly influenced by flow[5,34]. Studies have been carried out *in vivo* to describe the events that follow disruption of the contact-inhibited confluent endothelial cell monolayer. Since large areas of endothelial denudation are present over atherosclerotic plaques, many *in vivo* models consisted of producing large areas of aortic endothelial denudation using an inflated balloon catheter[35-38], by air-drying[39], or other means. The media was also inadvertently injured in these models systems[40]. The endothelial cells along the edge of the wound extended prominent lamellipodia, elongated and began to migrate into the area of injury. The endothelium migrated as a continuous sheet of cells with very few cells at the leading edge migrating as single cells[37]. This suggested that cell-to-cell contacts were maintained during the process of reendothelialization. There was subsequent endothelial cell proliferation associated with this migration which provided new cells to replace those that were removed[41].

Since large areas of endothelial denudation have not been shown to occur *in vivo*, model systems using controlled injury of the endothelium by a small fine surface, such as the tip of a thin nylon filament, have been employed to make very small denuding injuries which closely mimic naturally occurring injury[42-44]. These studies have shown that within 8-10 h, areas of denudation of 3-5 cells wide were repaired by the spreading and migration of endothelial cells adjacent to the wound. Proliferation was not required initially, although cell proliferation may have occurred later in order to restore the initial number of cells[43]. The rate of repair of small injuries appeared to be unaffected by hypertension or hyperlipidemia[45].

In Vitro Models

In the last 20 years, numerous tissue culture systems have been developed to study endothelial structure and function *in vitro*[46-48]. We have used the large *in vitro* wound model system to provide important information on the cellular processes involved in repair. We have used porcine aortic endothelial cells since the pig develops spontaneous atherosclerosis[49] and has a cardiovascular system similar to that of man[50]. In the large experimental wound model[51], porcine thoracic aortic endothelial cells migrate as a sheet of cells, although each individual cell migrates on its own within the advancing monolayer. The endothelial cells at the leading edge extrude lamellipodia, elongate and then migrate into the denuded area. This was similar to the findings of other *in vitro*[52,53] and *in vivo* studies[38]. By carrying out the wounding on confluent monolayers grown on glass coverslip, we have been able to study the cytoskeleton during repair. Quantitative measurements of rate of migration and cell proliferation can be easily made, live cell migration can be observed by time lapse cinemicrophotography[54], and cytoskeletal proteins can be localized using light and electron-microscopic immunocytochemical and fluorescent methods. In addition, Madri[55] has carefully studied the role of extracellular matrix components including laminin, fibronectin and type I and III and type IV collagen on endothelial cell proliferation and sheet migration in similar models.

We have developed a unique single cell wound model of endothelial injury[59]. To make the wound, a confluent monolayer grown on a glass coverslip is visualized under an inverted-phase microscope and a single endothelial cell is removed using gentle suction delivered by a micropipette attached to a micromanipulation system. All the endothelial cells around the periphery of the wound extrude lamellipodia into the area of denudation to close the wound within 30-40 minutes (Figure 1). There is no cell translocation nor is there cell proliferation in this model system. However, if more than 5 cells are removed, then cell translocation is also required to close the wound.

We have also characterized an aortic organ culture model to study reendothelialization[56]. Rectangular, full thickness pieces of porcine aorta that are not near branch sites are cut out of the thoracic aorta with a scalpel blade. Half of the endothelium is denuded lengthwise using a single gentle stroke of the scalpel blade in the direction of blood flow. The mode of repair is documented by scanning and transmission electron microscopy[56]. Pederson and Bowyer[57] also have reported on a rabbit aortic organ culture system in which small intimal wounds of defined width were created. The advantages of organ culture over cell culture include the presence of a natural substratum which is known to play an important role in endothelial cell migration[58], as well as the omission of a cell dispersion step which may lead to selection of particular cell phenotypes.

Regulation of Repair by Soluble Factors

A variety of soluble factors affect reendothelialization. TGF-ß delays endothelial repair during the first 24 hours following wounding[60] and fibroblast growth factor is required for the migration of bovine endothelial cells. Endogenously released bFGF not only promotes migration but also regulates the basal levels of plasminogen activator release and DNA synthesis[61]. This autocrine role for bFGF may be mediated by its release from endothelial cells through mechanically induced cell membrane disruption since bFGF does not have a signal sequence and thus lacks the structural features that are normally required for protein secretion[62]. Whether these factors have direct or indirect effects on the cytoskeleton is not known. It is of interest that platelet derived growth factor, which is secreted by endothelial cells[63] induces actin and vinculin reorganization and membrane ruffling in fibroblasts[64].

THE ENDOTHELIAL CELL CYTOSKELETON

The three major fibrous protein systems of the endothelial cell include microfilaments, microtubules and intermediate filaments[66] (Figure 2). These systems are dynamic and microfilaments and microtubules undergo rapid filament assembly and disassembly. They are regulated with respect to cross-linking and polymerization by many associated proteins[67,68]. Although referred to as the cell cytoskeleton these systems are distinct with respect to biochemical structure, immunological properties and function. There is, however, likely to be important interactions between these cytoskeletal systems[69-71].

Actin Microfilaments

Structure and Function. Actin, a contractile protein, is present in endothelial cells[72], as in most eucaryotic cells, in the filamentous form, F-actin, and the monomeric form, G-actin[73,74]. It is likely that the shift in equilibrium between the monomeric and the polymeric forms of actin is associated with many actin-mediated cell functions[75,76]. For example, when endothelial cells migrate, there is a shift in the ratio of G- to F-actin favoring G-actin while total actin remains unchanged[77].

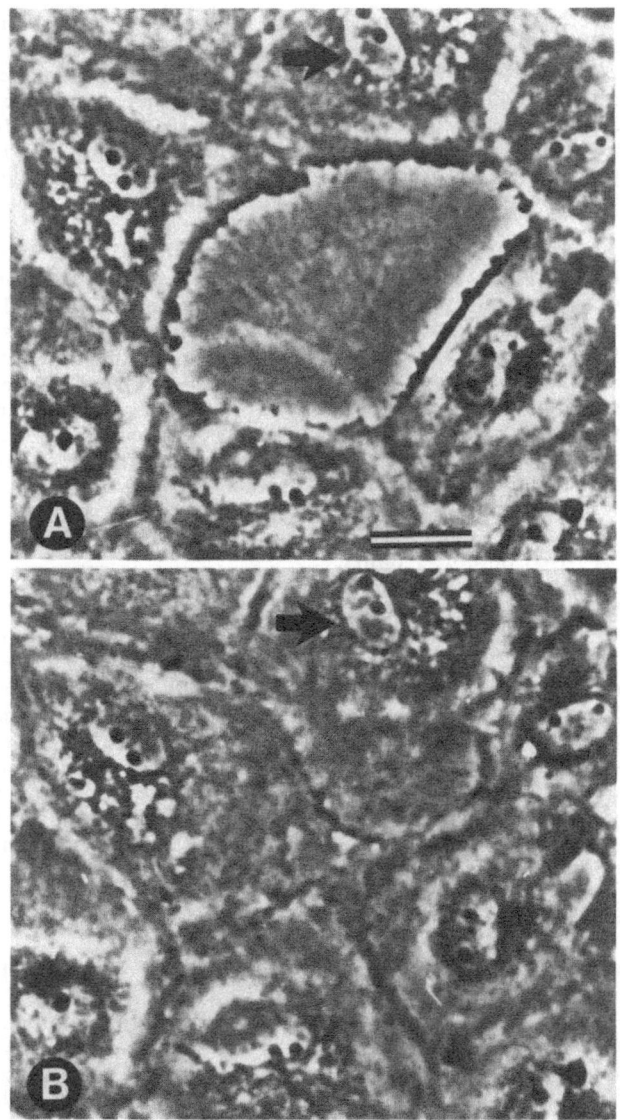

Figure 1. Closure of single cell wound made in a confluent endothelial monolayer of porcine aortic endothelial cell. (A) At time of wounding. (B) Wound almost completely closed by lamellipodia extrusion (x1600).

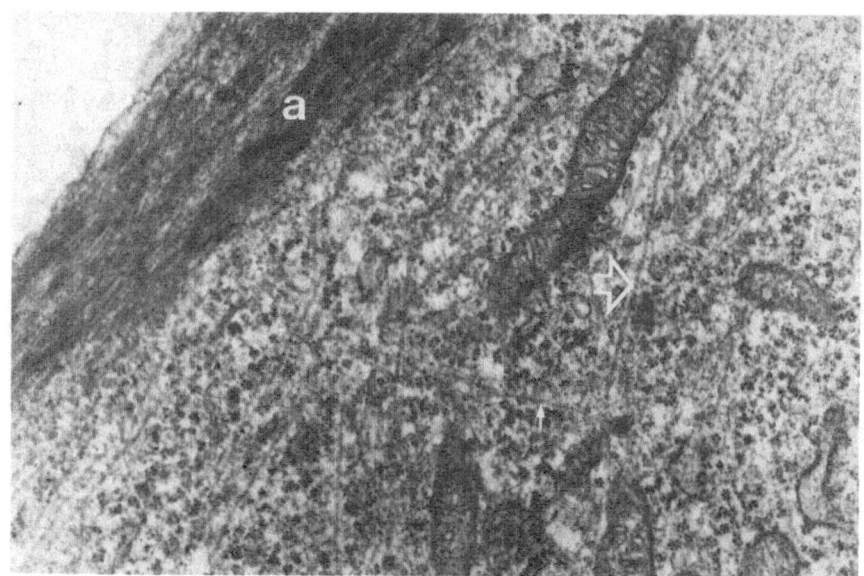

Figure 2. Transmission electron microscope photomicrograph showing actin filaments (a), microtubules (open arrow), and intermediate filaments (arrow). (x30,000)

Filamentous actin is organized into a diffuse network of short microfilaments[78] and into prominent microfilament bundles[79-81]. The diffuse network is present mainly in the cell cortex as well as in cellular processes such as lamellipodia[82]. The microfilament bundles include several different types[83], including the stress fibers. The stress fiber bundles are composed of actin filaments in parallel alignment with nonuniform polarity[84,85]. They were first identified and studied *in vitro* in a variety of cell types. In addition to actin, the stress fibers contain myosin, tropomyosin, and alpha actinin[86-89], and are thought to be contractile and under mechanical stress[90-92].

Microfilament bundles[93] have been considered important in providing the force of contraction for cell migration[94]. In addition, it is likely that at least some of these bundles function as long substrate adhesion complexes, especially in well spread non-migrating fibroblasts[95]. The term stress fibers has been applied to these ventral microfilament bundles since they are thought to be contractile and under mechanical stress, and to pull against a site of cell adhesion to the substratum[96]. Studies have also shown, however that microfilament bundles did not appear necessary for cell motility in some systems. In fact, when the bundles became more prominent, cells became extremely flattened on the substratum and were non-motile[97,98]. It is likely that there is an optimum number of stress fibers in a given cell to provide optimum conditions for migration.

Microfilament Bundles *In Vivo*. The actin microfilament bundles have been identified in vessel wall endothelial cells from a variety of locations using transmission electron microscopy[99,100] and, more recently, by *in situ* localization using immunofluorescence microscopy[101-103]. The actin microfilaments can be localized by using antibodies to actin as well as by derivatives of phallacidin and phalloidin such as 7-nitrobenz-2-oxa-1,1-diazole (NBD) phallacidin or rhodamine phalloidin. Phallacidin is a phallotoxin isolated from the amanita family of mushrooms that has a very high affinity for F-actin[104]. Actin microfilament bundles are located at the periphery of the cell as well as within the cell cortex. *In situ* staining has shown that these microfilaments contain myosin and alpha actinin. Recently, we have developed a technique in which perfusion-fixed aortas were stained *in situ* for F-actin by infusing rhodamine phalloidin via a peristaltic pump into the aortas at a slow flow rate. This new technique resulted in excellent visualization of branch points and allowed for a precise description of the actin microfilament bundles in endothelial cells along flow dividers[105,106]. The actin microfilaments reorganize in response to changes in hemodynamic shear stress[106a] and show modifications in regions of arteries prone to atherosclerotic lesion formation[107].

Microfilament Bundles *In Vitro*. We have shown that when endothelial cells form a confluent, contact inhibited monolayer, the periphery of the cell contains prominent circumferential microfilament bundles which we have termed the dense peripheral band (DPB) as well as shorter central microfilament bundles[108] (Figure 3). In low density culture, even in islands of endothelial cells where there is cell to cell contact, a DPB was not formed[109]. We have shown, using double immunofluorescence microscopy, that actin colocalizes with myosin, with tropomyosin, with alpha actinin and with vinculin within this DPB[109]. Although there were microtubules extending toward and into the DPB, there did not seem to be any preferential localization of microtubules within the band. Occasionally, microtubules ran parallel to the band along its inner aspect. As noted above, *in vivo* actin microfilaments are also distributed as peripheral and central bundles. In the thoracic and abdominal aorta, the peripheral actin is less prominent than in confluent cultures. The central bundles are similar but are more prominent in the abdominal aorta. They are generally oriented in the long axis of the cell. However, deviations of up to a 30° angle are often seen. At areas where there is very low shear stress, the cells are cobblestone and the morphology of the peripheral bands are similar to that seen *in vitro* (Figure 4).

Transmission electron microscopic examination of the DPB has shown that there were microfilaments which emanated from the band and extended into junctions which had cytoplasmic plaques. Often the junctions of adjacent cells had microfilaments extending into their respective plaques and the microfilaments appeared to be in alignment with each other[109]. Since they had actin microfilaments extending into them and since vinculin was present at the periphery of the endothelial cells associated with the DPB's, we postulate that these plaques are similar to adherens junctions[110]. In their ultrastructural study of aortic endothelium, Huttner *et al.*[111] showed that the endothelial

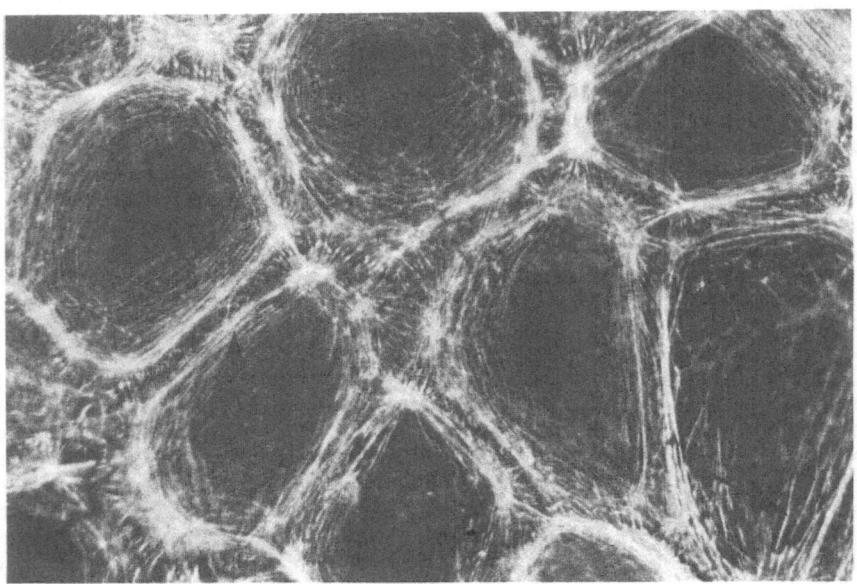

Figure 3. Fluorescence photomicrograph of endothelial cells showing actin microfilaments localized with rhodamine phalloidin. Note dense peripheral band (short arrow) and central microfilaments (long arrow). (x2000)

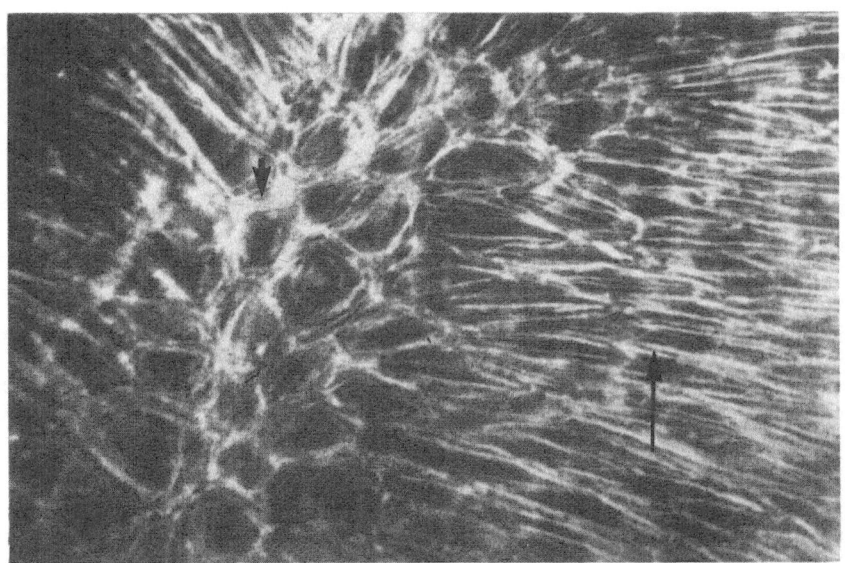

Figure 4. Fluorescent localization of endothelial actin microfilaments *in situ* at a flow divider in the rabbit aorta. Prominent bundles (long arrow) are present at areas of high shear and circumferential bundles (short arrow) are prominent in areas of low shear. The latter are similar to the DPB of confluent cultures. (x800)

cells are interconnected by gap and tight junctions. Adherens junctions between these cells were not described. Recently, monoclonal antibodies have been described which show a staining pattern of endothelium which suggests that the antigen is part of a junctional component[112].

Microtubules and Centrosomes

The centrosomes of endothelial cells consist of the paracentral paired centrioles and the amorphous material around them. The centrosome is a microtubule organizing center, a site capable of initiating the polymerization of microtubules from tubulin both *in vivo* and *in vitro*[113-116]. Microtubules emanate from the centrosomal area and are very prominent toward the center of the cell and less so at the periphery[66] (Figure 5a). The distribution of microtubules is similar in both low density and confluent cultures. Several studies have explored the role of microtubules[117,118] in cell migration. Badley *et al.*[117] compared the cytoskeleton in single migrating and stationary chick fibroblasts using immunofluorescence microscopy. They concluded that the distribution of microtubules is not altered significantly during the conversion from the migratory to the stationary state. There are, however, studies that show that coordinated movement in one direction requires the presence of microtubules and that movement is either reduced[118], inhibited[119], or can occur only randomly[120-122] when microtubules are disrupted, as in colchicine-treated cells. Malech *et al.*[123] showed that colchicine had no effect on random migration of human neutrophils; however, activated random migration was minimally decreased and directed migration was markedly inhibited. They also showed that the position of the centriole and its associated microtubules appear to be important in establishing the direction of migration of neutrophils. In other systems the centrosome does not appear to direct migration[124].

Intermediate Filaments

Endothelial cells contain intermediate filaments[125-127] (Figure 5b). Their physiological role in endothelial cell function is not known. It is believed that intermediate filaments have a mechanical role in cell function[127]. They are not, however, as dynamic as the microfilaments and microtubules. At the present time, these filaments have not been shown to have a role in endothelial motility. Antibodies to intermediate filaments injected into fibroblasts induce the filaments to coil around the nucleus; however, locomotion and cell morphology are unaltered[128,129].

CENTROSOME AND MICROTUBULES IN REENDOTHELIALIZATION

The function of the centrosome[130] during endothelial regeneration has been described using large wounds *in vitro*[131,132]. We have shown that endothelial cells migrating into the wound rapidly redistributed their centrosomes to the front of the cell between the nucleus and the leading lamellipodia (Figure 6). If the wound was treated first with colcemid to break down microtubules, redistribution did not occur. We have also shown that redistribution was independent of cell migration since wounds treated with cytochalasin B at concentrations which just inhibited migration still showed centrosomal redistribution. This redistribution, however, occurred more slowly than under normal conditions, suggesting that interactions between microtubules and microfilaments may play some role in enhancing centrosome redistribution[132].

The distribution of various cytoskeletal components may change when cells are removed from their *in situ* environment and grown *in vitro*[133]. We showed, however, that centrosome redistribution followed wounding not only in tissue culture but also in organ culture[134] as well as *in vivo*[135]. The orientation occurred most rapidly in the tissue culture model, suggesting that the subendothelial matrix and hemodynamic factors may

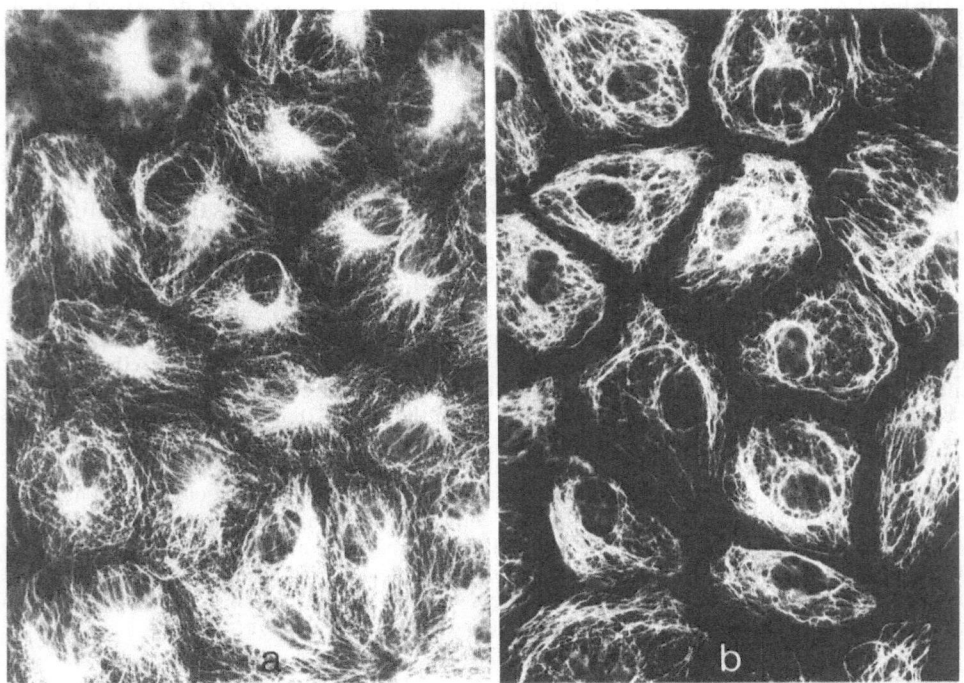

Figure 5. Immunofluorescent localization of endothelial cytoskeletal proteins. (a) Microtubules localized by an antibody to tubulin; (b) Intermediate filaments localized with an antibody to vimentin. (x1000)

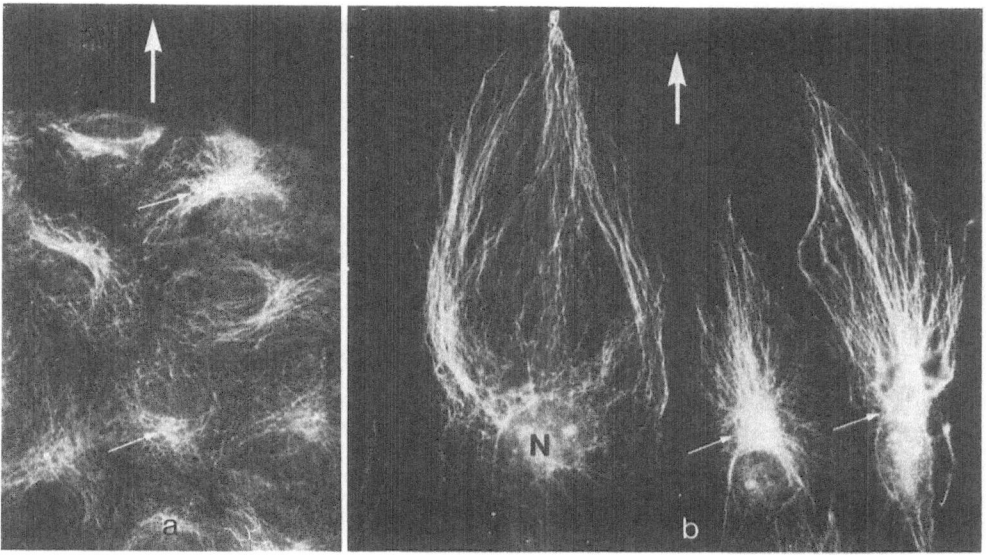

Figure 6. Migrating endothelial cells at (a) 0 and at (b) 4 hours following wounding of a confluent endothelial monolayer. Note the redistribution of the centrosomes (thin arrow) to the front of the cell. (Large arrow - direction of cell migration) (Modified, Gotlieb, A.I. *et al.*, J Cell Biol 96: 1266, 1983). (x2000)

regulate centrosomal distribution. Centrosomal redistribution in bovine aortic endothelial cells is enhanced by several factors including serum, multiplication stimulating factor, and insulin[136,137]. Although platelet derived growth factor had no effect on its own, it had a stimulatory synergistic effect with sub-effective doses of serum, insulin, and multiplication stimulating factor. These studies are important since they show that cytoskeletal events occurring during endothelial regeneration can be regulated by soluble external factors.

Centrosomal redistribution occurs in a variety of other cell systems under a variety of conditions[138-139]. In endothelial cells, it appears that reorientation is important in the initiation and regulation of migration and may thus act as a preprogrammed internal control mechanism. It has been suggested that, since the Golgi apparatus redistributes along with the centrosome, the function of centrosomal redistribution is to provide new membrane for extruding lamellipodia during cell migration[139].

What induces centrosomal redistribution? It appears to be triggered when the cell receives an external signal that migration is imminent. Within the confluent endothelial monolayer, we have seen centrosomal redistribution occurring well before the breakdown of the DPB in cells away from the leading edge[140]. These cells were not migrating, however, they received some signal from the cells in front of them which were migrating. The signal may be a physical stimulus such as might occur when the monolayer becomes less tightly packed, or a chemical signal passed through gap junctions[141,142].

The mechanism by which the centrosome moves toward the front of the cell is also not known. As noted previously, centrosomal redistribution requires intact microtubules and appears to be delayed somewhat if the microfilaments are broken down[131]. One possibility is that there are direct or indirect connections between the microtubules and the DPB, with the band anchoring the microtubules as the centrosome redistributes. Using time lapse cinemicrophotography, we have noted that in some cases the centrosome redistributed independently of any detectable nuclear movement. In other instances, however, it appeared that the centrosome and nucleus rotated together.

MICROFILAMENTS IN REENDOTHELIALIZATION

Large experimental wounds were used to study the changes in the F-actin cytoskeleton during reendothelialization. We found that four zones were present at the wound edge, each with characteristic features (Gotlieb *et al.*, 1984). Cells in the first two zones, the leading edge and elongated zone, showed absent or markedly reduced DPBs and prominent migration while the other two zones had intact DPBs and the cells did not migrate (Figure 7). Thus, the presence of the DPB was associated with a marked reduction of cell migration.

A comparison between the microfilament bundles that we observed in *in vitro* endothelial cells with those described in recent *in vivo* studies indicate many similarities.

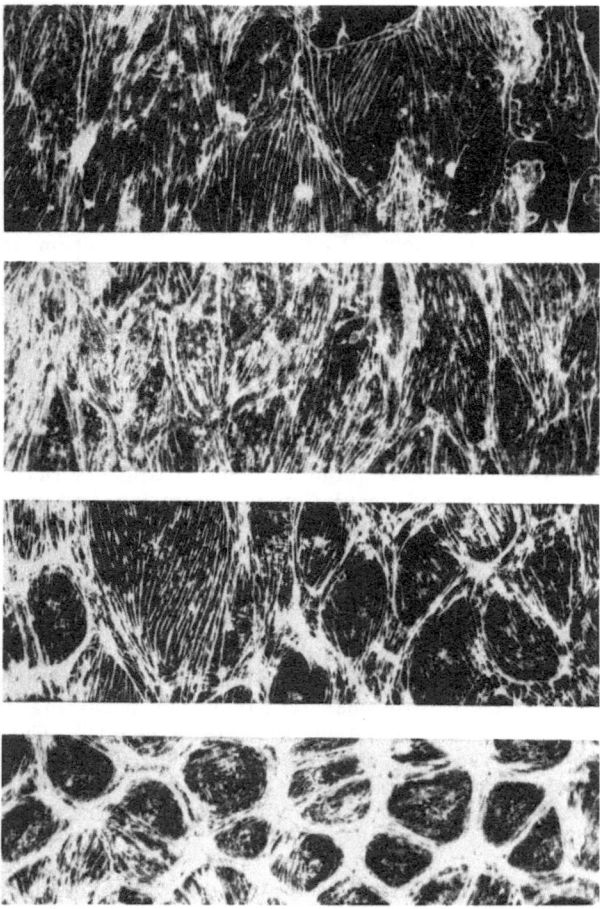

Figure 7. F-actin localized with rhodamine phalloidin in the endothelial cells at the edge of a wound 24 hours following wounding. Top panel is wound edge; bottom is non-migrating monolayer. (x1000)

Prominent microfilament bundles were present at the periphery of aortic endothelial cells from several species[143,144]. Central microfilament bundles were present in those endothelial cells that were located in areas which were believed to be under increased hemodynamic stress. *In vivo*, actin-containing microfilaments were reduced at the periphery of rabbit aortic endothelial cells following balloon denudation, while central microfilament bundles (termed stress fibers by the authors) increased in the cells migrating into the wound[144,145]. Since central bundles persisted long after reendothelialization had ceased, and since they were also normally present in areas of high shear force, the authors concluded that these fibers were not directly related to cell movement but instead were involved in adherence of the endothelial cell to the substratum. Endothelial cells both *in vivo*[3,5] and *in vitro*[146,147] are indeed responsive to hemodynamic shear stress and undergo changes in shape and orientation[4,5] and in cytoskeleton[4,146,147] in response to altered flow. *In vivo*, there is an increase of microfilament bundles (stress fibers) in the cytoplasm of regenerating endothelial cells as compared to normal resting endothelial cells[148]. These microfilament bundles are connected to membrane domains located at the abluminal aspect of endothelial cells. Actin stress fiber modification *in vivo* is also associated with compromised endothelial barrier function and coincides with the accumulation of low density lipoproteins in areas of intimal thickening[149].

Studies on the single cell wound model showed that, following removal of the cell, the part of the DPB facing the wound showed some splaying and became more prominent. Microfilaments emanated from the band into the lamellipodia. Wounds incubated with cytochalasin B at concentrations which caused loss of the DPB with sparing of the central microfilaments showed very little closure after a period of 6 hours. Complete closure occurred normally between 30-40 minutes. However, when the cytochalasin was washed out, the microfilaments of the DPB began to reappear immediately, and were associated with lamellipodia formation. Thus, reendothelialization occurring by spreading requires an intact DPB; however, the centrosome was not involved since centrosomal redistribution did not occur.

To characterize the full spectrum of cytoskeletal events occurring during small wound repair, the *in vitro* model was used to compare wounds of 1 to 4 cells with those of 6-12 cells. The former closed rapidly within an hour and the later closed from 3 to 6 hours. The repair process was observed by time-lapse cinemicrophotography. Using fluorescence and immunofluorescence microscopy, the cellular morphological events were correlated with the localization and distribution of actin microfilament bundles and vinculin plaques, and centrosomes and their associated microtubules. Single to 4 cell wounds were closed by cell spreading, while wounds 7 to 9 cells in size closed by initially spreading which was followed at approximately 1 hour after wounding by cell migration. These two processes showed different cytoskeletal patterns. Cell spreading occurred without centrosome relocation. However, centrosome distribution to the front of the cell occurred as the cell began to elongate and migrate. While the peripheral actin microfilament bundles (i.e., the dense peripheral band) remained intact during cell spreading, they broke down during migration and were associated with a reduction in peripheral vinculin plaque staining. Thus, the major events characterizing the closure of endothelial wounds were precise, followed a specific sequence, and were associated with specific cytoskeletal patterns which most likely were important in maintaining directionality of migration and reducing the adhesion of cells to their neighbors within the monolayer (Figure 8).

The DPB can also be regulated by external factors including thrombin, phorbol 12-myristate 13-acetate (TPA)[151], ethchlorvynol[152] and histamine[153]. These agents promote the reversible breakdown of the DPB in the intact confluent monolayer. We have reported that incubation of confluent monolayers of endothelial cells with thrombin resulted in the reversible loss of the DPB and was associated with a change in the shape of the cells from cobblestone to elongated (Figure 9). Associated with the DPB disruption, peripheral vinculin plaques were lost, only to reappear in association with the return of the DPB following wash out of thrombin. It is possible that thrombin may enhance the rate of repair by promoting rapid reversible reorganization of microfilaments similar to the pattern present in migrating cells.

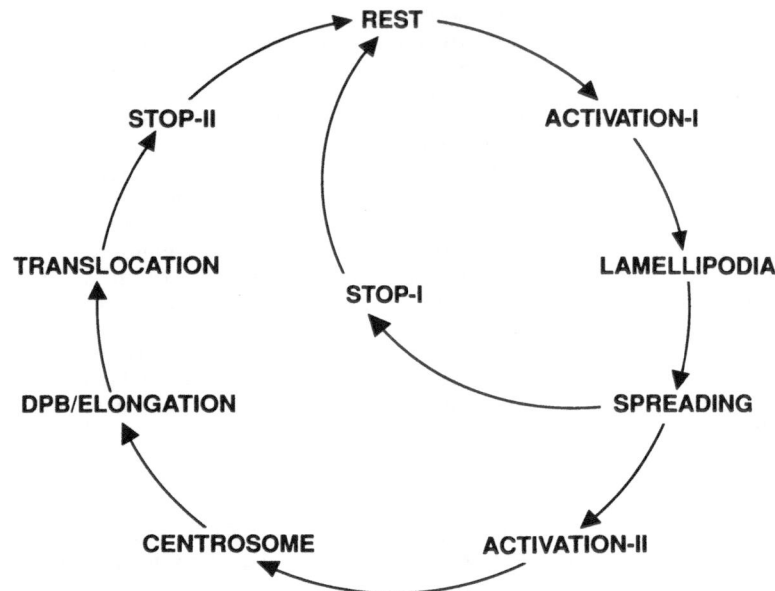

Figure 8. The closure of a small wound: The injury-repair cycle. Activation-I refers to that set of unknown cellular processes activated to regulate lamellipodia extension and spreading, while, Stop-I refers to the turning off of those processes; Activation-II refers to that set of unknown cellular processes activated to regulate centrosome redistribution, DPB reduction and loss, cell elongation, and finally translocation, while Stop II refers to the turning off of those processes. Repair is completed and the monolayer returns to rest.

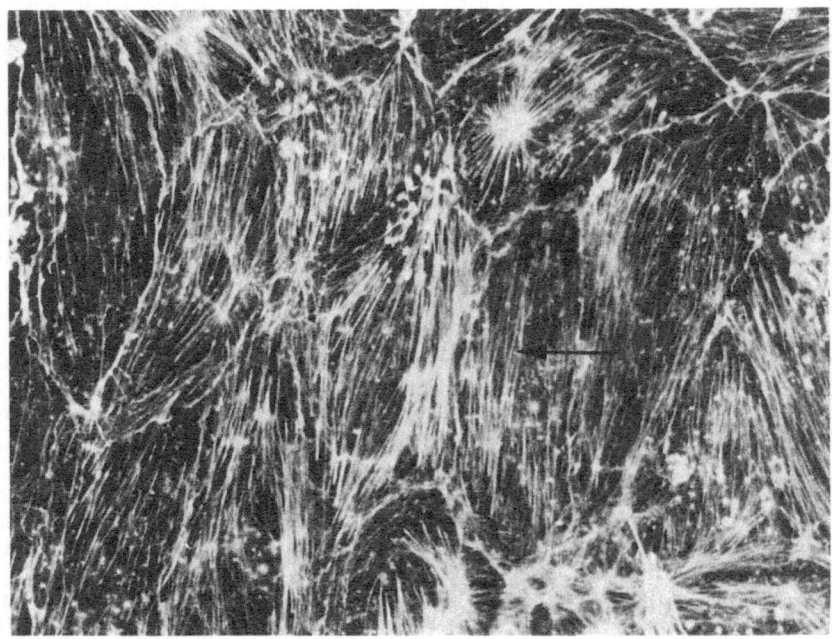

Figure 9. Fluorescent localization of actin microfilaments with rhodamine phalloidin in cultured endothelial cells treated with thrombin. Note that there is a loss of the DPB and an increase in central microfilament bundles (arrow). (x1500)

Wysolmerski et al.[152] showed that the cause of reversible pulmonary edema by ethchlorvynol may be related to the loss of the DPB which results in loss of endothelial integrity. Shasby et al. have proposed that the cytoskeleton is important in the regulation of endothelial permeability[154]. Recently, recombinant tumor necrosis factor and immune interferon caused human endothelial cells to rearrange their actin cytoskeleton so that the DPB was lost and central microfilament bundles became prominent[155]. Hyperoxia is associated with alterations in actin distribution of endothelial cells[156]. DPB become disrupted or lost and stress fibers are increased in number and thickness. Total actin remained the same while F-actin was increased. Thus, the DPB is sensitive to soluble factors which may in some cases, such as thrombin, be present at the site of endothelial injury. The nature of this effect and whether it is direct or indirect await further study.

REFERENCES

1. Ross, R. 1986. The pathogenesis of atherosclerosis, An update. N. Engl. J. Med. 314: 488-500.

2. Cotran, R.S., Gimbrone, M.A., Bevilacqua, M.P., Mandrick, D.L., Pober, J.S. 1986. Induction and detection of a human endothelial activation antigen in vivo. J. Exp. Med. 164: 661-666.

3. Nerem, R.M. and Cornhill, J.F. 1980. Hemodynamics and atherogenesis. Atherosclerosis 36: 151-157.

4. Dewey, C.F., Bussolari, S.R., Gimbrone, M.A. and Davies, P.F. 1981. The dynamic response of vascular endothelial cells to fluid shear stress. J. Biomech. Eng. 103: 177-185.

5. Langille, B.L. and Adamson, S.L. 1981. Relationship between blood flow direction and endothelial cell orientation at arterial branch sites in rabbits and mice. Circulation Res. 48: 481-488.

6. Kinlough-Rathbone, R.L., Packham, M.A. and Mustard, J.F. 1983. Vessel injury, platelet adherence, and platelet survival. Arteriosclerosis 3: 529-546.

7. Levy, R.I. 1981. Review: declining mortality in coronary heart disease. Arteriosclerosis 1: 312-325.

8. Stern, D.M., Drillings, M., Kisiel, W., et al. 1983. Activation of Factor IX bound to cultured bovine aortic endothelial cells. Proc. Natl. Acad. Sci. USA 31: 913-917.

9. Stern, D.M., Nawroth, P.P., Harris, K., Esmon, C.T. 1986. Cultured bovine aortic endothelial cells promote activated protein C-protein S-mediated inactivation of Factor Va. J. Biol. Chem. 261: 713-718.

10. Stern, D.M., Nawroth, P.P., Kisiel, W., et al. 1984. A coagulation pathway on bovine aortic segments leading to the generation of factor Va and thrombin. J. Clin. Invest. 74: 1910-1921.

11. Loskutoff, D.J. and Edgington, T.S. 1977. Synthesis of a fibrinolytic activator and inhibitor by endothelial cells, Proc. Natl. Acad. Sci. USA 74: 3903-3907.

12. Loskutoff, D.J., van Mourik, J.A., Erickson, L.A., Lawrence, D.A. 1983. Detection of an unusually stable fibrinolytic inhibitor produced by bovine endothelial cells. Proc. Natl. Acad. Sci. USA 80: 2956-2960.

13. Gerrity, R.C. 1981. The role of the monocyte in atherogenesis. II. Migration of foam cells from atherosclerotic lesions. Am. J. Path. 103: 191-200.

14. Faggiotto, A., Ross, R. and Hacker, L. 1984. Studies of hypercholesterolemia in the non-human primate. I. Changes that lead to fatty streak formation. Arteriosclerosis 4: 323-340.

14. Faggiotto, A. and Ross, R. 1984. Studies of hypercholesterolemia in the non-human primate. II. Fatty streak conversion to fibrous plaque. Arteriosclerosis 4: 341-356.

15. Bevilacqua, M.P., Pober, J.S., Wheller, M.E., Cotran, R.S. and Gimbrone, M.A. 1985. Interleukin-1 acts on cultured human vascular endothelium to increase adhesion of polymorphonuclear leukocytes, monocytes, and related leukocyte cell lines. J. Clin. Invest. 76: 2003-2011.

16. Jaffe, E.A. 1977. Endothelial cells and biology of factor VIII. N. Engl. J. Med. 296: 377-383.

17. Bevilacqua, M.P., Pober, J.S., Majeau, G.R., Cotran, R.S. and Gimbrone, M.A. 1984. Interleukin 1 (IL-1) induces biosynthesis and cell surface expression of procoagulants activity in human vascular endothelial cells. J. Exp. Med. 160: 618-623.

18. Stenflo, J. 1984. Structure and function of protein C. Semin. Thromb. Hemost. 10: 109-121.

19. Comp, D.C. and Esmon, C.T. 1981. Generation of fibrinolytic activity by infusion of activated protein C into dogs. J. Clin. Invest. 68: 1221-1228.

20. Weksler, B.B., Marcus, A.J. and Jaffe, E.A. 1977. Synthesis of prostaglandin I_2 (prostacyclin) by cultured human and bovine endothelial cells. Proc. Natl. Acad. Sci. USA 74: 3922-3926.

21. Libby, P., Ordovas, J.M., Auger, K.R., Robbins, H., Birinyi, L.K. and Dinarello, C.A. 1986. Endotoxin and tumor necrosis factor induce interleukin-1 gene expression in adult human vascular cells. Am. J. Path. 124: 179-186.

22. Pober, J.S. and Cotran, R.S. 1990. Cytokines and endothelial cell biology. Physiol. Rev. 70: 427-452.

23. Fox, P.L. and DiCorleto, P.E. 1984. Regulation of a production of a platelet derived growth factor-like protein by cultured bovine aortic endothelial cells. J. Cell Physiol. 121: 298-308.

24. Gospodarowicz, D. 1988. Molecular and developmental aspects of fibroblast growth factor, In: Biology of Growth Factors, Kudlow, J.E., MacLennan, D.H., Bernstein, A. and Gotlieb, A.I. (eds), Plenum, N.Y., pp. 23-39.

25. Ross, R., Glomset, J. and Harker, L. 1977. Response to injury and atherosclerosis. Am. J. Path. 86: 675-684.

26. Hansson, G.K. and Schwartz, S.M. 1983. Endothelial cell dysfunction without cell loss. In: Biochemical Interactions at the Endothelium, Cryer (ed), Elsevier Sciences Publishers, New York, pp. 343-361.

27. Taylor, K.E., Glagov, S. and Zarins, C.K. 1989. Presentation and structural adaptation of endothelium over experimental foam cell lesions: Quantative ultrastructural study, Arteriosclerosis 9: 881-894.

28. Ross, R., Glomset, J. and Harker, L. 1977. Response to injury and atherosclerosis. Am. J. Path. 86: 675-684.

29. Hansson, G.K. and Schwartz, S.M. 1983. Evidence of cell death in the vascular endothelium *in vivo* and *in vitro*. Am. J. Path. 112: 278-286.

30. Langille, B.L. 1984. Integrity of arterial endothelium following acute exposure to high shear stress. Biorheology 21: 333-346.

31. Langille, B.L., Reidy, M.A. and Klein, R.L. 1986. Injury and repair of endothelium at sites of flow disturbances near abdominal aortic coarctations in rabbits. Arteriosclerosis 6: 146-154.

32. Reidy, M.A. and Schwartz, S.M. 1983. Endothelial injury and regeneration. IV. Endotoxin: a nondenuding injury to aortic endothelium. Lab. Invest. 48: 25-34.

33. Schwartz, S.M. and Benditt, E.D. 1973. Cell replication in the aortic endothelium: A new method for study of the problem. Lab. Invest. 28: 699-707.

34. Flaherty, J.T., Peirce, J.E., Ferrans, V.J., Patel, D.J., Tucker, W.K. and Fry, D.L. 1972. Endothelial nuclear patterns in the canine arterial tree with particular reference to hemodynamic events. Circ. Res. 30: 23-33.

35. Baumgartner, H.R. 1963. Eine neue Methode zur Erzeugung von Thromben durch gezielte Ueberdehnung der Gefasswand. Z. ges. exp. Med. 137: 227-247.

36. Stemerman, M.B., Spaet, T.H., Pitlick, F., Cintron, J., Lejnioeks, I. and Tiell, M.L. 1977. Intimal healing: the pattern of reendothelialization and intimal thickening. Am. J. Path. 87: 125-137.

37. Schwartz, S.M., Haudenschild, C.C. and Eddy, E.M. 1978. Endothelial regeneration. I. Quantitative analysis of intimal stages of endothelial regeneration in rat aortic intima. Lab. Invest. 38: 568-579.

38. Haudenschild, C.C. and Schwartz, S.M. 1979. Endothelial regeneration. II. Restitution of endothelial continuity. Lab. Invest. 41: 407-418.

39. Fishman, J.A., Ryan, G.B. and Karnovsky, M.J. 1975. Endothelial regeneration in the rat carotid artery and the significance of endothelial denudation in the pathogenesis of myointimal thickening. Lab. Invest. 32: 339-351.

40. Reidy, M.A. and Silver, M. 1985. Endothelial regeneration. VII. Lack of intimal proliferation after defined injury to rat aorta. Am. J. Pathol. 118: 173-177.

41. Schwartz, S.M., Gajdusek, C.M., Reidy, M.A., Selden III, S.C. and Haudenschild, C.C. 1980. Maintenance of integrity in aortic endothelium. Fed. Proc. 39: 2618-2625.

42. Reidy, M.A. and Schwartz, S.M. Endothelial regeneration. III. Time course of intimal changes after small defined injury to rat aortic endothelium. Lab. Invest. 44: 301-306.

43. Ramsay, M.M., Walker, L.N. and Bowyer, D.E. 1982. Narrow superficial injury to rabbit aortic endothelium. Atherosclerosis 43: 233-243.

44. Hirsch, E.Z., Chisholm, G.M. and White, H. 1981. Endothelial regeneration and integrity in selectively denuded longitudinal tracks in thoracic aortas of rats. Fed. Proc. 40: 331.

45. Prescott, M.F. and Muller, K.R. 1983. Endothelial regeneration in hypertensive and genetically hypercholesterolemic rats. Arteriosclerosis 3: 206-214.

46. Jaffe, E.A., Nachman, R.L. and Becker, C.G. 1973. Culture of human endothelial cell derived from umbilical veins: identification of morphologic criteria. J. Clin. Invest. 52: 2745-2758.

47. Gimbrone, M.A., Cotran, R.S. and Folkman, J. 1974. Human vascular endothelial cells in culture: growth and DNA synthesis. J. Cell Biol. 60: 673-684.

48. Booyse, F.M., Sedlak, B.J. and Rafelson, M.E. 1975. Culture of arterial endothelial cells: characterization and growth of bovine aortic endothelial cells. Thromb. Diat. 34: 825-839.

49. Gottlieb, H. and Lalich, J.J. 1954. The occurrence of arteriosclerosis in the aorta of swine. Am. J. Path. 30: 851-853.

50. Scott, R.F., Thomas, W.A., Lee, W.M., Renier, J.M. and Florentin, R.A. 1979. Distribution of intimal smooth muscle cell masses and their relationship to early atherosclerosis in the abdominal aortas of young swine. Atherosclerosis 34: 291-301.

51. Sholley, M.M., Gimbrone, M.A. and Cotran, R.S. 1977. Cellular migration and replication in endothelial regeneration: A study using irradiated endothelial cultures, Lab. Invest. 36: 18-25.

52. Wall, R.T., Harker, L.A. and Striker, G.E. 1978. Human endothelial cell migration. Stimulation by released platelet factor. Lab Invest. 39: 523-529.

53. Thorgeirsson, G., Robertson Jr., A.L. and Cowan, D.H. 1979. Migration of human vascular endothelial and smooth muscle cells. Lab. Invest. 41: 51-62.

54. Gotlieb, A.I. and Spector, W. 1981. Migration into an experimental wound: a comparison of porcine aortic endothelial and smooth muscle cells and the effect of culture irradiation. Am. J. Path. 103: 271-282.

55. Madri, J.A., Pratt, B.M. and Yannariello-Brown, J. 1988. Matrix-driven cell size change modulates aortic endothelial cell proliferation and sheet migration. Am. J. Path. 132: 18-27.

56. Gotlieb, A.I. and Boden, P. 1984. Porcine aortic organ culture: A model to study the cellular response to vascular injury. In Vitro 20: 535-542.

57. Pederson, D.C. and Bowyer, D.E. 1985. Endothelial injury and healing *in vitro*: Studies using an organ culture system. Am. J. Path. 110: 264-272.

58. Buck, R.C. 1979. Contact guidance in the subendothelial space. Repair of rat aorta *in vitro*. Exp. Molec. Path. 31: 275-283.

59. Wong, M.K.K. and Gotlieb, A.I. 1984. *In vitro* reendothelialization of single cell wound: Role of microfilament bundles in rapid lamellipodia-mediated wound closure. Lab. Invest. 51: 75-81.

60. Heimark, R.L., Twardzik, D.R. and Schwartz, S. 1986. Inhibition of endothelial regeneration by type-beta transforming growth factor from platelet. Science 233: 1078-1080.

61. Sato, Y. and Rifkin, D.B. 1988. Autocrine activities of basic fibroblast growth factor: regulation of endothelial cell movement, plasminogen activator synthesis, and DNA synthesis. J. Cell Biol. 107: 1199-1205.

62. McNeil, P.L., Muthukrishnan, L., Warder, E. and D'Amore, P.A. 1989. Growth factors are released by mechanically wounded endothelial cells. J. Cell Biol. 109: 811-822.

63. DiCorleto, P.E., Gajdusek, C.M., Schwartz, S.M. and Ross, R. 1983. Biochemical properties of the endothelium-derived growth factor: comparison to other growth factors. J. Cell Physiol. 114: 339-345.

64. Herman, B. and Pledger, W.J. 1985. Platelet derived growth factor induced alteration in vinculin and actin distribution in BALB/C-3T3 cell. J. Cell Biol. 100:1031-1040.

65. Herman, B., Harrington, M.A., Olashaw, N.E. and Pledger, W.J. 1986. Identification of the cellular mechanisms responsible for platelet-derived growth factor induced alterations in cytoplasmic vinculin distribution. J. Cell Physiol. 126: 115-125.

66. Kalnins, V.I., Subrahmanyan, L. and Gotlieb, A.I. 1981. The reorganization of cytoskeletal fiber systems in spreading porcine endothelial cells in culture. Eur. J. Cell Biol. 24: 36-44.

67. Pollard, T.D. and Cooper J.A. 1988. Actin and actin binding proteins. A critical evaluation of mechanisms and functions. Ann. Rev. Biochem. 55: 987-1035.

68. Stossel, T.P., Chaponnier, C., Ezzel, R.M., Hartwig, J.H., Janmey, P.A., Kwiatkowski, D., Lind, S.E., Smith, D.B., Southwick, R.S., Yin, H.L. and Zaners, K.S. 1985. Nonmuscle actin-binding proteins. Ann. Rev. Cell Biol. 1: 353-402.

69. Singer, S.J., Ball, E.H., Geiger, B. and Chen, W.T. 1981. Immunolabeling studies of cytoskeletal associations in cultured cells. Symposium on Quantative Biology 46: 303-316.

70. Pollard, T.D., Selden, S.C. and Maupin, P. 1984. Interaction of actin filaments with microtubules. J. Cell Biol. 99: 33s-37s.

71. Euteneuer, U. and Schliwa, M. 1985. Evidence for an involvement of actin in the positioning and motility of centrosomes. J. Cell Biol. 101: 96-103.

72. Becker, C.G. and Murphy, G.E. 1969. Demonstration of contractile protein in endothelium and cells of the heart valve endocardium, intima, arteriosclerotic plaques and Aschoff bodies of rheumatic heart disease. Am. J. Path. 55: 1-.

73. Pollard, T.D. and Weihing, R.R. 1974. Actin and myosin and cell movement. CRC Crit. Rev. Biochem. 2:1.

74. Korn, E.D. 1983. Actin polymerization and its regulation by proteins from non-muscle cells. Physiol. Rev. 62: 672.

75. Howard, T.H. and Oresajo, C.O. 1985. The kinetics of chemotactic peptide-induced change in F-actin content, F-actin distribution, and the shape of neutrophils. J. Cell Biol. 101: 1078.

76. Gabbiani, G., Gabbiani, F., Heimark, R.L. and Schwartz, S.M. 1984. Organization of actin cytoskeleton during early endothelial regeneration in vitro. J. Cell Sci. 66: 39-50.

77. Willingham, M.C., Yamada, S.S., Davies, P.J.A., Rutherford, A.V., Gallo, M.G. and Pastan, I. 1981. Intracellular localization of actin in cultured fibroblasts by electron microscopic immunocytochemistry. J. Histochem. Cytochem. 29: 17-37.

78. Porter, K.R., Claude, A. and Fullam, E.F. 1945. A study of tissue culture cells by electron microscopy. J. Exp. Med. 81: 233.

79. Ishikawa, H., Bischoff, R. and Holtzer, H. 1969. Formation of arrowhead complexes with heavy meromyosin in a variety of cell types. J. Cell Biol. 43: 312-328.

80. Lazarides, E. and Weber, K. 1974. Actin antibody: the specific visualization of actin filaments in nonmuscle cells. Proc. Natl. Acad. Sci. USA 71: 2268-2272.

81. Small, J.V., Rinnethaler, G. and Hissen, H. 1981. Organization of actin meshworks in cultured cells: the leading edge. Symp. Quant. Biol. 46: 599-611.

82. Stossel, T.P. 1984. Contribution of actin to the structure of cytoplasmic matrix. J. Cell Biol. 99: 15s-21s.

83. Buckley, I.K. and Porter, K.R. 1967. Cytoplasmic fibrils in living cultured cells: a light and electron microscope study. Protoplasma 64: 349-380.

84. Byers, H.R. and Fujiwara, K.J. 1982. Stress fibers in cells in situ: Immunofluorescent visualization with anti-actin, anti-myosin, and anti-alpha-actinin. J. Cell Biol. 93: 804-811.

85. Gordon, W.E. 1978. Immunofluorescent and ultrastructural studies of "sarcomeric" units in stress fibers of cultured non-muscle cells. Exp. Cell Res. 117: 253-260.

86. Sanger, J.W., Sanger, J.M. and Jockusch, B.M. 1983. Differences in the stress fibers between fibroblast and epithelial cells. J. Cell Biol. 96: 961-969.

87. Lazarides, E. and Burridge, K. 1975. Alpha-actinin; immunofluorescent localization of a muscle structural protein in non-muscle cells. Cell 6: 289-298.

88. Weber, K. and Groeschel-Stewart, U. 1974. Antibody to myosin: the specific visualization of myosin containing filaments in non-muscle cells. Proc. Natl. Acad. Sci. USA 71: 4561-4564.

89. Lazarides, E. 1975. Tropomyosin antibody: the specific localization of tropomyosin in non muscle cells. J. Cell Biol. 65: 549-561.

90. Burridge, K. 1981. Are stress fibers contractile? Nature London 294: 691.

91. Burridge, K. and Connel, L. 1983. A cytoskeletal component concentrated in adhesion plaques and other sites of actin membrane inteactions. Cell Motility 516: 405-418.

92. Harris, A.K., Stopak, D. and Wild, P. 1981. Fibroblast traction as a mechanism for collagen morphogenesis. Nature London 290: 249.

93. Stossel, T.P. 1984. Contribution of actin to the structure of the cytoplasmic matrix. J. Cell Biol. 99: 15s-21s.

94. Kreis, T.E. and Birchmeier, W. 1980. Stress fiber sarcomeres of fibroblasts are contractile. Cell 22: 555-561.

95. Singer, I.I. 1982. Association of fibronectin and vinculin with focal contacts and stress fibers in stationary hamster fibroblasts. J. Cell Biol. 92: 398-408.

96. Harris, A.K., Wild, P. and Stopak, D. 1980. Silicone rubber substrata: A new wrinkle in the study of cell locomotion. Science 208: 177-179.

97. Bradley, R.A., Couchman, J.R. and Rees, D.A. 1980. Comparison of the cell cytoskeleton in migratory and stationary chick fibroblast. J. Muscle Res. Cell Motil. 1: 5-14.

98. Herman, I.M., Crisona, N.J. and Pollard, T.D. 1981. Relation between cell activity and the distribution of cytoplasmic actin and myosin. J. Cell Biol. 90: 84-91.

99. Gabbiani, G., Badonnel, M.C. and Roma, G. 1975. Cytoplasmic contractile apparatus in aortic endothelial cells of hypertensive rats. Lab. Invest. 32: 227-234.

100. Gabbiani, G., Elmer, G., Geulpa, C., Vallotton, M.B., Badonnel, M.C. and Huttner, I. 1979. Morphologic and functional changes of the aortic intima during experimental hypertension. Am. J. Path. 96: 339-342.

101. Wong, A.J., Pollard, T.D. and Herman, I.M. 1983. Actin filament stress fibers in vascular endothelial cells in vivo. Science 219: 867-869.

102. White, G.E., Gimbrone, M.A. and Fujiwara, K. 1983. Factors influencing the expression of stress fibers in vascular endothelial cells *in situ*. J. Cell Biol. 97: 416-424.

103. Rogers, K.A. and Kalnins, V.I. 1983. A method for examining the endothelial cytoskeleton *in situ* using immunofluorescence. J. Histochem. Cytochem. 31: 1217-1320.

104. Barak, L.S., Yocum, R.R., Nothnagel, E.A. and Webb, W.W. 1980. Fluorescence staining of the actin cytoskeleton in living cells with 7-nitrobenz-2-oxa-1,1-diazole-phallicidin, Proc. Natl. Acad. Sci. USA 77: 980-984.

105. Kim, D.W., Langille, B.L., Wong, M.K.K. and Gotlieb, A.I. 1989. Patterns of endothelial microfilament distribution in the rabbit aorta *in situ*. Circ. Res. 64: 21-31.

106. Kim, D.W., Gotlieb, A.I. and Langille, B.L. 1989. *In vivo* modulation of endothelial F-actin microfilaments by experimental alterations in shear stress. Arteriosclerosis 9: 439-445.

106a. Yost, J.C. and Herman, I.M. 1988. Age-related and site-specific adaptation of the arterial endothelial cytoskeleton during atherogenesis. Am. J. Path. 30: 595-604.

107. Gotlieb, A.I., Spector, W., Wong, M.K.K. and Lacey, C. 1984. *In vitro* reendothelialization: microfilament bundle reorganization in migrating porcine endothelial cells. Arteriosclerosis 4: 91-96.

108. Wong, M.K.K. and Gotlieb, A.I. 1984. *In vitro* reendothelialization of single cell wound: Role of microfilament bundles in rapid lamellipodia-mediated wound closure. Lab. Invest. 51: 75-81.

109. Wong, M.K.K. and Gotlieb, A.I. 1986. Endothelial cell monolayer integrity. I. Characterization of dense peripheral band of microfilaments. Arteriosclerosis 6: 212-219.

110. Geiger, B., Schmid, E. and Franke, W.W. 1983. Spatial distribution of proteins specific of desmosomes and adherens junctions in epithelial cells demonstrated by double immunofluorescence microscopy. Differentiation 23: 189-205.

111. Huttner, I., Boutet, M. and More, R.H. 1973. Studies on protein passage through arterial endothelium. Lab. Invest. 28: 672-677.

112. Muller, W.A., Ratti, C.M., McDonnell, S.L. and Cohn, Z.A. 1989. A human endothelial cell-restricted, externally disposed plasmalemmal protein enriched in intercellular junctions. J. Exp. Med. 170: 399-414.

113. Kirschner, M.N. 1978. Microtubule assembly in neucleation, Int. Rev. Cytol. 54: 1-71.

114. Osborn, M. and Weber, K. 1976. Cytoplasmic microtubules in tissue culture cells appear to grow from an organizing structure towards the plasma membrane. Proc. Natl. Acad. Sci. U.S.A. 73: 867-871.

115. Porter, K.R. 1966. Cytoplasmic microtubules and their functions. In: Principles of Biomolecular Organization, Wolstenholme, G.E.W. and O'Connor, M. (eds), Little Brown, and Company, Boston, pp. 308-356.

116. Raff, E.D. 1979. The control of microtubule assembly *in vivo*. Int. Rev. Cytol. 59: 1-96.

117. Badley, R.A., Couchman, J.R. and Rees, D.A. 1980. Comparison of the cell cyto-skeleton in migratory and stationary chick fibroblasts. J. Mus. Res. Cell Motility 1: 5-14.

118. Vasiliev, J.M. and Gelfand, I.M. 1976. Effects of colcemid on morphogenetic processes and locomotion of fibroblasts in cell motility. In: Cold Spring Harbor Conference on Cell Proliferation, Goldman, R., Pollard, T. and Rosenbaum, J. (eds), Cold Spring Harbor, N.Y., pp. 279-304.

119. Goldman, R.D. 1971. The role of three cytoplasmic fibers in BKH-21 cell motility. I. Microtubules and the effects of colchicine. J. Cell Biol. 51: 752-762.

120. Gail, M.H. and Boone, C.W. 1971. Effect of colcemid on fibroblast motility. Exp. Cell Res. 65: 221-227.

121. Cheung, H.T., Contarow, W.D. and Sundharadas, G. 1978. Colchicine and cytochalasin B (CB) effects on random movement, spreading and adhesion of mouse macrophages. Exp. Cell Res. 111: 95-103.

122. Bhisey, A.N. and Freed, J.J. 1971. Ameboid movement induced in cultured macrophages by colchicine or vinblastine. Exp. Cell Res. 64: 419-429.

123. Malech, H.L., Root, R.K. and Gallin, J.I. 1977. Structural analysis of human neutrophil migration. J. Cell Biol. 75: 666-693.

124. Sameshima, M., Imai, Y. and Hashimoto, Y. 1988. The position of the microtubule-organizing center relative to the nucleus is independent of the direction of cell migration in Dictyostelium discoideum. Cell Motility Cytoskeleton 9: 111-116.

125. Franke, W.W., Schmid, E., Osborn, M. and Weber, K. 1979. Intermediate-sized filaments of human endothelial cells. J. Cell Biol. 81: 570-580.

126. Blose, S.H. and Meltzer, D.I. 1981. Visualization of the 10 nm filament vimentin rings in vascular endothelial cells in situ: close resemblance to vimentin cytoskeletons found in monolayers in vitro. Exptl. Cell Res. 135: 299-309.

127. Lazarides, E. 1980. Intermediate filaments as mechanical integrators of cellular shape. Nature 283: 249-258.

128. Lin, J.J.C. and Feramisco, J.R. 1981. Disruption of the in vivo distribution of the intermediate filaments in fibroblasts through the microinjection of a specific monoclonal antibody. Cell 24: 185-193.

129. Gawlitta, W., Osborn, M. and Weber, K. 1981. Filaments induced by microinjection of vimentin specific antibody does not interfere with locomotion and mitosis. Eur. J. Cell Biol. 26: 83-90.

130. Albrecht-Buehler, G. 1977. Phagokinetic tracks of 3T3 cells: parallels between the orientation of track segments and cellular structures which contain actin or tubulin. Cell 12: 333-339.

131. Gotlieb, A.I., McBurnie-May, L., Subrahmanyan, L. and Kalnins, V.I. 1981. Distribution of microtubule organizing centers in migrating sheets of endothelial cells. J. Cell Biol. 91: 589-594.

132. Gotlieb, A.I., Subrahmanyan, L. and Kalnins, V.I. 1983. Microtubule organizing centers and cell migration. Effect of inhibition of migration and microtubule disruption in endothelial culture. J. Cell Biol. 96: 1266-1272.

133. Rogers, K.A. and Kalnins, V.I. 1983. Comparison of the cytoskeleton in aortic endothelial cells in situ and in vitro. Lab. Invest. 49: 650-654.

134. Rogers, K.A., Boden, P., Kalnins, V.I. and Gotlieb, A.I. 1986. The distribution of centrosomes in endothelial cells of non-wounded and wounded aortic organ cultures. Cell Tissue Res. 243: 223-277.

135. Rogers, K.A. McKee, N. and Kalnins, V.I. 1985. The preferential orientation of centrioles towards the heart in endothelial cells of major blood vessels is reestablished following reversal of a segment. Proc. Natl. Acad. Sci. USA 82: 3272-3276.

136. Mascardo, R.N. and Sherline, P. 1984. Insulin and multiplication-stimulating activity induce a very rapid centrosomal orientation response to wounding in endothelial cell monolayers. Diabetes 33: 1099-1105.

137. Mascardo, R.N. 1988. The effects of hyperglycemia on the directed migration of wounded endothelial cell monolayers. Metabolism 37: 378-385.

138. Kupfer, A., Louvard, D. and Singer, S.J. 1982. Polarization of the Golgi apparatus and the microtubule organizing center in cultured fibroblasts at an edge of an experimental wound. Proc. Natl. Acad. Sci. USA 79: 2603-2607.

139. Kupfer, A., Dennert G. and Singer, S.J. 1983. Polarization of the Golgi apparatus and the microtubule-organizing center within cloned natural killer cells bound to their targets. Proc. Natl. Acad. Sci. USA 80: 7224-7228.

140. Gotlieb, A.I. and Wong, M.K.K. 1988. Current concepts on the role of the endothelial cytoskeleton in endothelial integrity, repair, and dysfunction, In: Endothelial Cells, Volume II, Ryan, U. (ed), CRC Press Inc., Boca Raton, FL, pp. 81-101.

141. Larson, D.M. and Sheridan, J.D. 1982. Intercellular junctions and transfer of small molecules in primary vascular endothelial cultures. J. Cell Biol. 92: 183-191.

142. Larson, D.M. and Haudenschild, C.C. 1988. Junctional transfer in wounded cultures of bovine aortic endothelial cells. Lab. Invest. 59: 373-379.

143. Wong, A.J., Pollard, T.D. and Herman, I.M. 1983. Actin filament stress fibers in vascular endothelial cells in vivo. Science 219: 867-869.

144. Gabbiani, G., Gabbiani, F., Lombardi, D. and Schwartz, S.M. 1983. Organization of actin cytoskeleton in normal and regeneration arterial endothelial cells. Proc. Natl. Acad. Sci. USA 80: 2361-2364.

145. Rogers, K.A., Sandig, M., McKee, N. and Kalnins, V.I. 1989. The distribution of microfilament bundles in rabbit endothelial cells in the intact aorta and during wound healing in situ. Biochem. Cell Biol. 67: 553-562.

146. Franke, R.P., Grafe, M., Schmittler, H., Seiffge, D., Mittermayer, C. and Drenckhahn, D. 1984. Induction of human vascular endothelial stress fibers by fliud shear stress. Nature 307: 648-649.

147. Herman, I.M., Brant, A.M., Warty, V.S., Bonaccorso, J., Klein, E.C., Kormos, R.L. and Borovetz, H.S. 1987. Hemodynamics and the vascular endothelial cytoskeleton. J. Cell Biol. 105: 291-302.

148. Huttner I., Walker, C. and Gabbiani, G. 1985. Aortic endothelial cell during regeneration. Remodeling of cell junctions, stress fibers, and stress fiber-membrane attachment domains. Lab. Invest. 53: 287-302.

149. Baja, F., Blatter, M.C., James, R.W., Pometta, D. and Gabbiani, G. 1989. Actin stress fiber content of regenerated endothelial cells correlates with intramural retention of intermediate plus low density lipoproteins in rat aorta after balloon injury. Atherosclerosis 76: 181-191.

150. Wong, M.K.K. and Gotlieb, A.I. 1988. The reorganization of microfilaments centrosomes, and microtubules during in vitro small wound reendothelialization. J. Cell Biol. 107: 1777-1783.

151. Wong, M.K.K. and Gotlieb, A.I. 1990. Endothelial monolayer integrity. Perturbation of F-actin filaments and the dense peripheral band-vinculin network. Arteriosclerosis 10: 76-84.

152. Wysolmerski, R., Lagunoff, D. and Dahms, T. 1984. Ethchlorvynol-induced pulmonary edema in rats. An ultrastructural study. Am. J. Path. 115: 447-457.

153. Wysolmerski, R.B. and Lagunoff, D. 1988. Inhibition of endothelial cell retraction by ATP depletion. Am. J. Path. 132: 28-37.

154. Shasby, D.M., Shasby, S.S., Sullivan J.M. and Peach, M.J. 1982. Role of endothelial cell cytoskeleton in control of endothelial permeability. Circ. Res. 51: 657-661.

155. Stolpen, A.H., Guinan, E.C., Fiers, W. and Pober, J.S. 1986. Recombinant tumor necrosis factor and immune interferon act singly and in combination to reorganize human vascular endothelial cell monolayers. Am. J. Path. 123: 16-24.

156. Phillips, P.G., Higgins, P.J., Malik, A.B. and Tsan, M.-F. 1988. Effect of hyperoxia on the cytoarchitecture of cultured endothelial cells. Am. J. Path. 132: 59-72.

THROMBIN-STIMULATED ENDOTHELIAL CELL FUNCTIONS:

MONOCYTE ADHESION AND PDGF PRODUCTION

Paul E. DiCorleto, Carol de la Motte
and Ravi Shankar

Department of Vascular Cell Biology
and Atherosclerosis
Cleveland Clinic Research Institute
Cleveland, OH 44195

INTRODUCTION

In the past decade there has been much interest in the concept of an activated or injured endothelium that exhibits properties distinct from healthy, adult endothelium. The activated state of the endothelial cell may result from the action of cytokines, as recently reviewed in detail by Pober and Cotran[1]. Alternatively, injury to the endothelium may shorten endothelial cell lifetime, causing increased turnover in specific regions of the artery. This may, in turn, lead to the expression of genes which are suppressed under physiological rather than pathological conditions, even in the absence of exogenous stimulators. Extrapolation of *in vitro* findings has suggested to us a possible role of dysfunctional endothelium in the development of the atherosclerotic plaque[2]. Activated endothelial cells may (1) express binding sites for monocytes and perhaps secrete monocyte activators; (2) secrete oxygen free radicals that can modify nearby low density lipoprotein (LDL); (3) act as a procoagulant rather than an anti-coagulant surface, and (4) synthesize and secrete a PDGF-like protein and/or other mitogens and chemoattractants for medial smooth muscle cells.

We have recently focussed our attention on the role of thrombin as a possible generator of the activated state of the endothelium. Thrombin is a coagulation system protease, as well as a potent platelet aggregating substance, which would be expected to be present at sites of vascular injury. Thrombin is known to bind to specific receptor sites on endothelial cells and to stimulate diverse functions in these cells, examples of which are an increase in prostacyclin production[3,4], the release of von Willebrand factor[5], an increase in expression of the platelet-derived growth factor genes[6,7] and an increase in neutrophil binding to the endothelial cell surface[8,9]. We have recently reported that thrombin is also capable of stimulating monocyte adhesion to the endothelium *in vitro*[10].

We have been interested in defining the intracellular pathways employed by thrombin in its induction of monocyte adhesion and PDGF production. This protease is known to activate multiple signalling systems. Thrombin causes a transient increase in cytosolic calcium in human arterial and umbilical vein endothelial cells[11,12]. The production of specific inositol phosphates is also increased in endothelial cells in response to this protease[13,14], and increased levels of diacylglycerol have been observed in thrombin-treated fibroblasts[15,16]. The role of these various signalling systems in thrombin-induced monocyte adhesion and PDGF production has not been well-established.

Atherosclerosis, Edited by A. I. Gotlieb *et al.*
Plenum Press, New York, 1991

Adhesion of Monocytes to the Endothelium

The monocyte-derived macrophage has been implicated as a participant in several aspects of atherosclerotic plaque development. Macrophages evolve into foam cells through unregulated ingestion of lipid. They also have the capacity to produce growth factor(s) for vascular smooth muscle cells and to generate cytotoxic factors for neighboring cells leading to necrosis. The initial event in the interaction of a monocyte with the vessel wall is its attachment to the endothelium. The adherence of mononuclear cells to the endothelial cells of large vessels during the early stages of experimentally-induced atherosclerosis has been observed in multiple animal models[17-20]. In each of these systems a distinctly focal adhesion of monocytes was observed as an early event following cholesterol feeding. The mechanism by which hypercholesterolemia causes increased adherence of monocytes to the endothelium and the reason why monocytes adhere to specific regions of the vessel wall remain unknown.

Experiments with cultured endothelial cells have led to several conclusions about monocyte adhesion that appear to be applicable to multiple vascular beds in several species. First, the expression of monocyte binding sites on the endothelial cell surface is low under "control" or "physiological" conditions, i.e., when the endothelial cells are confluent, quiescent, and unstimulated by exogenous agents. Secondly, when treated with specific exogenous agents[10,21-23] or when stimulated to migrate and divide in response to an *in vitro* "wound"[24], endothelial cells exhibit a greatly increased ability to bind monocytes. Thirdly, though the above mentioned stimulators also induce increased neutrophil adhesion to endothelial cells, the cell surface binding sites on the endothelial cells may be specific for one or the other of these two leukocytes[25]. Thus, monocyte and neutrophil adhesion to endothelial cells may be regulated through distinct pathways and induced differentially by certain, as yet undefined, exogenous agents.

Thrombin-induced Monocyte Adhesion

Thrombin has been shown to stimulate the adhesion of neutrophils to vascular endothelial cells in a rapid process that may involve the generation of platelet-activating factor[8]. We tested whether α-thrombin was also capable of altering adhesion of peripheral blood monocytes and the human monocytic cell line, U937 cells, to cultured endothelial cells[10]. Human, bovine and porcine endothelial cells, exhibited a 3 to 13-fold increase in adhesion of monocytic cells when treated with bovine α-thrombin. A half-maximal stimulation of monocytic cell adhesion to porcine endothelial cells occurred at 15 units/ml. The thrombin stimulation occurred in the presence and absence of serum suggesting that thrombin neither requires association with a serum protein to be active nor is it inhibited in its action by serum proteins. A concentration-response curve for thrombin stimulation of human aortic endothelial cells to bind U937 cells and produce PDGF is shown in Figure 1.

The time required for thrombin to induce maximal monocytic cell adhesion was found to be substantially longer (6 h) than the time reported by others to be required for thrombin-induced neutrophil binding (5 to 60 min)[8,9]. The stimulated level of monocyte binding to porcine aortic endothelial cells persisted through 24 h; whereas, thrombin-stimulated neutrophil binding has been reported to decay within hours of treatment[8,9]. The time required to attain maximal monocytic cell adhesion in response to thrombin was also longer than that required by other activators of monocytic cell adhesion; in our system interleukin-1, lipopolysaccharide and phorbol esters induced maximal adhesion in 4 h (Figure 2). We investigated the possibility that thrombin was acting indirectly through the induction of an autocrine factor which induced monocyte adhesion sites. Conditioned media from thrombin-treated endothelial cells did not stimulate control endothelial cells which were specifically blocked from thrombin stimulation by prior treatment with heparin; thus, we were unable to detect an autocrine inducer, though the possibility still exists.

The proteolytic activity of thrombin is known to be required for many of its induced changes in platelet and endothelial cell function. We tested whether this enzymatic activity was required for stimulation of monocytic cell adhesion to

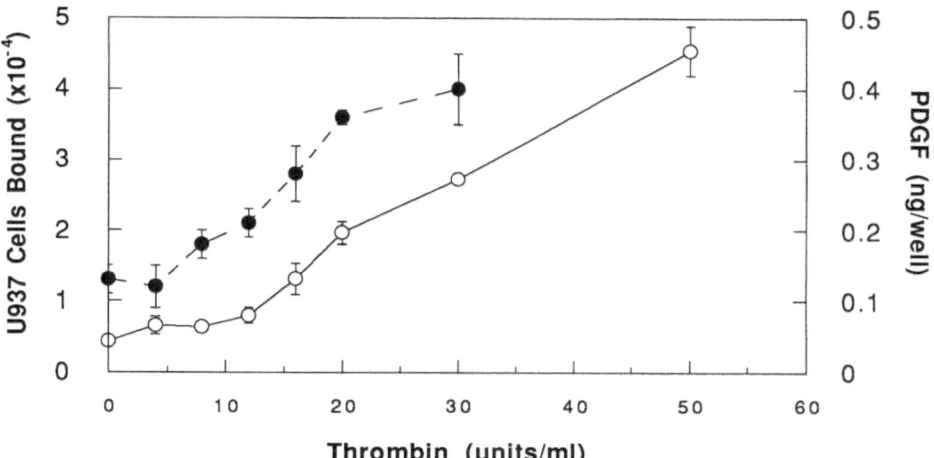

Figure 1. Thrombin stimulation of U937 cell adhesion to, and PDGF production by, human aortic EC. Confluent human aortic EC were incubated with varying concentrations of bovine thrombin in MCDB 107 + 5% FBS and incubated for 6 h at 37°C before performing the U937 cell adhesion assay (o) or 8 h at 37°C before removing the supernatant for PDGF quantitation (●). Monocyte adhesion and PDGF production were determined as described[37].

endothelial cells[6]. Phenylmethylsulfonyl fluoride (PMSF)-treated thrombin exhibited a stimulatory activity that was proportionate to its reduced level of proteolytic activity and γ-thrombin, which is catalytically inactive, did not stimulate monocyte adhesion suggesting the requirement for a proteolytic cleavage in the process of cell signalling. Similar results have been obtained by others for thrombin-induced PDGFc production[6]. One must consider the possibility, however, that the PMSF-treated thrombin is unable to interact with its specific cell surface receptor and that the inhibition of thrombin action is due to this rather than the absence of proteolytic activity.

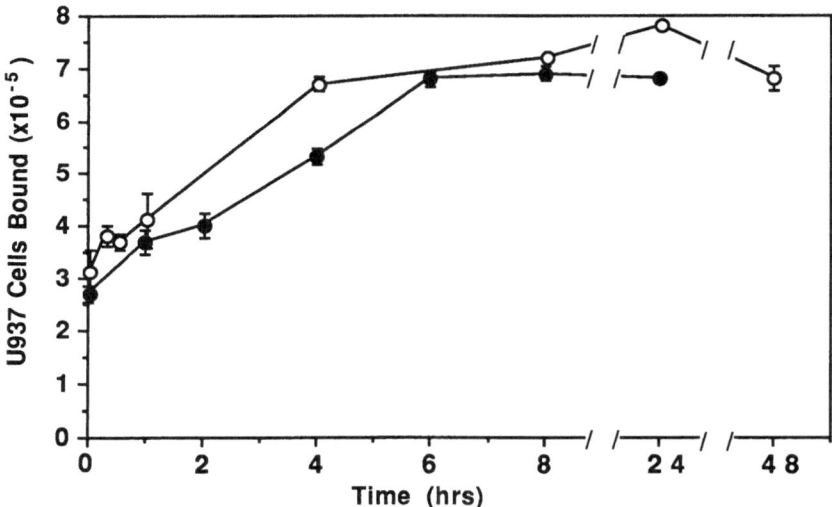

Figure 2. Time course of thrombin and PMA stimulation of EC to bind U937 cells. Confluent porcine aortic EC were incubated with bovine thrombin (●) (25 units/ml) or PMA (o) (10 nM) in DME/F12 + 5% FBS at staggered times before assay for U937 cell adhesion. Methods are as described[37]. This figure has been previously published[10].

Since platelet-activating factor (PAF) production by endothelial cells is stimulated by thrombin and since PAF has been implicated in stimulated neutrophil adhesion to endothelial cells[8,26], we investigated the effect of PAF on monocytic cell adhesion to endothelial cells. PAF, even at pharmacological concentrations (10^{-8} - 10^{-6} M) had no effect on the ability of endothelial cells to bind monocytic cells[10]. Thrombin is known as a potent stimulator of prostanoid production by endothelial cells[3,4]. Two other agents which stimulate cyclooxygenase activity, arachidonate (400 μM) and calcium ionophore A23187 (0.6 μM), had no effect on monocyte binding to endothelial cells and acetylsalicylic acid (100 mg/ml), which inhibits cyclooxygenase activity, did not affect the basal or thrombin-stimulated level of adhesion of monocytic cells to endothelial cells[10].

Protein Kinase C and Thrombin-Induced Monocyte Adhesion

Phorbol esters have been shown to cause increased adhesion of leukocytes to endothelial cells[23,27], suggesting a role for protein kinase C (PKC) in the stimulatory response. We tested the ability of PMA, as well as other known PKC activators, to stimulate monocytic cell adhesion to endothelial cells in our system. PMA, phorbol dibutyrate, and 1-oleoyl-2-acetyl-*rac*-glycerol stimulated endothelial cells to bind enhanced numbers of U937 cells; whereas, the non-activating phorbol had no effect in the assay. PMA stimulation was half-maximal at 1 nM and maximal at 10 nM, which is consistent with activation of PKC[10]. The time required for a PMA response was consistent with a role for PKC in thrombin stimulation of endothelial cells.

We investigated the effect of the PKC inhibitor H7, and its structural analogue HA1004, which is a more potent inhibitor of cAMP- and cGMP-dependent protein kinases. The inhibitor H7 caused nearly complete inhibition of not only PMA and thrombin-stimulated adhesion of U937 cell adhesion to endothelial cells but also adhesion stimulated by IL-1, PMA, and LPS[10]. HA1004, on the other hand, had no effect on the stimulation of monocyte adhesion by any of the activators. PKC activation therefore appears to represent a common step in the induction of monocyte binding sites by very different stimuli. The ability of PMA and diacylglycerol to stimulate monocyte binding to endothelial cells also suggests that inositol phosphate production and release of intracellular calcium stores, which are induced in endothelial cells in response to thrombin, are not involved in the monocyte adhesion pathway. Brock and Capasso have demonstrated in the same endothelial cell culture system used in our studies that PKC activators block both the rapid, transient increase in cytosolic calcium and the rise in inositol phosphate levels that are generated in response to thrombin[28]. The PKC inhibitor had no effect on binding of monocytes to sparse, unstimulated endothelial cells or to endothelial cells responding to an *in vitro* wound.

Thrombin-Stimulated PDGF Production by Endothelial Cells

Endothelial cells were the first vascular cells (other than the platelet) shown to secrete PDGF[29]. Endothelial cells cultured from all major vessels, and from all species examined to date, secrete PDGF into their medium. The process is constitutive, that is, exogenous stimulation of the cells is not required; however, there is evidence that factors related to arterial injury further stimulate the release of the growth factor. Several agents that mortally injure cultured bovine aortic endothelial cells, including bacterial lipopolysaccharide, cause the release of a large burst of PDGF, presumably from dying cells[30]. Certain members of the coagulation cascade, including thrombin and Factor Xa, increase the rate of synthesis of PDGF by cultured endothelial cells by up to 10-fold[6,31]. The only transmembrane signal that has been shown to influence thrombin-stimulated PDGF production is the activation of adenyl cyclase[7,32-34]. However, this pathway is inhibitory for thrombin-induced PDGF production and no evidence exists for the activation of adenyl cyclase by thrombin.

We have initiated studies to define the intracellular pathways employed by thrombin to stimulate PDGF production with a goal of determining whether this cellular endpoint is differentially   regulated from thrombin-induced monocyte adhesion. Higher levels of thrombin are required to stimulate monocyte adhesion than PDGF production in replicate cultures of human aortic endothelial cells (Figure 1). This consistent finding with multiple isolates of endothelial cells would suggest that different thrombin receptor classes or different coupling systems are required for the stimulation of the two cellular functions. Secondly, we have observed independent isolates of human aortic endothelial cells which are stimulated by thrombin to induce only one of the two endothelial cells functions which is further evidence that the thrombin receptor can be uncoupled from either pathway.

Activation of the Na^+/H^+ antiporter and phospholipase C through receptor-G protein coupling are two important transcellular signals elicited by thrombin in several cell lines including endothelial cells. We have examined whether thrombin-induced PDGF production by human aortic endothelial cells or monocyte adhesion to these endothelial cells is regulated by a G protein and/or the Na^+/H^+ antiporter. This antiporter is a non-electrogenic transport system which is widespread in the plasmalemma of eukaryotic cells and plays a pivotal role in cellular regulation through modulation of intracellular pH (pH_i) and cell volume[35]. Thrombin is believed to cause an increase in pH_i in endothelial cells via the activation of this antiporter which is considered a mediator of some of thrombin's cellular effects such as platelet-activating factor production[36]. Intracellular pH measurements with the fluorescent dye 2'7' bis (carboxyethyl) - 5(6') - carboxyfluorescein not only substantiate this claim but also show that thrombin-induced increases in pH_i can be blocked either by amiloride and its analogues or by excluding extracellular Na^+ [35,36]. Amiloride and its analogues are believed to be competitive inhibitors with respect to extracellular Na^+ [35]. We have found that inhibition of the Na^+/H^+ antiporter by either amiloride and its analogues or by depletion of extracellular sodium completely blocks thrombin induction of PDGF production without affecting thrombin-induced monocyte adhesion[37,38] (Figure 3).

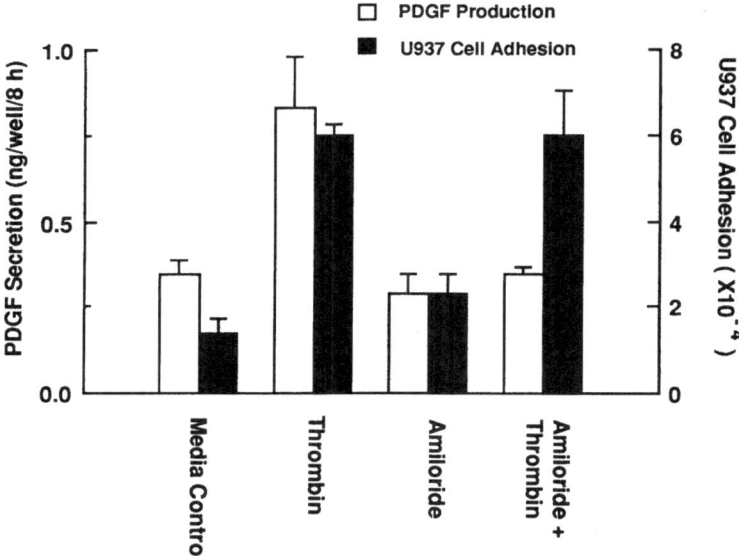

Figure 3. Effect of amiloride on basal and thrombin-stimulated PDGF production by, and monocyte adhesion to, human aortic endothelial cells. Confluent human aortic EC were preincubated with or without amiloride (100 μM) for 30 min at 37°C. Appropriate concentrations of bovine α-thrombin (12 U/ml for PDGF stimulation and 30 U/ml for monocyte adhesion) were added to the wells and the incubation continued for a further 6 h for monocyte adhesion and 8 h for PDGF production. Monocyte adhesion and PDGF production were determined as described[37].

Many macromolecular agonists achieve their cellular effects through coupling cell surface receptors to second messenger systems via a G protein. The activation of adenyl cyclase, phospholipase C and ion channels are some examples of such G protein-coupled effector systems. Evidence for receptor-coupled G protein activation is emerging for many agonist-induced endothelial cell functions. Furthermore, the susceptibility of these functions to specific toxins indicates that distinct G proteins are associated with different agonists. While bradykinin-induced phosphoinositide turnover in endothelial cells is insensitive to pertussis toxin[39], both ATP-stimulated prostacyclin production[40] and histamine-stimulated phosphatidylinositol turnover[39] are pertussis toxin-sensitive. Brock and Capasso[41] have recently demonstrated that thrombin stimulates phosphatidylinositol turnover through a pertussis toxin-insensitive G protein in permeabilized human endothelial cells. While our results have demonstrated that PDGF production and monocyte adhesion are differentially regulated by a Na^+/H^+ antiporter dependent-process, thrombin stimulation of both processes appeared to involve a pertussis toxin-insensitive G protein[37,38]. Thrombin induced a substantial increase in 35S-GTPγS binding to human aortic endothelial cell membranes compared to untreated membranes. In addition, GTPγS, when added to a suboptimal concentration of thrombin, induced both monocyte adhesion and PDGF production to levels similar to those at maximal stimulation by thrombin. This stimulatory effect of GTPγS on PDGF production appeared to be transcriptionally regulated as indicated by an increase in the steady state mRNA levels of both PDGF-A and -B chains.

SUMMARY

The endothelium is susceptible to stimulation by multiple humoral and paracrine factors including cytokines and lipopolysaccharide. In addition, the coagulation system protease thrombin, a molecule of physiological and pathological relevance at sites of injury, has the capacity to trigger dramatic changes in endothelial cell function. Two thrombin-stimulated processes which are of direct pertinence to the development of the atherosclerotic plaque are the expression of PDGF genes and the expression of leukocyte adhesion molecules. We have observed that a pertussis toxin-insensitive G protein is important in transducing signals involved in thrombin-stimulated PDGF production and monocyte adhesion in human aortic endothelial cells. The activation of a thrombin-stimulated amiloride-sensitive Na^+/H^+ antiporter, on the other hand, plays a role only in PDGF production. The differential regulation of multiple signals generated in endothelial cells by a single agonist may govern the specific response of the cell to its environment. For example, endothelial cells may preferentially respond to thrombin *in vivo* by increasing monocyte adhesion sites under circumstances when the pH_i may not be conducive for the activation of Na^+/H^+ antiporter. Under different environmental conditions PDGF gene expression may accompany the induction of monocyte adhesion sites. Thus, specific paracrine functions of the endothelial cells may be activated temporally to catalyze such processes as wound healing, inflammation and atherosclerosis.

ACKNOWLEDGEMENTS

These studies were supported in part by NIH grants HL29582 and HL34727, and by a post doctoral fellowship grant to R.S. from the American Heart Association, Northeast Ohio Affiliate. P.E.D. is the recipient of a Research Career Development Award (HL1561) from the NIH.

REFERENCES

1.   Pober, J.S. and Cotran, R.S. 1990. Cytokines and endothelial cell biology. Physiol. Rev. 70: 427-451.

2.   DiCorleto, P.E. and Chisolm, G.M. 1986. Participation of the endothelium in the development of the atherosclerotic plaque. Prog. Lipid Res. 25: 365-374.

3.   Weksler, B.B., Ley C.W. and Jaffe, E.A. 1978. Stimulation of endothelial cell prostacyclin production by thrombin, trypsin and the ionophore A23187. J. Clin. Invest. 62: 923-930.

4.   Glassberg, M.K., Bern, M.M., Coughlin, S.R., Haudenschild, C.C., Hoyer, L.W., Antoniades, H.N. and Zetter, B.R. 1982. Cultured endothelial cells derived from the human iliac arteries. In Vitro 10: 859-866.

5.   de Groot, P.G., Gonsalves, M.D., Loesberg, C., van Buul-Wortelboer, M.F., van Aken, W.G. and van Mourik, J.A. 1984. Thrombin-induced release of von Willebrand factor from endothelial cells is mediated by phospholipid methylation. J. Biol. Chem. 259: 13329-13333.

6.   Harlan, J.M., Thompson, P.J., Ross, R.R. and Bowen-Pope, D.F. 1986. α-Thrombin induces release of platelet-derived growth factor-like molecules by cultured human endothelial cells. J. Cell Biol. 103: 1129-1133.

7.   Daniel, T.O., Gibbs, V.C., Milfay, D.F. and Williams, L.T. 1987. Agents that increase cAMP accumulation block endothelial c-sis induction by thrombin and transforming growth factor ß. J. Biol. Chem. 262: 11893-11896.

8.   Zimmerman, G.A., McIntyre, T.M. and Prescott, S.M. 1985. Thrombin stimulates the adherence of neutrophils to human endothelial cells in vitro. J. Clin. Invest. 76: 2235-2246.

9.   Bizios, R., Lai, L.C., Cooper, J.A., Del Vecchio, P.J. and Malik, A.B. 1988. Thrombin-induced adherence of neutrophils to cultured endothelial monolayers: Increased endothelial adhesiveness. J. Cell Physiol. 134: 275-280.

10.   DiCorleto, P.E. and de la Motte, C.A. 1989. Thrombin causes increased monocytic cell adhesion to endothelial cells through a protein kinase C-dependent pathway. Biochem. J. 264: 71-77.

11.   Hallam, T.J., Pearson, J.D. and Needham, L.A. 1988. Thrombin-stimulated elevation of human endothelial cell cytoplasmic free calcium concentration causes prostacyclin production. Biochem. J. 251: 243-249.

12.   Ryan, U.S., Avdonin, P.V., Posin, E.Y.A., Popov, E.G., Danilo, S.M. and Tkachuk, V.A. 1988. Influence of vasoactive agents on cytoplasmic free calcium concentration in INDO-1 loaded vascular endothelial cells. J. Appl. Physiol. 65: 2221-2227.

13.   Pollock, W.K., Wreggett, K.A. and Irvine, R.F. 1988. Inositol phosphate production and Ca^{2+} mobilization in human umbilical vein endothelial cells stimulated by thrombin and histamine. Biochem. J. 256: 371-376.

14.   Jaffe, E.A., Grulich, J., Weksler, B.B., Hampel, G. and Watanabe, K. 1987. Correlation between thrombin-induced prostacyclin production and inositol triphosphate and cytosolic free calcium levels in cultured human endothelial cells. J. Biol. Chem. 262: 8557-8565.

15.   Raben, D.M., Yasuda, K. and Cunningham, D.D. 1987. Modulation of thrombin-stimulated lipid responses in cultured fibroblasts. Evidence for two coupling mechanisms. Biochemistry 26: 2759-2765.

16.   Wright, T.M., Rangan, L.A., Shin, H.S. and Raben, D.M. 1988. Kinetic analysis of 1,2 diacylglycerol mass levels in cultured fibroblasts. J. Biol. Chem. 263: 9374-9380.

17.   Gerrity, R.G. 1981. The role of the monocyte in atherogenesis. I. Transition of blood-borne monocytes into foam cells in fatty lesions. Am. J. Pathol. 103: 181-200.

18.　Taylor, R.G. and Lewis, J.C. 1986. Endothelial cell proliferation and monocyte adhesion to atherosclerotic lesions in White Carneau pigeons. Am. J. Pathol. 125: 152-160.

19.　Joris, I.T., Zand, J.J., Nunnari, F.J., Krolikowski, F.J. and Majno, G. 1983. Studies on the pathogenesis of atherosclerosis. I. Adhesion and emigration of mononuclear cells in the aorta of hypercholesterolemic rats. Am. J. Pathol. 113: 341-358.

20.　Faggioto, A., Ross, R. and Harker, L. 1984. Studies of hypercholesterolemia in non-human primates. I. Changes that lead to fatty streak formation. Arteriosclerosis 4: 323-340.

21.　Bevilacqua, M.P., Pober, J.S., Wheeler, M.E., Cotran, R.S. and Gimbrone, M.A. 1985. Interleukin-1 acts on cultured human vascular endothelium to increase the adhesion of polymorphonuclear leukocytes, monocytes and related leukocyte cell lines. J. Clin. Invest. 76: 2003-2011.

22.　Broudy, V.C., Harlan, J.M. and Adamson, J.W. 1987. Disparate effects of tumor necrosis factor-α/cachectin and tumor necrosis factor-ß/lymphotoxin on hematopoietic growth factor production and neutrophil adhesion molecule expression by cultured human endothelial cells. J. Immunol. 138: 4298-4302.

23.　Goerdt, S., Zwadlo, G., Schlegel, R., Hagemeier, H.-H. and Sorg, C. 1987. Characterization and expression kinetics of an endothelial cell activation antigen present in vivo only in acute inflammatory tissues. Expl. Cell Biol. 55: 117-126.

24.　DiCorleto, P.E. and de la Motte, C.A. 1985. Characterization of the adhesion of the human monocytic cell line U937 to cultured endothelial cells. J. Clin. Invest. 75: 1153-1161.

25.　Bevilacqua, M.P., Stengelin, S., Gimbrone, M.A. and Seed, B. 1989. Endothelial leukocyte adhesion molecule 1: An inducible receptor for neutrophils related to complement regulatory proteins and lectins. Science 243: 1160-1165.

26.　Prescott, S.M., Zimmerman, G.A. and McIntyre, T.M. 1984. Human endothelial cells in culture produce platelet-activating factor (1-alkyl-2-acetyl-sn-glycero-3-phosphocholine) when stimulated with thrombin. Proc. Natl. Acad. Sci. USA 81: 3534-353.

27.　Schleimer, R.P. and Rutledge, B.K. 1986. Cultured human vascular endothelial cells acquire adhesiveness for neutrophils after stimulation with interleukin 1, endotoxin and tumor-promoting phorbol diesters. J. Immunol. 136: 649-654.

28.　Brock, T.A. and Capasso, E.A. 1988. Thrombin and histamine activate phospholipase C in human endothelial cells via a phorbol ester-sensitive pathway. J. Cell Physiol. 136: 54-62.

29.　DiCorleto, P.E. and Bowen-Pope, D.F. 1983. Cultured endothelial cells produce a platelet-derived growth factor-like protein. Proc. Natl. Acad. Sci. USA 80: 1919-1923.

30.　Fox, P.L. and DiCorleto, P.E. 1984. Regulation of production of a platelet-derived growth factor-like protein by cultured bovine aortic endothelial cells. J. Cell Phys. 121: 298-308.

31.　Gajdusek, C., Carbon, S., Ross, R., Nawroth, P. and Stern, D. 1986. Activation of coagulation releases endothelial cell mitogens. J. Cell Biol. 103: 419-428.

32.　Starksen, N.F., Harsh, G.R., Gibbs, V.C. and Williams, L.T. 1987. Regulated expression of the platelet-derived growth factor-A chain gene in microvascular endothelial cells. J. Biol. Chem. 262: 14381-14384.

33.   Kavanaugh, W., Harsh, G.R., Starksen, N.F., Rocco, C.M. and Williams, L.T. 1988. Transcriptional regulation of the A- and B-chain genes of platelet-derived growth factor in microvascular endothelial cells. J. Biol. Chem. 263: 8470-8472.

34.   Daniel, T.O. and Fen, Z. 1988. Distinct pathways mediate transcriptional regulation of platelet-derived growth factor B/c-sis expression. J. Biol. Chem. 263: 19815-19820.

35.   Grinstein, S., Rotin, D. and Mason, M.J. 1989. Na^+/H^+ exchange and growth factor-induced cytosolic pH changes. Role in cellular proliferation. Biochim. Biophys. Acta 988: 73-97.

36.   Ghigo, D., Bussolino, F., Garbarino, G., Heller, R., Turrini, F., Pescarmona, G., Cragoe, Jr. E.J., Pegoraro, L. and Bosia, A. 1988. Role of Na^+/H^+ exchange in thrombin-induced platelet-activating factor production by human endothelial cells. J. Biol. Chem. 263: 19437-19446.

37.   Shankar, R., de la Motte, C.A. and DiCorleto, P.E. Thrombin stimulates platelet-derived growth factor production and monocyte adhesion through distinct intracellular pathways in human aortic endothelial cells. (manuscript submitted)

38.   Shankar, R. and DiCorleto, P.E. 1989. Thrombin stimulates PDGFc production by human aortic endothelial cells through activation of a G protein and the Na^+/H^+ antiporter. J. Cell Biol. 109: 152a.

39.   Voyno-Yasenetskaya, T.A., Panchenko, M.P., Nupenko, E.V., Rybin, V.O. and Tkachuk, V.A. 1989. Guanine nucleotide-dependent, pertussis toxin-insensitive regulation of phosphoinositide turnover by bradykinin in bovine pulmonary artery endothelial cells. FEBS Lett 259: 67-70.

40.   Pirotton, S., Erneux, C. and Boeynaems, J.M. 1987. Dual role of GTP-binding proteins in the control of endothelial prostacyclin. Biochem. Biophys. Res. Commun. 147: 1113-1120.

41.   Brock, T.A. and Capasso, E.L. 1989. GTPᵧS increases thrombin-mediated inositol trisphosphate accumulation in permeabilized human endothelial cells. Am. Rev. Respir. Dis. 140: 1121-1125.

SECTION 3

EXTRACELLULAR MATRIX INTERACTIONS IN ATHEROGENESIS

DYNAMIC RESPONSES OF COLLAGEN AND ELASTIN

TO VESSEL WALL PERTURBATION

F.W. Keeley

Division of Cardiovascular Research
Research Institute
The Hospital for Sick Children
Toronto, Canada, M5G 1X8

INTRODUCTION

The concept that mechanical forces can influence the composition and structure of tissues is common to many areas of physiology and developmental biology. Metabolism in many tissues is affected by physical stress, and the tissue response is generally anabolic in nature. Perhaps the most familiar example of such a response is the effect of exercise on skeletal muscle[1]. However, analogous responses in other tissues have also been described. These tissues include cardiac muscle[2-5], bone[6,7] and cartilage[8,9]. Our laboratory has been investigating the nature and mechanism of the response of large arteries to such physical stresses.

In 1973, in a review on the subject of the blood vessel wall, Harvey Wolinsky pointed out that "virtually any perturbation of the vessel wall results in damage to or proliferation of medial and sub-endothelial muscle cells and in accumulation of fibrillar and mucopolysaccharide elements"[10]. Wolinsky listed several types of vascular perturbations that had been investigated, including freezing, heating, X-irradiation, inflation by balloon catheter, the presence of incisions or sutures, various drugs and hormones, and pathological situations such as hyperlipidemia and hypertension.

Although such perturbations can take many forms, the three types of mechanical stress or trauma which have been investigated most thoroughly are injuries to the endothelial cell layer of the vessel wall, clearly an important factor in the development of atherosclerosis, shear stress resulting from altered flow rates and flow patterns in vessels, and tangential stress generated within the vessel wall by distension, the type of stress which is elevated in the pathological situation of hypertension. In this discussion, I will be focussing on the nature of the response of the arterial wall to such tangential wall stresses.

The tangential wall stress, which is a measure of the load exerted on the vessel wall by a distending pressure, is defined by LaPlace's law as it applies to a cylindrical vessel of finite wall thickness:

$$\text{Tangential Wall Stress} = \frac{\text{Blood Pressure x Internal Radius}}{\text{Wall Thickness}}$$

From this relationship it is clear that the tangential wall stress can be elevated by increasing either pressure or internal radius of the vessel, but that this increased wall stress can be counteracted by an increase in wall thickness, which will restore the stress level in the vessel towards normal levels. Several years ago, Wolinsky and Glagov

Atherosclerosis, Edited by A. I. Gotlieb *et al.*
Plenum Press, New York, 1991

demonstrated this balance between blood pressure and radius on the one hand and wall thickness on the other[11]. Comparing the dimensions and physical forces acting on thoracic aortas of adult animals ranging in size from the mouse to the human, they observed that wall thickness in these species varied such that the stress experienced by the wall of the vessel was remarkably constant. This suggested that, at least in the aorta, there was an optimum wall stress level, determined by the nature and architecture of the components of the wall, which was maintained through adjustments in wall thickness.

Two constituents of the arterial wall which are vital to its physical integrity and important in its ability to withstand such tangential stresses are the connective tissue proteins, collagen and elastin. In large vessels such as the aorta, these connective tissue proteins not only make up the majority of the wall mass, but also to a great extent determine the passive physical properties of the vessel[12]. Collagen is a family of proteins with a widespread distribution. In general, collagen is important for the physical integrity of tissues, imparting properties such as tensile strength and resistance to compression. In contrast, elastin has a more restricted distribution. However, elastin is present in high proportions in tissues such as the aorta (up to 50% of dry weight), certain elastic ligaments (up to 80% of dry weight), and the parenchyma of the lung (15-20% of dry weight). In these tissues, elastin is responsible for the properties of extensibility and elastic recoil, characteristics which are particularly important for their normal physiological function.

Once assembled into their extracellular, cross-linked matrices, collagen and elastin are both very stable proteins. Collagen has been reported to have a half-life of 60-90 days[13], and most data suggest that arterial elastin does not turn over at all under normal circumstances[14]. Indeed, measurements of the turnover of elastin usually indicate a half-life longer than the life span of the experimental animal. As might be expected for such a stable protein, most of the synthesis and accumulation of arterial elastin takes place during development and growth of the vessel. In the aorta of the growing chick, elastin accumulation is rapid only for the first 4-6 weeks after hatching, after which time little further elastin accumulation takes place. Synthesis of elastin peaks at about 2 weeks after hatching, then falls rapidly to very low levels[15,16]. This developmental pattern of elastin synthesis in the chick aorta is also reflected in steady state levels of mRNA for elastin measured either by *in vitro* translation or using cDNA probes. Similar patterns of arterial development are also seen in other species[17,18].

Suggestions have been made that the rapid post-natal accumulation of connective tissue proteins in blood vessels may be related to physical stresses due to the increase in blood pressure that normally takes place over this period of development[19,20]. However, if the timing of the post-natal accumulation of elastin in the aorta of the chick is compared to developmental blood pressure changes and the calculated wall stress in the tissue (Figure 1), it is clear that elastin accumulation correlates better with increasing wall stress, which takes into account both blood pressure and the diameter of the vessel, than with increasing blood pressure alone. This observation suggests that changes in the calibre of the vessel are also important for post-natal accumulation of arterial connective tissues.

RESPONSES OF VASCULAR ELASTIN AND COLLAGEN DURING DEVELOPMENT OF HYPERTENSION

Many investigations of the effects of physical stresses on blood vessels have used the pathological situation of hypertension as a model system. For over a hundred years it has been known that one of the consequences of hypertension is a marked thickening of the walls of blood vessels. This thickening involves both large and small vessels, and includes substantial increases in absolute amounts of elastin and collagen, although the proportion of these proteins by weight in the vessel wall is unchanged[21,22]. Indeed, Cleary and Moont[23] have shown that, in spite of increases in the total contents of these proteins, the decreasing elastin/collagen gradient moving down the aorta is not altered in hypertension.

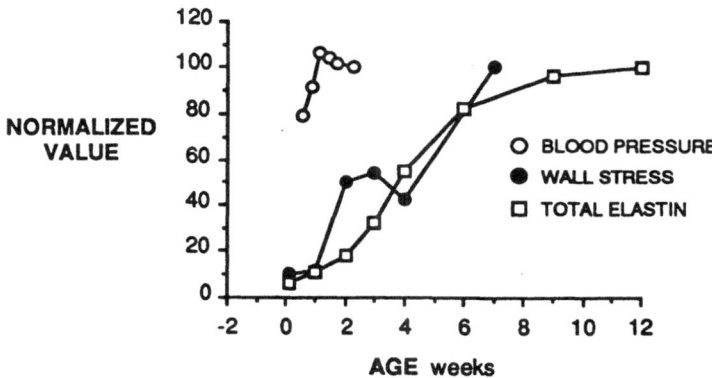

Figure 1. Blood pressure, total accumulated insoluble elastin and calculated wall stress in chicken aorta as a function of age. Data were normalized to the oldest data point in each set. Blood pressure data are from Gardner et al.[20]. Wall stress was calculated using LaPlace's Law. Vessel radius and wall thickness were measured from aortas fixed under pressure.

Most observations of connective tissue changes in vessels in response to hypertension had been made on tissues well after hypertension and its consequences were established. Using several rat models of hypertension, including the DOCA-salt model[24], the renal clip or Goldblatt model, and the Dahl salt-sensitive rat model, we have investigated the effects of developing hypertension on the production of elastin and collagen in arteries.

Methodology for the determination of elastin and collagen synthesis and total accumulated elastin and collagen is essentially identical to that described previously[24,25]. The technique exploits the fact that insoluble elastin contains no methionine residues and thus resists digestion by cyanogen bromide. As a result, the insoluble residue remaining after thorough treatment of these vascular tissues with cyanogen bromide is essentially pure elastin. This can be confirmed by amino acid analysis. The weight of the residue after cyanogen bromide treatment is taken as total accumulated insoluble elastin. If this tissue has been previously incubated in the presence of radioactive amino acids, then the radioactivity incorporated into this residue is a measure of newly synthesized insoluble elastin. The material solubilized by the cyanogen bromide treatment is a mixture of collagen and all other proteins in the vascular wall. However, collagen is the only protein in this mixture that will contain significant amounts of hydroxyproline. Thus, the hydroxyproline content of the extract is a measure of the total accumulated collagen in the tissue, and radioactive hydroxyproline is representative of newly synthesized collagen. Radioactive and total hydroxyproline in the extract can be quantitated by amino acid analysis or using a modification of the method of Kivirikko et al.[26].

A typical response is that seen for the renal clip model (Figure 2). Systolic blood pressure, measured by tail cuff, rises within a few days of clipping and reaches a plateau by about 15 days, with the final blood pressure level dependent on the size of the clip. Coincident with this rise in blood pressure is an increase in the total insoluble elastin content of the arterial segment, and an increase in synthesis of insoluble elastin which peaks at about 12 days and then falls back towards control levels. This responsiveness is confined neither to elastin nor to the aorta. On the contrary, early and sensitive responses are also seen for collagen accumulation and in smaller arterial vessels (Figure 3).

The general characteristics of this in vivo response to developing systemic hypertension can be summarized as follows:

1. The aorta is highly sensitive to increases in blood pressure, with increased elastin accumulation appearing almost as soon as elevations in pressure can be detected.

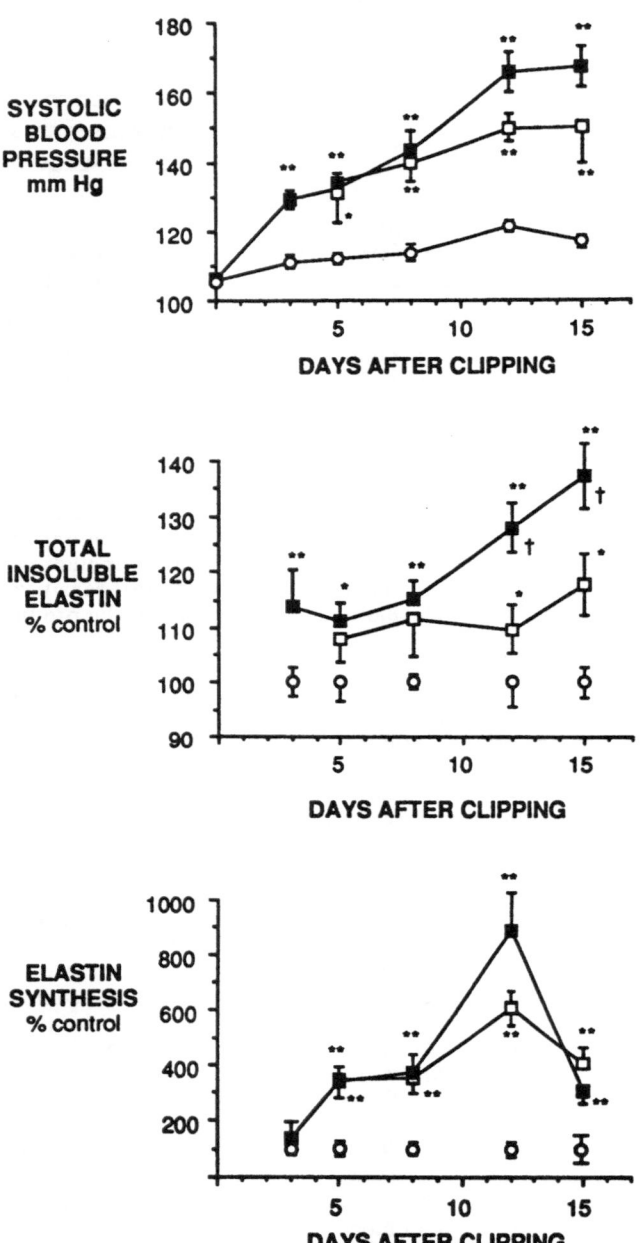

Figure 2. Time course of the changes in blood pressure, total elastin and elastin synthesis in the wall of the thoracic aorta of rats made hypertensive by the renal clip method. Total elastin and elastin synthesis are expressed as a percent of age-matched, sham-operated controls. Open circles are sham-operated animals. Open squares are animals made hypertensive with a clip of internal width 0.25 mm. Closed squares are animals made hypertensive with a clip of internal width 0.20 mm. Single and double asterisks refer to p<0.05 and p<0.01, respectively, using an unpaired student t-test.

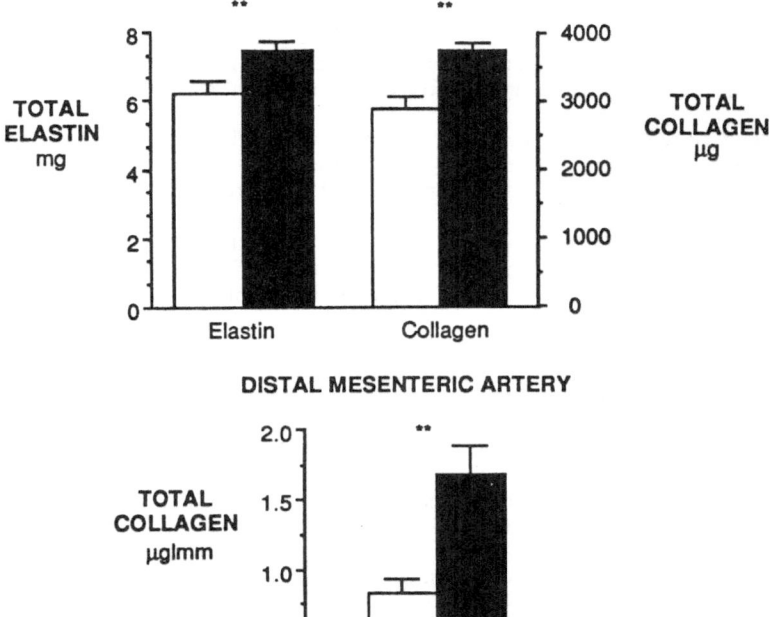

Figure 3. Effect of renal clip hypertension on vascular elastin and collagen in thoracic aorta and distal mesenteric artery two weeks after rats were made hypertensive with a clip of internal width 0.25 mm. Double asterisks refer to p<0.01 using an unpaired student t-test. Open bars are sham-operated animals. Filled bars are renal clipped animals.

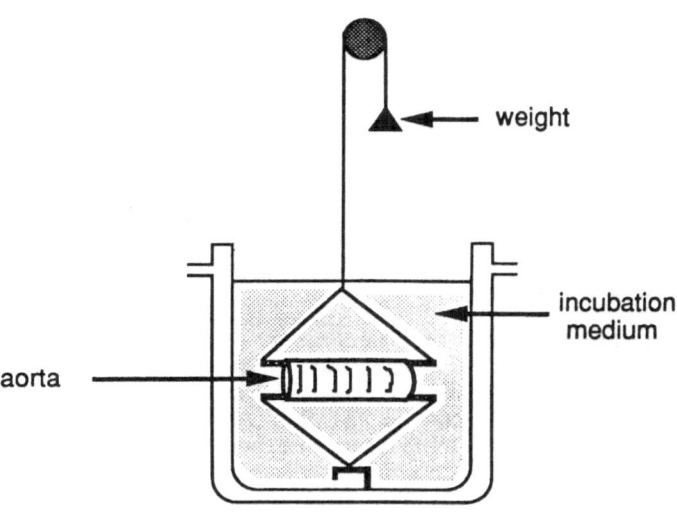

Figure 4. Schematic diagram of the *in vitro* "hanging" model for studying the effects of wall stress on elastin production in aortic tissue.

We have reported a similar sensitivity of pulmonary arteries to pulmonary hypertension induced by monocrotaline[25].

2. In all cases, and from the earliest stages, the response includes large and coordinated increases in collagen and elastin production such that proportions of these proteins relative to each other and relative to total vessel weight remain unchanged.

3. The system behaves as if the response were driven by increased wall stress. That is, when sufficient new matrix has accumulated in the vessel wall to restore normal wall stress, production of collagen and elastin returns to normal levels.

4. The response does not appear to involve any significant hyperplasia of aortic cells. Thus, at least at early stages in the development of hypertension, the increase in synthesis of these connective tissue proteins takes place in the absence of cell proliferation.

It appears therefore that the response of the aorta to increased wall stress, whether as part of a normal developmental process or as a result of the pathological stresses of hypertension, is similar and involves a coordinated and proportional increase in elastin and collagen production, with little or no proliferation of cells.

RESPONSE OF VASCULAR ELASTIN AND COLLAGEN TO *IN VITRO* MODELS OF INCREASED WALL STRESS

In order to investigate the mechanism of this response in more detail, we have developed *in vitro* organ culture models which respond to increased wall stress with increased production of elastin and collagen. Similar models have previously been described in the literature for a variety of tissues and cells[8,27-32]. A schematic representation of one of the organ culture models used by us is shown in Figure 4. Thoracic aortas from 7 day old chickens are suspended between two stainless steel rods in an incubation bath. Knowing the dimensions of the vessel and the force applied, the stress on the vessel wall can be calculated, and the response of connective tissue protein production measured.

The relationship between wall stress and elastin production derived from this model is shown in Figure 5. Similar results are seen for collagen production. It is clear that even in this *in vitro* model, in the absence of neurohumoral influences or circulating factors, connective tissue protein production remains sensitive to increased wall stress, responding quickly and in a graded fashion. The maximum effect is seen at a stress equivalent to an *in vivo* mean arterial pressure of approximately 180 mm Hg, and the response appears to diminish at wall stresses equivalent to much higher pressures.

The threshold for the response of the vessel wall to static stress in this *in vitro* model is low, with the first indications of increased elastin production appearing at an equivalent mean arterial pressure of between 90 and 140 mm Hg. This is remarkably similar to the threshold seen in *in vivo* models of hypertension in which vessels are exposed to increased cyclic stresses, suggesting that mean arterial pressure may be the important factor in triggering the response. As was the case in *in vivo* models of hypertension, the response *in vitro* also appears to involve a coordinated effect on both collagen and elastin production.

Although endothelial factors may modulate this response to increased wall stress, an intact endothelium is clearly not required for the response, since scanning electron microscopy has shown that the manipulations required for mounting the vessels on the stainless steel rods effectively remove the endothelium. The fact that an intact endothelium was not necessary for the response allowed the use of a simplified organ culture model (Figure 6). In this case the vessel is threaded onto polypropylene tubing of an appropriate size to stretch the walls of the vessel. Tubing of a size which fills the lumen but does not stretch the vessel wall can be used as a control. Although the wall stress level to which the vessel is subjected cannot be either varied or readily calculated

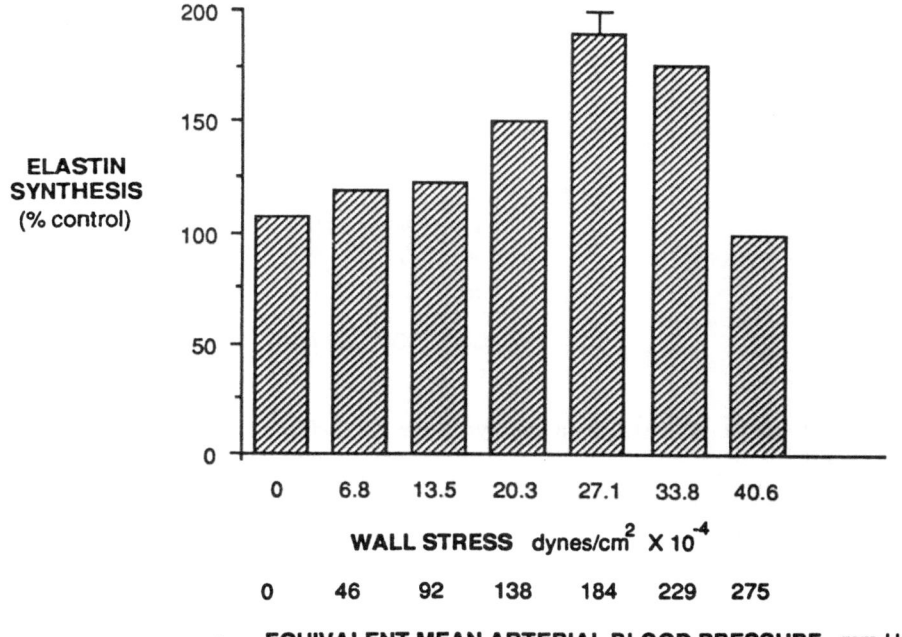

Figure 5. Effect of wall stress on the synthesis of elastin in 7 day old chick aorta using the "hanging" model. Elastin synthesis is expressed as radioactivity of radiolabelled amino acids incorporated into insoluble elastin of stressed vessels as a percent of that incorporated into control vessels incubated in the same bath. The equivalent mean arterial blood pressure was calculated using the internal radius and wall thickness of the vessel. Stress levels of 0 represent vessels in which rods were inserted, but not suspended.

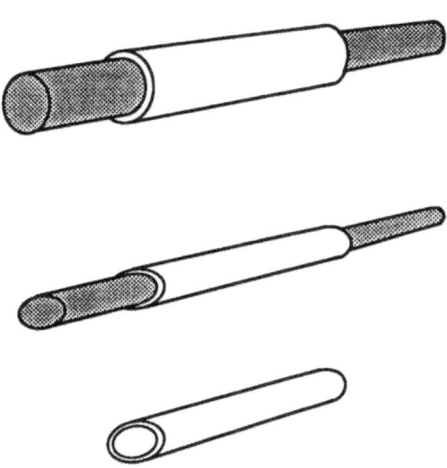

Figure 6. Schematic diagram of the *in vitro* "tubing" model for studying the effects of wall stress on elastin production in aortic tissue.

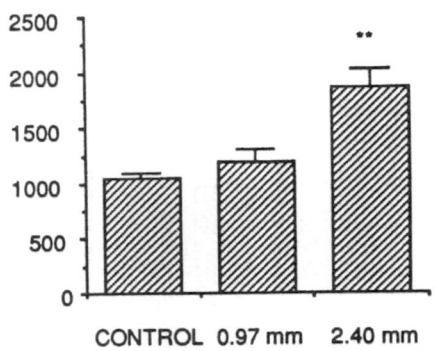

Figure 7. Effect of wall stress on the synthesis of elastin in 1 day old chick aorta
using the "tubing" model. Synthesis is expressed as radioactivity incorporated into
insoluble elastin. The control vessel had no tubing inserted. Tubing of diameter
0.97 mm, which filled the lumen but did not distend the vessel, was used as an
additional control. The double asterisk refers to p<0.01, using an unpaired student
t-test.

using this model, it does have the advantage of simplicity. In addition, a symmetrical
stress is applied to the vessel wall in contrast to the asymmetrical stress produced by the
hanging model. Results using this tubing model are similar to those obtained with the
hanging model (Figure 7). In addition to its low threshold, the response is also rapid,
and can be seen within 4 hours of application of the increased stress (Figure 8).

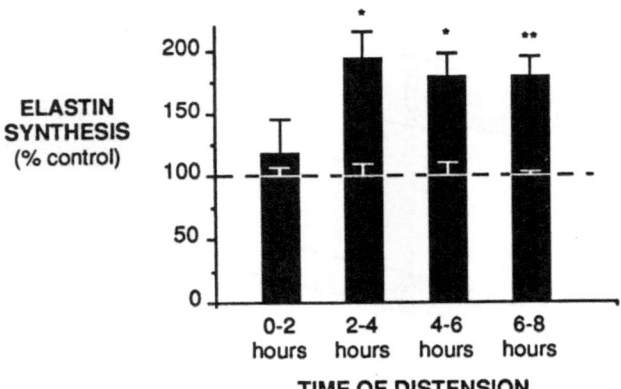

Figure 8. Time course of the response of elastin synthesis to wall stress using the
"tubing" model. Vessels were incubated in medium containing ^{14}C-proline for the
period indicated and then chased for 3 hours in unlabelled medium. Synthesis is
expressed relative to synthesis by control vessels in the same incubation bath.
Single and double asterisks refer to p<0.05 and p<0.01, respectively, using an
unpaired student t-test.

POSSIBLE MECHANISMS FOR THE CONNECTIVE TISSUE RESPONSE
TO WALL STRESS

Increased tangential stress in vessels is likely manifested as an increased distention or strain on the cells in the vessel wall. However, the level of strain that a cell in a vessel wall will "feel" for a given wall stress can be affected by many factors. First, when a distending force is applied to a matrix, cells attached to that matrix will experience a strain only as long as they remain adherent to the matrix. Distension of the matrix sufficient to detach the cells will relieve that strain. Thus, the response to stress may depend on the strength of adhesion of the cells to the matrix. Secondly, cells actively dividing will detach or loosen their adherence to the matrix and reattach after dividing, presumably resuming their unstrained conformation in the process. Normally cells in arteries are slow to turnover[33,34], but it is at least theoretically possible that cells might "escape" from increased strain through an increased rate of division.

The strain experienced by a cell for a given level of stress will also depend on the stiffness of the matrix to which the cell is attached, a property which will be influenced by relative contents and arrangements of elastin, collagen, proteoglycans and other matrix components. Furthermore, the matrix may not be symmetrical in its stiffness properties, nor the cell symmetrical in its sensitivity to strain, so that the orientation

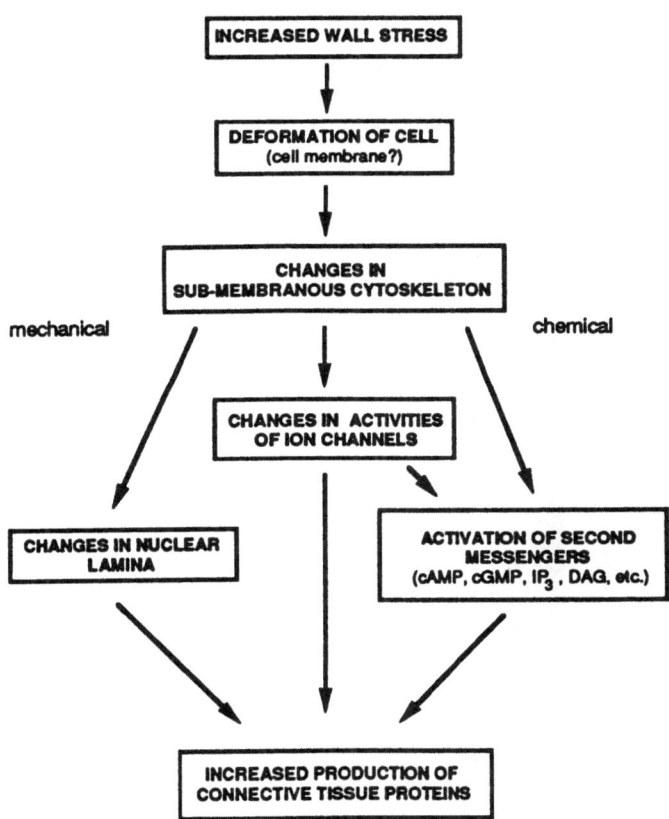

Figure 9. Hypothetical scheme representing possible mechanisms for the transduction of elevated vascular wall stress into increased production of elastin and collagen by arterial cells.

of the cell relative to the matrix and to the direction of the stress may also be important factors. In addition, these vessels are not passive, elastic tubes. Clearly the contractile state of the cell, influenced by both neurohumoral and circulating factors, may also influence the extent of cell strain produced by a given stress.

Assuming that the driving force for the connective tissue response is cell stretch, what could be the mechanism by which this physical distension is transduced into a coordinated increase in collagen and elastin production in the vessel wall? Several processes could be involved (Figure 9). For example, distension of the cell may result in physical changes in its submembranous cytoskeleton. This influence may be transmitted either directly to the nuclear membrane and, hence, to the nuclear lamina

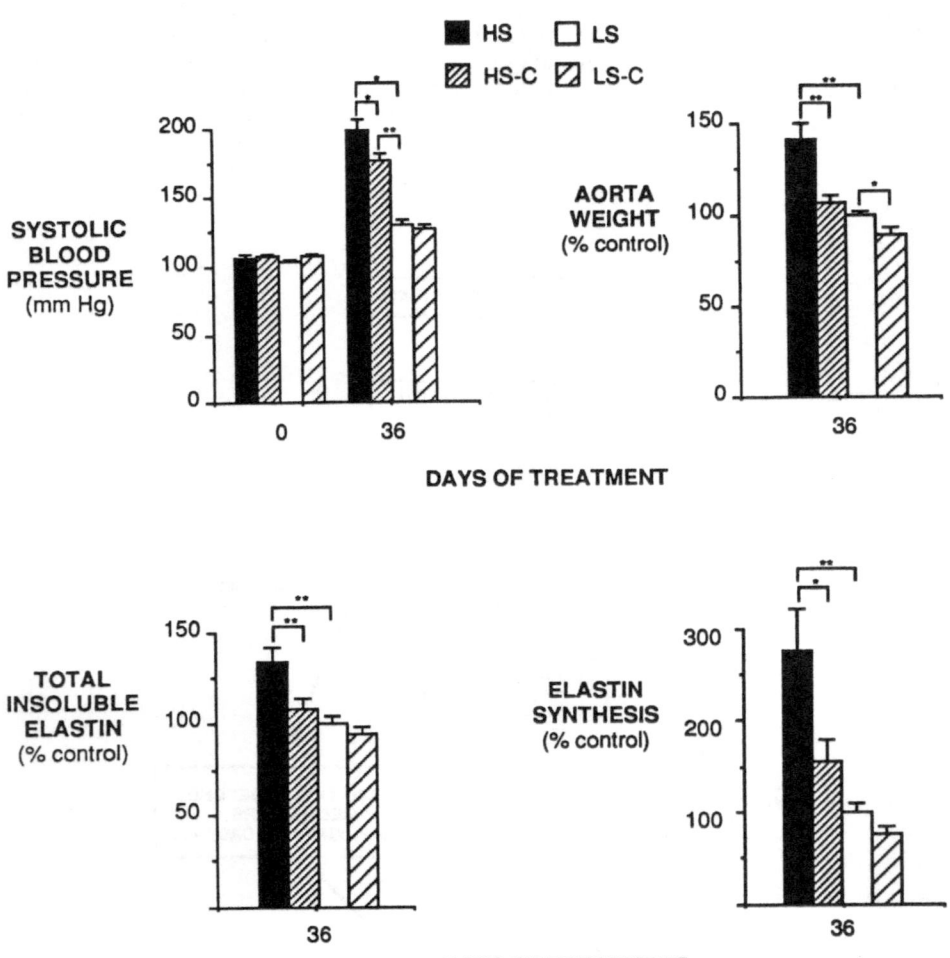

Figure 10. Effect of colchicine (0.2 mg/kg body weight, daily intraperitoneal) on the hypertensive and hypertrophic response in aorta of Dahl salt sensitive rats made hypertensive with a high salt diet. Treatment groups were: HS-C (high salt with colchicine treatment); HS (high salt); LS-C (low salt with colchicine treatment); LS (low salt). Total elastin and elastin synthesis are expressed as a percent of age-matched controls. Single and double asterisks refer to p<0.05 and p<0.01, respectively, using an unpaired student t-test.

by a mechanical pathway involving other elements of the cytoskeleton, or through already identified mediators of signal transduction within the cell. Additionally or alternatively, physical forces on cell membranes are known to affect ion channel activities[35], raising the possibility that such a mechanism may be involved either directly or indirectly in the response. The net effect is hypertrophy of the cell and increased production of collagen and elastin.

If such a scheme has any relationship to reality, then the cytoskeleton may be central to the process of transduction of physical forces. Interactions between some cytoskeletal elements and submembranous components of signal transduction pathways have been postulated[36-39]. If activation of signal transduction systems can result in changes in cell shape, is it possible that the converse is also true? That is, can changes in cell shape influence signal transduction in such a way as to alter connective tissue production by cells?

At the present time, evidence for a role for the cytoskeleton is slim but tantalizing. Using the *in vivo*, Dahl salt-sensitive rat model of hypertension, we have shown that treatment with colchicine to disrupt microtubules prior to and during development of hypertension essentially abolishes the hypertrophic response of the vessel wall, without affecting the increase in blood pressure (Figure 10). Furthermore, in colchicine-treated hypertensive animals, elastin synthesis is decreased to normal levels. Although other explanations are possible, this may suggest that microtubular elements are important for the perception of increased strain by the cell. Steady state levels of mRNA for elastin approximately doubled in vessels subjected to increased wall stress using the *in vitro* organ culture system (Figure 11). Significantly, this response was abolished in the presence of colchicine, giving support to the hypothesis that an intact cytoskeleton may be necessary for the perception of increased wall stress by arterial cells.

All of this is not meant to imply that mechanical stresses are the only important considerations with regard either to vascular development or in response to pathological influences, nor that elastin and collagen are the only significant elements of the arterial wall. Certainly many other factors, both interacting and independent, must also be considered. Nevertheless, the undoubted effects of these physical forces and the central role of these connective tissue proteins in the physical properties and physiological functions of arteries suggests that they are important factors in the regulation of vascular responses in development and pathology.

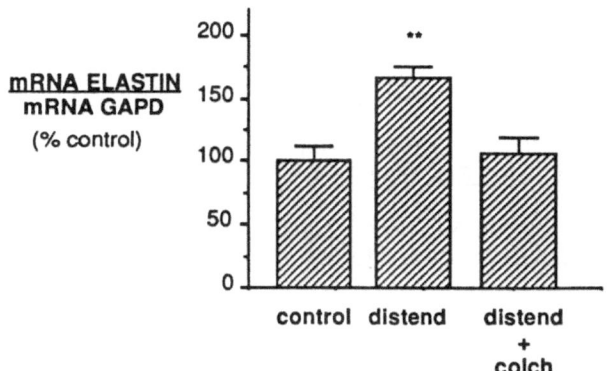

Figure 11. Effect of colchicine (0.1 mM) on steady state levels of elastin mRNA in chick aorta subjected to increased wall stress. Message levels are expressed relative to steady state levels of message for glyceraldehyde-3-phosphate dehydrogenase (GAPD).

ACKNOWLEDGEMENTS

This work was supported by the Heart and Stroke Foundation of Ontario and by the Physicians' Services Incorporated Foundation.

REFERENCES

1. Vandenburgh, H.H., Hatfaludy, S., Karlisch, P. and Shansky, J. 1989. Skeletal muscle growth is stimulated by intermittent stretch-relaxation in tissue culture. Am J. Physiol. 256: C674-C682.

2. Schreiber, S.S., Rothschild, M.A., Evans, C., Reff, F. and Oratz, M. 1975. The effect of pressure or flow stress on right ventricular protein synthesis in the face of constant and restricted coronary flow. J. Clin. Invest. 55: 1-11.

3. Schwartz, K., de la Bastie, D., Bouveret, P., Oliviero, P., Alonso, S. and Buckingham, M. 1986. Alpha-skeletal muscle actin mRNA's accumulate in hypertrophied adult rat hearts. Circ. Res. 59: 551-555.

4. Kent, R.L., Hoober, J.K. and Cooper, G. 1989. Load responsiveness of protein synthesis in adult myocardium: Role of cardiac deformation linked to sodium influx. Circ. Res. 64: 74-85.

5. Kira, Y., Kochel, P.J., Gordon, E.E. and Morgan, H.E. 1984. Aortic perfusion pressure as a determinant of cardiac protein synthesis. Amer. J. Physiol. 246: C247-C258.

6. Silbermann, M., von der Mark, K., van Menxel, M. and Reznick, A.Z. 1987. Effect of short term physical stress on DNA and collagen synthesis in the femur of young and old mice. Gerontology 33: 49-56.

7. Sandy, J.R., Meghji, S., Farndale, R.W. and Meikle, M.C. 1989. Dual elevation of cyclic AMP and inositol phosphates in response to mechanical deformation of murine osteoblasts. Biochim. Biophys. Acta. 1010: 265-269.

8. de Witt, M.T., Handley, C.J., Oakes, B.W. and Lowther, D.A. 1984. *In vitro* response of chondrocytes to mechanical loading. The effect of short term mechanical tension. Connect. Tiss. Res. 12: 97-109.

9. Vasan, N. 1983. Effects of physical stress on the synthesis and degradation of cartilage matrix. Connect. Tiss. Res. 12: 49-58.

10. Wolinsky, H. 1973. Mesenchymal response of the blood vessel wall. A potential avenue for understanding and treating atherosclerosis. Circ. Res. 32: 543-549.

11. Wolinsky, H. and Glagov, S. 1967. A lamellar unit of aortic medial structure and function in mammals. Circ. Res. 20: 99-111.

12. Roach, M.R. and Burton, A.C. 1957. The reason for the shape of the distensibility curves of arteries. Can. J. Biochem. Physiol. 35: 681-690.

13. Nissen, R., Cardinale, G.J. and Udenfriend, S. 1978. Increased turnover of arterial collagen in hypertensive rats. Proc. Nat. Acad. Sci. USA 75: 451-453.

14. Lefevre, M. and Rucker, R.B. 1980. Aorta elastin turnover in normal and hypercholesterolemic Japanese quail. Biochim. Biophys. Acta. 630: 519-529.

15. Keeley, F.W. and Johnson, D.J. 1985. The effect of induced atherosclerosis on the synthesis of elastin in chick aortic tissue. Atherosclerosis 54: 311-319.

16. Keeley, F.W. 1979. The synthesis of soluble and insoluble elastin in chicken aorta as a function of development and age. Effect of a high cholesterol diet. Can. J. Biochem. 57: 1273-1280.

17. Looker, T. and Berry, C.L. 1972. The growth and development of the rat aorta. II. Changes in nucleic acid and scleroprotein content. J. Anatomy 113: 17-34.

18. McCloskey, D.I. and Cleary, E.G. 1974. Chemical composition of the rabbit aorta during development. Circ. Res. 34: 828-835.

19. Leung, D.Y.M., Glagov, S. and Mathews, M.B. 1977. Elastin and collagen accumulation in rabbit ascending aorta and pulmonary trunk during postnatal growth. Circ. Res. 41: 316-323.

20. Gardner, R., Heng, H., Penner, M., Sedgwick, C. and Rucker, R. 1984. Elastin accumulation in the chick aorta. Effect of 6-hydroxydopamine and deoxycorticosterone acetate. Res. Commun. Chemical Pathol. Pharmacol. 43: 251-264.

21. Wolinsky, H. 1970. Response of the rat aortic media to hypertension: Morphological and chemical studies. Circ. Res. 26: 507-522.

22. Wolinsky, H. 1971. Effects of hypertension and its reversal on the thoracic aorta of male and female rats. Morphological and chemical studies. Circ. Res. 28: 622-637.

23. Cleary, E.G. and Moont, M. 1976. Hypertension in weanling rabbits. Adv. Exper. Med. Biol. 79: 477-490.

24. Keeley, F.W. and Johnson, D.J. 1986. The effect of developing hypertension on the synthesis and accumulation of elastin in the aorta of the rat. Can. J. Biochem. Cell. Biol. 64: 38-43.

25. Todorovich-Hunter, L., Johnson, D.J., Ranger, P., Keeley, F.W. and Rabinovitch, M.R. 1988. Altered elastin and collagen synthesis associated with progressive pulmonary hypertension induced by monocrotaline. A biochemical and ultrastructural study. Lab. Invest. 58: 184-195.

26. Kivirikko, K.I., Laitinen, O. and Prockop, D.J. 1967. Modifications of a specific assay for hydroxyproline in urine. Anal. Biochem. 19: 249-255.

27. Leung, D.Y.M., Glagov, S. and Mathews, M.B. 1976. Cyclic stretching stimulates synthesis of matrix components by arterial smooth muscle cells in vitro. Science 191: 475-477.

28. Kollros, P.R., Bates, S.R., Mathews, M.B., Horwitz, A.L. and Glagov, S. 1987. Cyclic AMP inhibits increased collagen production by cyclically stretched smooth muscle cells. Lab. Invest. 56: 410-417.

29. Sumpio, B.E. and Banes, A. 1988. Response to porcine aortic smooth muscle cells to cyclic tensional deformation in culture. J. Surg. Res. 44: 696-701.

30. Sumpio, B.E., Banes, A.J., Link, W.G. and Johnson, G. 1988. Enhanced collagen production by smooth muscle cells during repetitive mechanical stretching. Arch. Surgery 123: 1233-1236.

31. Sutcliffe, M.C. and Davidson, J.M. 1988. Increased tropoelastin production by porcine aorta smooth muscle cells stretched during in vitro culture. Collagen Rel. Res. 8: 538.

32. Tozzi, C.A., Poiani, G.J., Harangozo, A.M., Boyd, C.J. and Riley, D.J. 1988. Pulmonary vascular endothelial cells modulate stretch-induced DNA and connective tissue synthesis in rat pulmonary artery segments. Chest 93: 169S-170S.

33. Berry, C.L., Looker, T. and Germain, J. 1972. The growth and development of the rat aorta. I. Morphological aspects. J. Anatomy 113: 1-16.

34. Schwartz, S.M., Campbell, G.R. and Campbell JH. 1986. Replication of smooth muscle cells in vascular disease. Circ. Res. 58: 427-444.

35. Kirber, M.T., Singer, J.J. and Walsh, Jr. J.V. 1987. Stretch-activated channels in freshly dissociated smooth muscle cells. Biophys. J. 51: 252A.

36. Rasenick, M.M., Stein, P.J. and Bitensky, M.W. 1981. The regulatory subunit of adenylate cyclase interacts with cytoskeletal components. Nature 294: 560-562.

37. Sugrue, S.P. and Hay, E.D. 1986. The identification of extracellular matrix (ECM) binding sites on the basal surface of embryonic corneal epithelium and the effect of ECM binding on epithelial collagen production. J. Cell. Biol. 102: 1907-1916.

38. Omann, G.M., Allen, R.A., Bokoch, G.M., Painter, R.G., Traynor, A.E. and Sklar, L.A. 1987. Signal transduction and cytoskeletal activation in the neutrophil. Physiol. Rev. 67: 285-322.

39. Nakano, T., Hanasaki, K. and Arita, H. 1989. Possible involvement of cytoskeleton in collagen-stimulated activation of phospholipases in human platelets. J. Biol. Chem. 264: 5400-5406.

DYNAMIC INTERACTION OF PROTEOGLYCANS

Thomas N. Wight

Department of Pathology
University of Washington
Seattle, WA 98195

Proteoglycans are complex macromolecules which consist of a core glycoprotein backbone to which one or more glycosaminoglycan chains are attached through O-glycosidic linkages to serine residues. Glycosaminoglycans are unbranched chains of repeating disaccharide units in which one of the monosaccharides is an amino sugar and the other is invariably a hexuronic acid. Usually one type of glycosaminoglycan predominates on a single core protein giving rise to four main families: chondroitin sulfate proteoglycan (CS-PG), dermatan sulfate proteoglycan (DS-PG), heparan sulfate proteoglycan (HS-PG), and keratan sulfate proteoglycan (KS-PG) (1,2). An example of the basic structure of a typical proteoglycan is shown in Figure 1.

The diversity of proteoglycans largely derives from the number of different protein cores within specific proteoglycan families and from the polydispersity produced by a large variety of post translational modifications required to construct the final molecule (1,2). An example of the structural diversity of proteoglycans found in blood vessels is given in Figure 2.

Although proteoglycans constitute a minor component of vascular tissue (2-5% by dry weight), a number of studies over the years have demonstrated that these macromolecules are of enormous importance in influencing such arterial properties as viscoelasticity, permeability, lipid accumulation, hemostasis, and thrombosis (see reviews 3-5). The principal proteoglycans in arterial tissue are CS-PGs, DS-PGs, and HS-PGs and these different families are present in different locations within the vascular wall. Immunocytochemical studies have shown that a large CS-PG ($\approx 1.2 \times 10^6$ Mr) is present principally in the interstitial matrix of blood vessels while a small DS-PG ($\approx$ 120-180 kD) is present on the periphery of collagen fibrils within the interstitial matrix. Heparan sulfate proteoglycans are present in arterial basement membranes and are closely associated with elastic fibers (see review 6). These locations are documented in Figure 3.

A hallmark of early and late atherosclerosis is the accumulation of proteoglycans in the intimal lesions of atherosclerosis. Following experimental injury, CS-PG accumulates dramatically in fibromuscular intimal thickenings (Figure 4) while DS-PG accumulates in late atherosclerotic lesions that are enriched in collagen (Figure 5). Such accumulation may predispose the arterial wall to lipid accumulation, calcification and thrombosis by virtue of the ability of proteoglycans to interact with component molecules involved in these processes. An example of this possibility comes from studies using the balloon injury model of experimental atherosclerosis. Following such injury, proteoglycans accumulate in the regions of endothelial regrowth (Figure 6A). If animals are put on a high fat diet, lipid accumulates in the regions enriched in proteoglycans

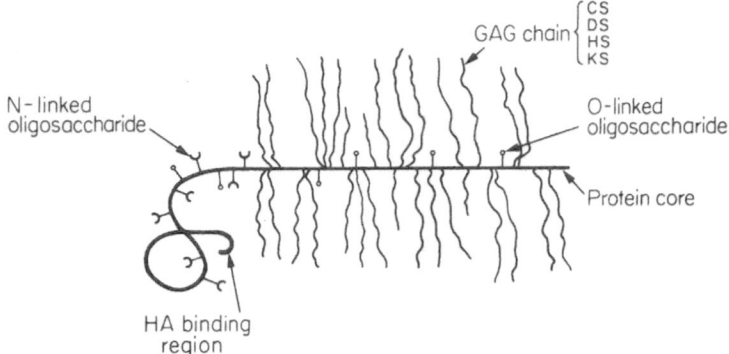

Figure 1. A schematic model of the proposed structure of a typical proteoglycan monomer. The molecule consists of a central protein core to which are attached a variable number of glycosaminoglycan (GAG) side chains and various proportions of O-linked and N-linked oligosaccharides. Usually, one type of GAG chain is associated with a single protein core. The N-terminal end of the protein core may possess the capacity to interact with hyaluronic acid (HA-binding region), while other protein cores may contain hydrophobic regions that facilitate their insertion into membranes. CS = chondroitin sulfate; DS = dermatan sulfate; HS = heparan sulfate; KS = keratan sulfate. (Reproduced with permission from Singer et al. *Recent advances in hematology*, 1985; 4: 1-24.)

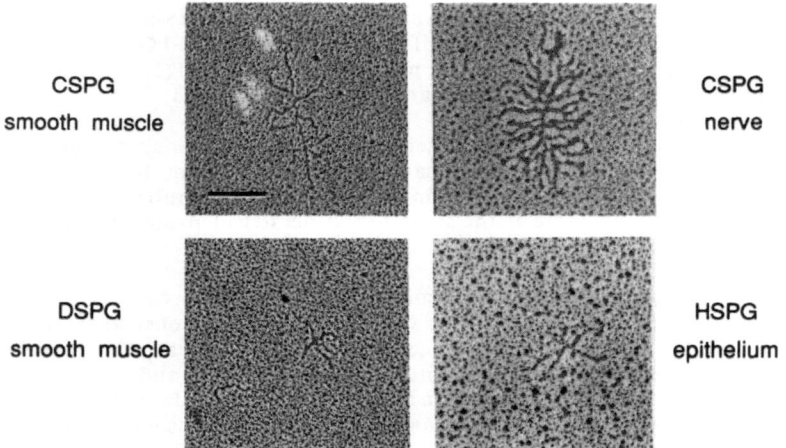

Figure 2. Electron micrographs of purified proteoglycan monomers prepared from different non-cartilaginous sources. The chondroitin sulfate proteoglycan (CSPG) prepared from arterial smooth muscle cell cultures consists of a central core of ≈ 275 nm to which are attached ≈ 10 side chains averaging 70 nm in length. Another form of CSPG is present in the nerve terminal of electric organ (Carlson, S. and Wight, T.N. *J Cell Biol*, 1987; 105: 3075-3086). This molecule possesses a central core of ≈ 345 nm with 20 to 25 side projections averaging 113 nm in length. Smaller proteoglycans containing dermatan sulfate (DSPG) also are synthesized by arterial smooth muscle cells. These molecules possess a central core of ≈ 100 nm and only one or two side projections of varying length (40 to 70 nm). Proteoglycans containing heparan sulfate (HSPG) vary enormously in size. The HSPG presented in this figure has been isolated from the plasma membrane of epithelial cells (Rapraeger et al. In: Wight, T.N. and Mecham, R.P., eds. Biology of proteoglycans. New York: Academic Press, 1987: 129-154). It consists of a central of ≈ 130 nm with 3 to 4 side chains of varying length (50 to 70 nm), X 83,000. Bar = 0.1 μm.

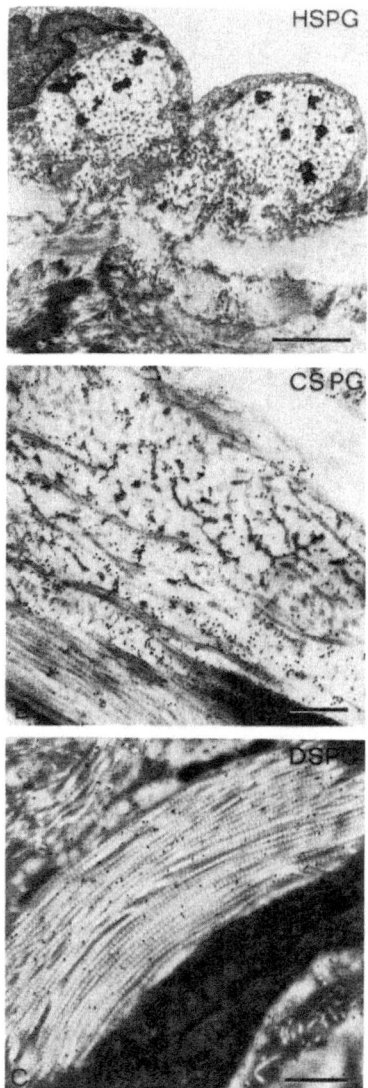

Figure 3. Electron micrographs demonstrating specific locations of various families of aortic proteoglycans. A. Segments of fixed rat aorta were incubated with ^{125}I-anti-HSPG antisera and were labeled with antibody that localized to the basement membrane beneath endothelial cells. x 8500. Bar = 2 μm. B. Thin sections of rabbit aorta that were lightly "etched" and immunostained with a monoclonal antibody against aortic CSPG and secondary antibody tagged with immunogold. Immunogold is confined to strands in the interstitial matrix. x 25,000. Bar = 0.5 μm. C. Thin section of human superficial artery immunogold stained with antisera against DSPG. Note specific location of gold particles along collagen fibrils. x 27,000. Bar = 0.5 μm.

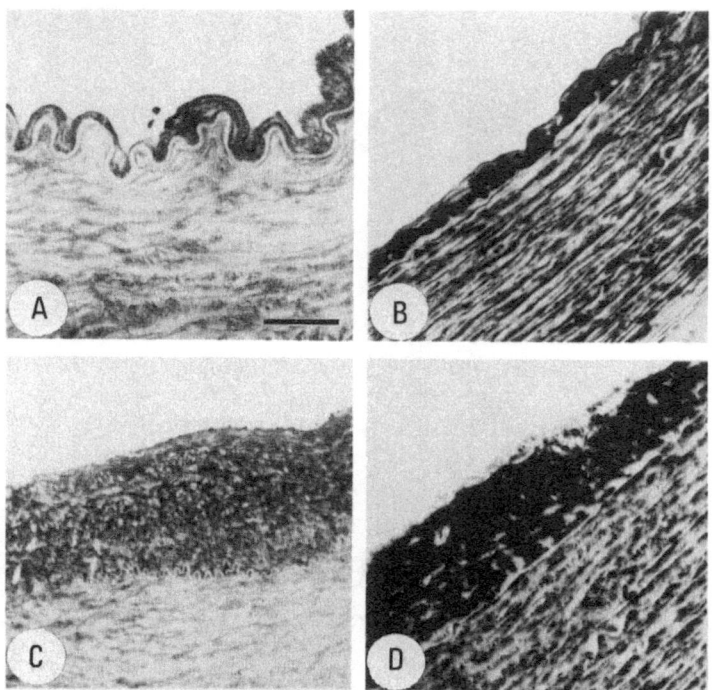

Figure 4. Light micrograph illustrating that the narrow intima of a normal blood vessel stains more intensely with (A) alcian blue and (B) a monoclonal antibody against aortic CSPG than the underlying medial layer. Vessels undergoing intimal hyperplasia (early atherosclerosis) possess thickened intimas that also stain intensely with alcian blue (C) and a monoclonal antibody against aortic CSPG (D). x 281. Reproduced with permission from Wight et al. In: Wight, T.N. and Mecham, R.P., eds. *Biology of Proteoglycans*. Orlando, FL: Academic Press, 1987: 267-300. Bar = 50 µm.

(Figure 6B). Such studies raise the interesting possibility that regenerating endothelium contributes to the proteoglycan composition of the arterial wall and in some way influences the metabolism of proteoglycans and lipids present in this region of the blood vessel. The ability of proteoglycans to interact with lipoproteins is well documented and this interaction may be one mechanism by which lipoproteins become trapped during lesion formation (see review 4).

In an attempt to understand why proteoglycans accumulate in the vessel wall, we have turned to cell culture to investigate factors that regulate proteoglycan metabolism by vascular cells. Vascular endothelial and smooth muscle cells synthesize a variety of different proteoglycans (reviewed in 6). Such proteoglycans can usually be identified in culture by extracting sulfate-labelled macromolecules in the different culture compartments using strong dissociative solvents with detergent, and separating proteoglycans from glycoproteins by ion exchange chromatography, molecular sieve chromatography, and/or SDS PAGE. Using this approach, we have found that endothelial cells synthesize both heparan and dermatan/chondroitin proteoglycans. The heparan sulfate proteoglycans from endothelial cells contain molecular species which have a variety of functions such as potentiating the inactivation of thrombin by binding to antithrombin III (11), binding to lipoprotein lipase (12), and inhibiting smooth muscle cells from proliferating (11,13). The dermatan/chondroitin class appear to consist of a specific class of small proteoglycan ($\approx$ 200-240 kD) recently named biglycan (14,15), which may serve as a precursor for some of the other forms of DS-PGs found in endothelial cell cultures (Kinsella and Wight, unpublished observations). The

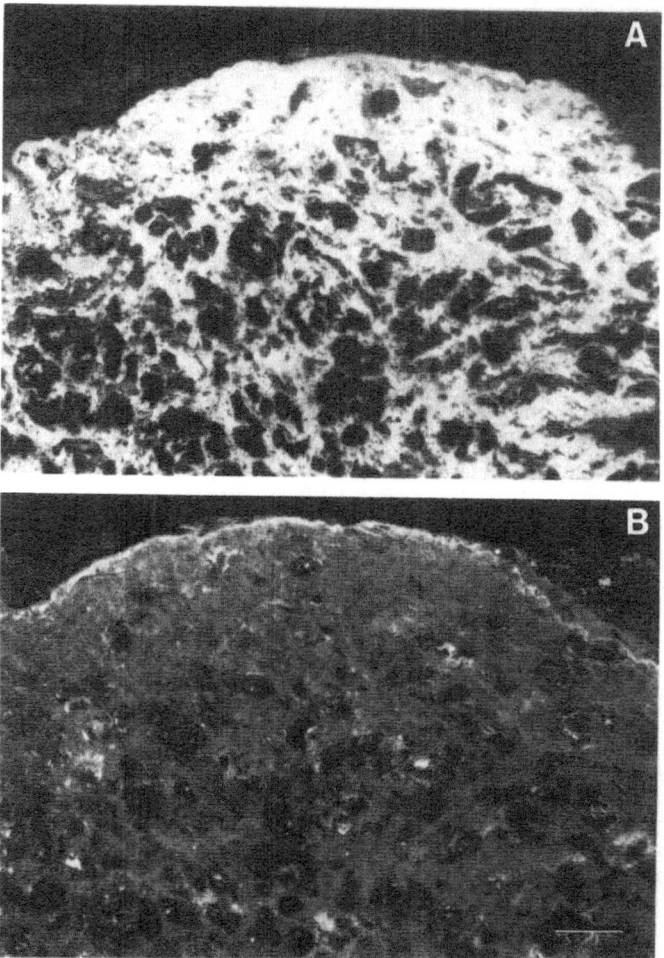

Figure 5. Sections of advanced human atherosclerotic lesions of the superficial femoral artery immunostained with anti-DSPG antisera (A) and antifibronectin antisera (B). Note the intense immunofluorescence demonstrated with anti-DSPG antisera. x 394. Bar = 20 μm.

role for this class of proteoglycan is not known, but it is of interest that these cells do not contain mRNA for the other small DS-PG (125-160 kD), decorin (15), which is synthesized by most other cells, including smooth muscle cells, and which appears to bind fibrillar collagen (reviewed in 6).

Proteoglycan metabolism by endothelial cells is influenced by events related to both development and disease. For example, proteoglycan synthesis is markedly stimulated when endothelial cells are stimulated to migrate (16) (Figure 7). Analysis of proteoglycans synthesized by migrating endothelial cells indicates a shift toward the deposition of the DS/CS-PG class rather than the HS-PG class. Autoradiographic analysis of the cultures revealed that migrating cells exhibited increased radiolabeling with 35S-sulfate. Inhibition of DNA synthesis did not interfere with this proteoglycan metabolic modulation or with the migratory behavior of the endothelial cells. It may be that HS-PG promotes cell attachment in this system while CS/DS-PG destabilizes attachment sites and facilitates migration. Whether these proteoglycan changes are required for endothelial cell movement needs further study.

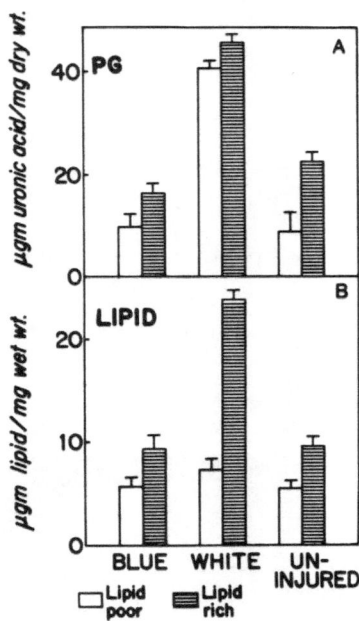

Figure 6. Correlation of lipid and proteoglycan concentration in de-endothelialized and re-endothelialized aortas in rabbits on normal and lipid-rich diets. Note that the re-endothelialized unstained (white) regions contained the greatest concentration of proteoglycan (PG) and lipid in the fat-fed animals when compared to the de-endothelialized, Evans blue-stained (blue) regions. The finding that the white regions contained significantly more PG than the blue regions in animals on regular diets suggests that this increase in PG may predispose the white region to lipid accumulation.

Unlike endothelial cells that synthesize predominantly HS-PGs, vascular smooth muscle cells synthesize principally CS-PGs and DS-PGs (reviewed in 6). In general, these proteoglycans resemble the bulk of the proteoglycans present in intact arteries. The major proteoglycan synthesized by arterial smooth muscle cells is a large CS-PG ($\approx$ 1.2 x 10^6 D) which is present throughout the interstitial matrix of blood vessels. These cells also synthesize two forms of the smaller DS-PGs: one form that contains one dermatan sulfate chain and a 45 kd core glycoprotein (decorin) and a larger form ($\approx$ 225 kd) which contains two dermatan sulfate chains with the same size core protein as the smaller form. In addition to these extracellular matrix forms, smooth muscle cells also possess at least two membrane associated forms: one that contains chondroitin sulfate and one that contains heparan sulfate (Aulinskas, Schönherr and Wight, unpublished observations).

Two key events in the development of the atherosclerotic plaque are the proliferation of arterial smooth muscle cells and the deposition of components of the extracellular matrix. An important question is whether these two events are related. Proteoglycan synthesis occurs principally during the G1 phase of the cell cycle when quiescent arterial smooth muscle cells are stimulated to divide (17); however, it is not clear whether cell proliferation is a prerequisite for elevated proteoglycan synthesis by arterial smooth muscle cells. To address this question, we have evaluated whether platelet derived growth factor (PDGF), which stimulates arterial smooth muscle cell proliferation, and transforming growth factor B (TGF-ß) which inhibits these cells from dividing, differentially affect proteoglycan metabolism (Figure 8). Both PDGF and TGF-ß stimulated proteoglycan synthesis by these cells and both appeared to preferentially affect the large CS-PG interstitial matrix class (18). Both of these peptides influenced the post-translational processing of the proteoglycans. For example, the CS-PG isolated from both the PDGF and TGF-ß treated cultures contained CS chains that were considerably longer than found in the CS-PG from quiescent cultures. Interestingly, the CS chains from the growth stimulated cultures had a 6-sulfate to 4-sulfate disaccharide ratio which was almost twice that of the CS chains in the growth inhibited cultures (i.e., TGF-ß). These observations suggest that proliferating arterial smooth muscle cells are synthesizing a CS-PG enriched in C-6 sulfate. The functional significance of these differences needs further study.

Another proteoglycan change associated with arterial smooth muscle cell proliferation appears to involve a membrane associated class of CS-PG. Thus, we found that an antibody directed against a CS-PG present on the surface of melanoma cells (19) exhibits elevated binding to arterial smooth muscle that has been stimulated to divide. A similar antibody immunostains proliferative lesions in blood vessels (20) so it may be that this CS-PG is specifically expressed when arterial smooth muscle cells proliferate. Whether the expression of this specific form of CS-PG is required for cell proliferation is currently under investigation.

Finally, heparin, a highly sulfated glycosaminoglycan, influences vascular wall structure by inhibiting smooth muscle cell proliferation (11). However, little is known as to whether heparin also influences the extracellular matrix. Thus, we have examined the collagen, elastin and proteoglycan content of experimentally injured vessels following a two week administration of heparin (Figure 9). Heparin infusion markedly decreased intimal thickening (21) and quantitative electron microscopy of the lesions

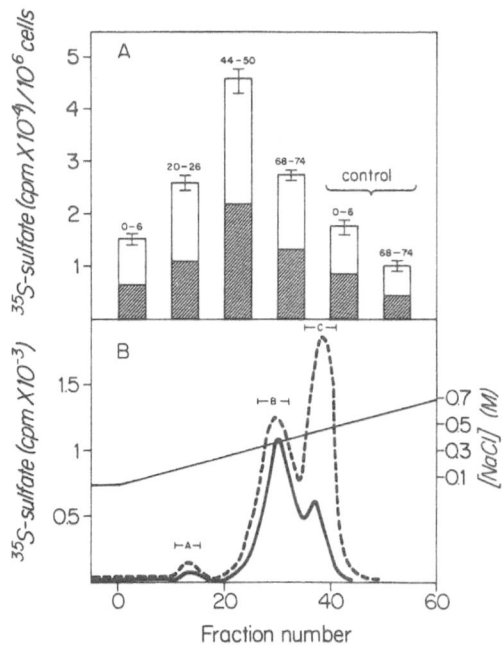

Figure 7. A. Time course of ^{35}S-sulfate incorporation into proteoglycan after multiscratch wounding of confluent endothelial monolayer cultures. Maximum incorporation (proteoglycan synthesis) is observed at 44 to 50 hours at the time when endothelial cell migration is maximal. Shaded areas represent proportion of total radioactivity present in the cell layer. B. DEAE-Sephacel ion exchange chromatography of ^{35}SO$_4$-labeled cell layer extracts from confluent (—) and wounded (- - -) cultures 48 hours after wounding. Note the large increase in the proportion of radioactivity associated with DSPG containing peak C.

revealed a marked decrease in the collagen and elastin content and an increase in the proteoglycan content throughout the intima, when compared to the intima from controls infused with saline instead of heparin. Immunohistochemistry demonstrated decreased staining of the intima in the heparin group with elastin antibodies but increased staining with antibodies against CS-PG and DS-PG. Such studies emphasize the need to consider the nature of the ECM when such therapeutic compounds are administered since such large changes in ECM components influence the viscoelastic properties of the vessel wall. The mechanism(s) responsible for heparin's effect on fibrous proteins and proteoglycans is under investigation. Recent studies have demonstrated that heparin stimulates proteoglycan synthesis while decreasing collagen synthesis by arterial smooth muscle cells *in vitro* (22).

In summary, this review has focused on the role of proteoglycans in vascular wall biology. There are a number of ways that proteoglycans can influence key events in both vascular wall development and disease. We know that a number of factors contribute to alterations in vascular proteoglycans and such alterations may in turn regulate several key processes fundamental to the development of the atherosclerotic plaque. We have just begun to unravel several fascinating facts about these complex molecules and the future promises to be both exciting and informative as we continue to explore the importance of proteoglycans in atherosclerosis.

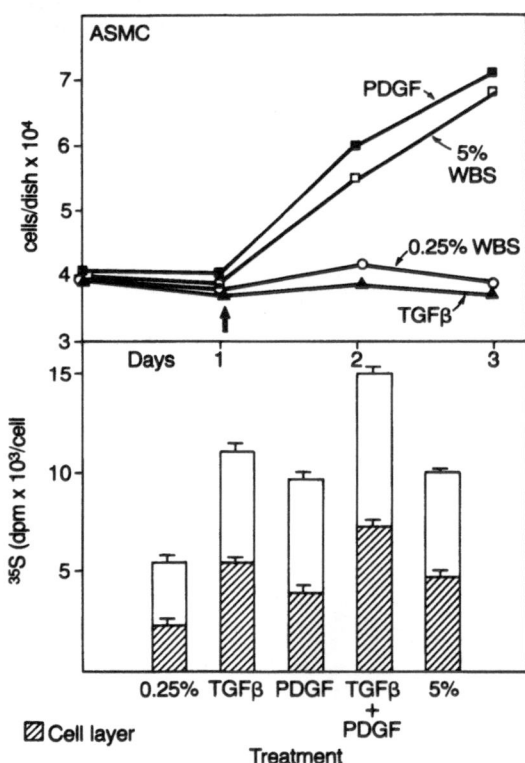

Figure 8. Top Panel. Growth curve of arterial smooth muscle cells. Cells were made quiescent in medium containing 0.25% serum for 48 hours and stimulated with either PDGF (10 ng/ml), TGF-ß (1 ng/ml) or 5% serum. Cell numbers were determined in duplicate on days 1, 2, and 3 following stimulation. Bottom Panel. Synthesis and secretion of proteoglycans by arterial smooth muscle cells in the presence of TGF-ß and PDGF. The incorporation of sulfate into proteoglycan was determined by CPC precipitation. Cell layer (hatched) and medium (white) samples are represented. Labelling was done for the first 24 hours following addition of the growth factors.

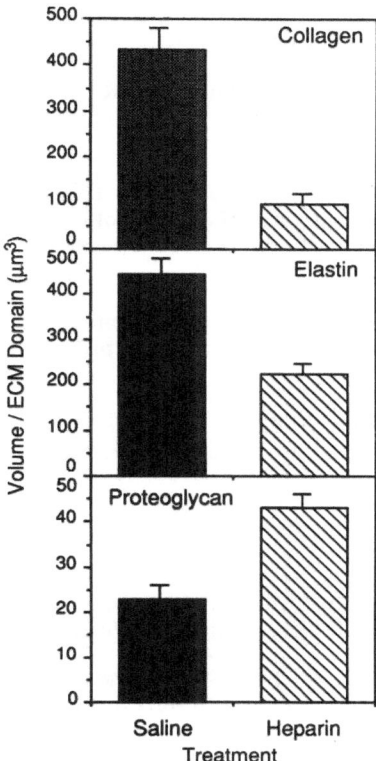

Figure 9. The volume of collagen, elastin and proteoglycan as determined by morphometry in experimentally injured blood vessels from saline or heparin treated animals. Animals (rats) were subjected to balloon catheterization and then infused with either saline or heparin for two weeks. The volume of collagen, elastin and proteoglycan was determined by ultrastructural stereology (21).

ACKNOWLEDGEMENTS

I would like to thank the many colleagues who have provided the data to include in this review. These include Drs. Michael Kinsella, Michael Lark, Tet-Kin Yeo, Alan Snow, Elke Schöenherr and Hannu Järväläinen. This work has been supported in part by a NIH Grant HL 18645. Special thanks to Ms. Kerri Wight and Mrs. Barbara Kovacich for the typing of this manuscript.

REFERENCES

1. Berenson, G.S., Radhakrishnamurthy, B., Srinivasan, S.R., Vijayagopal, P., Dalferes, E.R., Jr. and Sharma, C.   1984.   Recent advances in molecular pathology. Carbohydrate-protein macromolecules and arterial wall integrity - A role in atherogenesis. Exp. Mol. Pathol. 41: 267-287.

2. Camejo, G. 1982. The interaction of lipids and lipoproteins with the intercellular matrix of arterial tissue: Its possible role in atherogenesis. Adv. Lipid Res. 19: 1-53.

3. Castellot, J.J., Jr., Addonizio, M.L., Rosenberg, R., and Karnovsky, M.J. 1981. Cultured endothelial cells produce a heparinlike inhibitor of smooth muscle growth. J. Cell Biol. 90: 372-379.

4. Falcone, D.J., Hajjar, D.P. and Minick, C.R. 1980. Enhancement of cholesterol and cholesteryl ester accumulation in re-endothelialized aorta. Am. J. Pathol. 99: 81-104.

5. Falcone, D.J., Hajjar, D.P. and Minick, C.R. 1984. Lipoprotein and albumin accumulation in reendothelialized and deendothelialized aorta. Am. J. Pathol. 114: 112-120.

6. Fisher, L.W., Termine, J.D. and Young, M.F. 1989. Deduced protein sequences of bone small proteoglycan I (Biglycan) shows homology with proteoglycan II (Decorin) and several non-connective tissue proteins in a variety of species. J. Biol. Chem. 264: 4571-4576.

7. Garrigues, H.J., Lark, M.W., Lara, S., Hellstrom, I., Hellstrom, K.E. and Wight, T.N. 1986. The melanoma proteoglycan: restricted expression on microspikes. J. Cell Biol. 103: 1699-1710.

8. Hascall, V.C. and Hascall, G.K. 1981. Proteoglycans, in: Cell Biology of Extracellular Matrix, E.D. Hay, ed., Plenum Press, New York.

9. Hassel, J.R., Kimura, J.H. and Hascall, V.C. 1986. Proteoglycan core protein families. Annu. Rev. Biochem. 55: 539-567.

10. Järväläinen, H., Kinsella, M.G., Sandell, L.J. and Wight, T.N. 1989. The small dermatan sulfate proteoglycan II (PG II) is expressed by bovine arterial smooth muscle cells but not by endothelial cells. J. Cell Biol. 109: 233a.

11. Kinsella, M.G. and Wight, T.N. 1986. Modulation of sulfated proteoglycan synthesis by bovine aortic endothelial cells during migration. J. Cell Biol. 102: 679-687.

12. Klinger, M.M., Margolis, R.U. and Margolis, R.K. 1985. Isolation and characterization of the heparan sulfate proteoglycans of brain. Use of affinity chromatography on lipoprotein lipase agarose. J. Biol. Chem. 260: 4082-4090.

13. Marcum, J.A., Reilly, C.F. and Rosenberg, R.D. 1987. Heparan sulfate species and blood vessel wall function, in: Biology of Extracellular Matrix: Biology of Proteoglycans, T.N. Wight and R.P. Mecham, eds., Academic Press, Orlando, FL.

14. McEvoy, L.M. and Bumol, T.F. 1988. Biosynthesis and localization of the core glycoprotein of chondroitin sulfate in primate aorta. J. Cell Biol. 107: 156a.

15. Richardson, M., Ihnatowycz, I. and Moore, S. 1980. Glycosaminoglycan distribution in rabbit aortic wall following balloon catheter deendothelialization. An ultrastructural study. Lab. Invest. 43: 509-516.

16. Schöenherr, E., Sandell, L.J. and Wight, T.N. 1989. Differential effect of PDGF and TGFß on proteoglycan and DNA synthesis by cultured arterial smooth muscle cells and chondrocytes. J. Cell Biol. 109: 234a.

17. Snow, A.D., Bolender, R.P., Wight, T.N. and Clowes, A.W. 1990. Heparin modulates the composition of the extracellular matrix domain surrounding arterial smooth muscle cells. Amer. J. Pathol. in press.

18. Tan, E.M.L., Levine, E., Sorger, P., Unger, G.A., Hacobian, N., Planick, B. and Iozzo, R.V. 1989. Heparin and endothelial cell growth factor modulate collagen and proteoglycan production in human smooth muscle cells. Biochem. Biophys. Res. Comm. 163: 84-92.

19. Wight, T.N., Curwen, K.D., Litrenta, M.M., Alonso, D.R. and Minick, C.R. 1983. Effect of endothelium on glycosaminoglycan accumulation in injured rabbit aorta. Am. J. Pathol. 113: 156-164.

20. Wight, T.N., Potter-Perigo, S. and Aulinskas, T. 1989. Proteoglycans and vascular cell proliferation. <u>Am. Rev. Resp. Dis.</u> 140: 1132-1135.

21. Wight, T.N. 1988. Cell biology of arterial proteoglycans. <u>Arteriosclerosis</u> 1: 1-39.

22. Wight, T.N. 1980. Vessel proteoglycans and thrombogenesis, <u>in</u>: Progress in hemostasis and thrombosis, T.H. Spaet, ed., Grune and Stratton, New York.

ENDOTHELIAL CELL - EXTRACELLULAR MATRIX INTERACTIONS:

MODULATION OF VASCULAR CELL PHENOTYPE

BY MATRIX COMPONENTS AND SOLUBLE FACTORS

Joseph A. Madri

Department of Pathology
Yale University, School of Medicine
New Haven, CT 06510

ABSTRACT

The vessel wall is comprised of several different cell populations residing in and on a complex extracellular matrix. Each of the resident cell types has diverse functions and morphologies and each has a role in repair processes following injury. Responses to injury vary depending upon the extent and type of the injury and the vascular beds affected. Briefly, large vessel endothelial cells respond to denudation injury by sheet migration and proliferation. This is in contrast to the invasion into and migration through soft tissues and tube formation with lumen formation exhibited by microvascular endothelial cells in response to injury. Vascular smooth muscle cells of larger vessels respond to intimal injury by migration from the arterial media into the intima, followed by proliferation and matrix biosynthesis, ultimately giving rise to intimal thickening. Microvascular pericyte response to injury is less well-documented. Specifically, responses to injury by these vascular cell types appears to be modulated, in part, by the composition and organization of the surrounding matrix as well as by the various platelet factors and cytokines found at sites of injury. These observations suggest that extracellular matrix and soluble factors modulate each other's effects on local vascular cell populations following injury.

INTRODUCTION

The lumen of the vascular system is lined by mitotically quiescent, metabolically active endothelial cells which, in addition to having a broad range of metabolic activities, provide a non-thrombogenic surface for blood flow. Beneath the luminal endothelium, smooth muscle cells are found in the media of large vessels, and pericytes are found in close association with the endothelial cells of microvascular beds. The smooth muscle cells and pericytes are known to play major roles in maintaining vessel wall integrity, being responsible for the maintenance of the connective tissues of the vessel wall as well as controlling vascular tone[7]. The various vascular cell populations (large and small vessel derived endothelial, pericyte and smooth muscle cells) have been found to respond to injury in specific ways, depending upon the vascular bed(s) and the cell type(s) affected. In muscular and elastic arteries, following denudation injury evoked by angioplasty, endarterectomy or autologous or synthetic grafting, the endothelial cells bordering the affected area will exhibit rapid sheet migration over the exposed extracellular matrix and proliferate in an attempt to reconstitute the normal continuous endothelial cell lining[17,19]. In contrast, the medial smooth muscle cells of large and medium-sized vessels respond to intimal injury by migrating into the intima, where they proliferate and synthesize matrix components, which results in the formation of a thickened intima which narrows the vessel lumen[32]. Following soft tissue injury or in response to a variety of angiogenic factors, microvascular endothelial cells respond by releasing themselves from the constraints of their investing basement membranes. Following this they migrate and proliferate in the surrounding three-dimensional interstitial stroma and ultimately form new microvessels[21]. The role(s) of pericytes (smooth muscle cell analogs which are found surrounding microvessels following new vessel formation) during the reparative response is less well-studied and their origin(s) (endothelial, undifferentiated mesenchymal-fibroblastic or vascular smooth muscle cell) is still a matter of controversy[12,21].

In this paper we will discuss the modulation of large vessel endothelial and smooth muscle cell responses to *in vivo* denudation and *in vitro* injury by components of the extracellular matrix and selected soluble factors, including platelet-derived growth factor (PDGF) and transforming growth factors ß1 and ß2 (TGF-ß1 and TGF-ß2).

METHODS AND MATERIALS

Cells: Bovine aortic endothelial (BAEC) and bovine aortic smooth muscle (BASMC) cells were isolated, cultured and characterized as described[14,16].

Matrices: Collagen types I, III, IV and V, laminin and fibronectin were isolated, purified and characterized as described[15,19,20]. Bacteriologic plastic culture dishes were coated with matrix components[19].

Cell Proliferation: Proliferative rates were determined by cell counting using a Coulter Counter and by quantitating cellular DNA[18,25].

Migration Assay: BAEC and BASMC migration rates were quantitated using a previously described assay utilizing a teflon fence to initiate migration[28]. Migration areas and migration rates were determined using a Macintosh Plus computer.

Immunofluorescence: Immunofluorescence of BAEC, BASMC monolayer cultures were performed as described[2,35].

Electron Microscopy: Electron microscopy was performed using standard procedures[25].

ELISA: Quantitative assays of matrix components were performed using an ELISA assay developed in the laboratory[6,16,23].

Northern Blot Analyses: Quantitation of mRNA levels of selected matrix components and cell surface matrix binding proteins was performed by Northern blotting techniques[12,23].

Growth Factors: TGF-ß1 and TGF-ß2 were generous gifts of Drs. Anita Roberts and Michael Sporn, Laboratory of Chemoprevention, NCI, NIH, Bethesda, MD. PDGF-AA, AB and BB were purchased from Collaborative Research, Inc., Boston, MA.

RESULTS AND DISCUSSION

The Effects of Extracellular Matrix Components and Soluble Factors
on Bovine Aortic Endothelial Cell Migration *In Vitro* and *In Vivo*

Mechanical, viral or chemical injury to the endothelium, usually elicits platelet aggregation and release of growth factors and vasoactive agents, including PDGF, TGF-ß1, serotonin, norepinephrine and histamine, which promote vascular smooth muscle cell chemotaxis, migration and proliferation. These responses to injury have been proposed as key events in the development of arteriosclerosis[24]. In addition, many of the currently available invasive treatments of large and medium vessel occlusive disease (angioplasty, autologous and synthetic grafting, endarterectomy, atherectomy and laser ablation) also result in significant de-endothelialization, exposure of matrix

128

components and medial injury of vessel segments, and the resultant platelet adhesion, aggregation and release of soluble factors and "activation" of local medial smooth muscle cells are thought to be associated with the development of post-therapy stenosis that is noted in up to 50% of the patient populations studied[27]. During the past several years we have investigated the role of matrix components and selected soluble factors in the modulation of endothelial and vascular smooth muscle cell attachment, proliferation and migration[4,20,22,24].

It is now widely accepted that matrix components have a significant effect on aortic endothelial cell migration *in vitro*. Using an *in vitro* migration assay that was developed in our laboratory, we have demonstrated that interstitial collagens elicit the most rapid migration rate, basement membrane components elicit intermediate migratory rates, while fibronectin was found to elicit the lowest migration rate[13,19,20,28,29]. Additionally, we have found that large vessel endothelial cell migratory behavior on these substrates could be modulated by altering the amounts of particular matrix components coated on the culture dishes and migratory behavior could be correlated with the organizational patterns and dynamics of selected cytoskeletal components such as fodrin[28], protein 4.1[13], vinculin and α-actinin[22]. Specifically, bovine aortic endothelial cells (BAEC) cultured on substrates that elicit high to intermediate migratory rates, including collagen types I, III, IV and laminin, were also noted to reorganize the cortical cytoskeletal components fodrin and protein 4.1 following stimuli to migrate. In contrast, substrates which elicit low migratory rates, such as fibronectin, do not allow reorganization of these cytoskeletal components following stimuli to migrate[13,28]. These data are consistent with the concept that the existing and possibly newly synthesized extracellular matrix modulates cellular behavior by affecting the organization and dynamics of the cytoskeleton presumably via cell surface matrix binding proteins[19,20,22,24,28].

We have recently shown that both integrin and non-integrin matrix binding proteins participate in *in vitro* BAEC adhesion and migration events[2,35]. Immunoprecipitation and immunoblot analyses revealed that BAEC express at least two different integrin heterodimers, ß1 and ß3 class integrins. Attachment assays using RGD-containing peptides documented dose-dependent RGD sensitivity on fibrinogen, laminin and fibronectin substrates. Immunofluorescence studies of BAEC ß1 and ß3 integrins revealed rapid matrix protein-specific organization within one hour of plating. ß-1 integrin molecules were observed to become organized in linear stress fiber-type patterns on fibronectin and laminin substrates. In contrast, ß3 integrins were noted to organize into punctate patterns on a fibrinogen substratum at this early time point. The non-integrin laminin binding protein LB69 was not observed to organize on any of these matrix components at this early time point[2]. However, at later time points, LB69 was noted to organize following plating, suggesting a spatiotemporal segregation of integrin and non-integrin binding proteins during adhesive events. Supporting this concept of a spatiotemporal segregation of integrin and non-integrin binding proteins during adhesive events in BAEC are the findings that antibodies directed against both integrin and RGD-containing peptides inhibit BAEC attachment and spreading, while antibodies directed against LB69 as well as YIGSR-containing peptides only affect BAEC spreading. In addition, selective disorganization of ß1 integrins (but not LB69) are noted following RGD-containing peptide inhibition of BAEC sheet migration on laminin substrates. Additional studies have demonstrated that, while the underlying matrix elicits specific patterns of integrin organization of BAEC, there are no changes in the sizes of cell surface pools of ß1 or ß3 integrins. In contrast, selected soluble factors (TGF-ß1 and PDGF) were observed to modulate (increase) sizes of cell surface pools of BAEC ß1 or ß3 integrins and BASMC ß3 integrins but do not alter integrin organization in these two cell types *in vitro*[3]. These data suggest a complex modulation of vascular cell behavior occurring, in part, through a coordination of soluble factor and extracellular matrix protein regulation of extracellular matrix integrin expression and organization[2,3,23].

In addition, we have found that, in a rat carotid balloon de-endothelialization model in which there is incomplete re-endothelialization of the de-endothelialized area, there is increased fibronectin and TGF-ß staining throughout the media and luminal surface of the chronically de-endothelialized region of the vessel[23]. These *in vivo* findings correlate well with *in vitro* studies in which BAEC migration on a type I collagen coating was <u>inhibited</u> and BASMC migration was <u>enhanced</u> by: A) the addition of soluble fibronectin to the culture media; B) coatings of increasing fibronectin concentration and C) TGF-ß1 treatment of migrating cells in which there is a significant increase in fibronectin mRNA and protein levels[24]. Thus, both *in vivo* and *in vitro*, the composition of the existing and newly synthesized underlying extracellular matrix, as well as the presence of soluble factors (TGF-ß1), appear to have profound effects upon the migratory abilities of large vessel endothelial and medial smooth muscle cells. Additional studies using this *in vivo* model have revealed distinct differences in the localizational patterns of ß1 or ß3 integrins in the neointimal and luminal regions of both the chronically de-endothelialized and re-endothelialized areas[3]. Specifically, in the chronically de-endothelialized area, neointimal smooth muscle cells exhibited increased staining for ß3 integrins but no changes in ß1 integrins compared with normal and post-injury medial smooth muscle cells. This observation is consistent with the effects of TGF-ß1 and PDGF on BASMC noted *in vitro*. In contrast, in the re-endothelialized area, the neointimal smooth muscle cells nearer the lumen display less intense staining for ß3 integrins (as well as for fibrinogen) than those deeper in the neointima. These observations support the concept of endothelial cell modulation of smooth muscle phenotype.

In addition to these particular matrix- and soluble factor-mediated effects, BAEC have been found to be responsive to a variety of soluble factors including those obtained from platelet releasates. In the past, several investigators have demonstrated effects on motility, proliferation, morphology and surface protein expression of large vessel endothelial cells when they have been incubated with a variety of soluble factors including PDGF, TGF-ß1, serotonin, histamine and a variety of cytokines[1,4,5,8,10,31,34]. In recent studies, the platelet factors histamine, serotonin, norepinephrine, PDGF and TGF-ß1 were all found to decrease BAEC sheet migration while having markedly different effects on BAEC proliferation. In addition, inhibition of proliferation with mitomycin C treatment did not alter the effects of these added factors, supporting the concept that the processes of migration and proliferation are not linked[4]. It is important to recognize that platelets are not the sole cellular source for any of these factors and it is conceivable that the physiologic sources of these factors may be non-platelet mesenchymal cells resident in the vessel wall as well as circulating cells.

Preliminary studies investigating possible mechanisms by which extracellular matrix components and TGF-ß1 mediate changes in BAEC migratory behavior have revealed that both matrix composition and soluble factors modulate levels of BAEC plasminogen activator and plasminogen activator inhibitor-1, and that inhibitors of serine proteases decrease migration rates. Specifically, plasminogen activator activity levels were found to be decreased by TGF-ß1 treatment. In contrast, plasminogen activator inhibitor-1 levels were increased by treatment with TGF-ß1 (L. Bell & J. Madri, *in preparation*). These data are consistent with the concept that changes in this protease-protease inhibitor system may also play a role in modulating migratory behavior.

<u>The Effects of Extracellular Matrix and Soluble Factors on Bovine Aortic Smooth Muscle Cell Migration *In Vitro* and *In Vivo*</u>

In addition to the pivotal roles played by the overlying endothelial cells in modulating the responses to vascular injury and repair, medial smooth muscle cells also have an important role in the response to injury and subsequent repair processes. Vascular smooth muscle response to injury is one of migration from the media into the intima, where they proliferate and synthesize extracellular matrix. This response (driven by mononuclear cell and platelet factors, loss of overlying endothelial cells and autocrine/paracrine factors produced by the smooth muscle cells themselves as well as by dysfunctional endothelium at or near the site of injury) is thought to lead to intimal thickening and ultimately atherosclerosis[27,31]. It is known that vascular smooth muscle cell migration and proliferation are modulated, in part, by the composition of the surrounding extracellular matrix and platelet factors (such as PDGF and TGF-ß) and the migratory and proliferative responses of this cell type contribute to atherogenesis[4,9,32]. In recent studies we have found that the platelet factors histamine, serotonin, norepinephrine, PDGF and TGF-ß1 all increase bovine and rat aortic medial smooth muscle cell migration, while having markedly different effects on BASMC proliferation. In addition, as noted in experiments using BAEC, inhibition of proliferation with mitomycin C treatment did not alter the effects of these added factors, again supporting the concept that the processes of migration and proliferation are not linked[4]. Further studies have revealed that, in addition to increasing BASMC migratory rate, TGF-ß1 treatment increases fibronectin mRNA and protein levels in confluent and migrating BASMC cultures and that TGF-ß1 and PDGF treatment of cultured BASMC elicits an increase in the cell surface pool size of ß3 integrins. These findings support the concept that TGF-ß1 may be eliciting its effects on migration, in part, via the selective modulation of matrix component(s) and cell surface matrix receptors including the integrins[3,24].

Thus, it appears that selected matrix components (fibronectin) and soluble factors (TGF-ß1) elicit opposite effects in aortic endothelial and smooth muscle cells, enhancing migration in one case (BASMC) while inhibiting migration in the other (BAEC). These effects may, in part, lead to the development and progression of atherosclerotic lesions. In addition, TGF-ß1 appears to have its effects on these vascular cells, in part, by modulating the matrix synthetic profiles of the cells, eliciting increases in fibronectin mRNA and protein levels in both aortic endothelial and smooth muscle cells as well as by selectively modulating the sizes of surface pools of ß1 and ß3 integrins in these cell types. Furthermore, as noted for BAEC, both matrix composition and soluble factors (TGF-ß1) modulate plasminogen activator and plasminogen activator inhibitor-1 levels in BASMC. In contrast to BAEC, BASMC plasminogen activator levels are decreased in cells plated and migrating on a fibronectin substrate compared to type I collagen and increased in response to TGF-ß1. Conversely, BASMC levels of plasminogen activator inhibitor-1 are increased following TGF-ß1 treatment (L. Bell & J.A. Madri, *in preparation*). These data, taken together with our observations in BAEC cultures and in the rat carotid balloon de-endothelialization model which we use, are consistent with the concept that modulation of vascular cell migratory behavior is complex and involves several mechanisms including protease-protease inhibitor systems, changes in selected matrix component synthesis and organization, changes in expression and organization of cell surface matrix binding proteins (integrins and non-integrins) and cytoskeletal reorganization[2,3,16,18,22,23,24].

TGF-ß Isoforms as Modulators of Large Vessel Vascular Cell Behavior

Transforming growth factors ß are thought to play important regulatory roles in many biological processes including organ and tissue development, wound healing and arteriosclerosis. TGF-ß1 is one member of a family of polypeptides including the heterodimer TGF-ß1.2, the homodimer TGF-ß2 and the more recently discovered isoforms TGF-ß3, TGF-ß4 and TGF-ß5[24]. TGF-ß1 and TGF-ß2 are the most intensely studied members of this family and have been found to have similar, if not identical, activities in many mesenchymal cell types[33]. However, recent studies suggest that they elicit different responses in selected cell types[30]. Previous studies on large vessel endothelial cells have revealed that TGF-ß2 had no effect on proliferation[11]. In light of these data and the observation that TGF-ß1 and TGF-ß2 are present in differing amounts in a variety of cell types[33], it is likely that these TGF-ß isoforms may prove to be important in modulating the behavior of vascular cells in the processes of wound healing, angiogenesis and atherosclerosis[26]. In recent studies we have confirmed the observation that TGF-ß2 has no effect on BAEC proliferation and demonstrated that TGF-ß2 does not compete at all with TGF-ß1 in proliferation assays and only modestly in migration assays. TGF-ß2 also does not compete with TGF-ß1 for receptor binding using morphological assays (competitive binding with biotinylated TGF-ß1 followed by avidin-gold localization), FACS analyses (competitive binding with biotinylated TGF-ß1 followed by avidin-FITC fluorescence analyses), and radioligand binding assays (competitive binding with ^{125}I-TGF-ß1 followed by crosslinking, SDS-PAGE and autoradiography)[26]. In contrast, we have observed that the proliferation of cultured bovine aortic medial smooth muscle cells is inhibited by TGF-ß2 to the same extent as

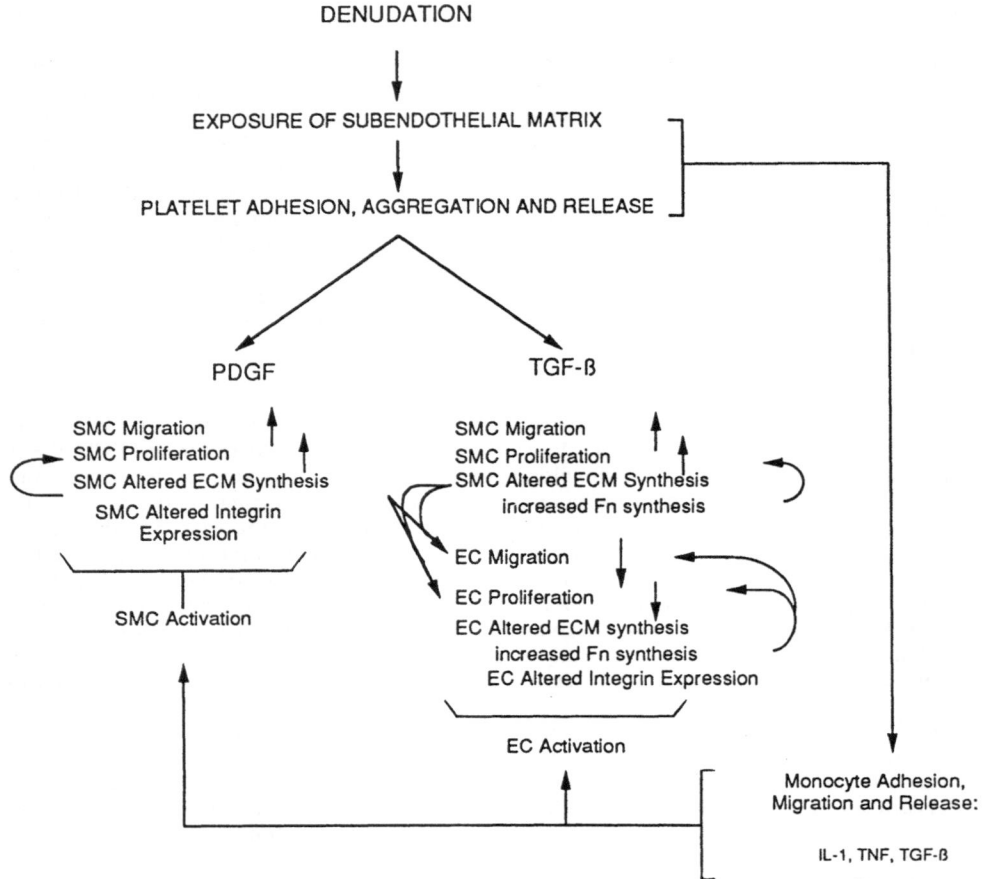

Figure 1. Schematic representation of the modulation of medial smooth muscle cell (SMC) and endothelial cell (EC) behavior following denudation injury by soluble factors (PDGF and TGF-ß1) and extracellular matrix components (Fn). In this scheme soluble factors elicit changes in matrix production and integrin expression which, in turn, modulate cell migration and proliferation.

by TGF-ß1, while smooth muscle cell migration was unaffected by TGF-ß2 at all concentrations tested (0.05 to 5.0 ng/ml)[26]. In addition, ^{125}I-TGF-ß1 crosslinking analyses revealed differences in the Type I to Type II TGF-ß receptor ratios in these two vascular cell types, namely BAEC 1:1 and BASMC 3:1. Whether these differences in receptor type ratio mediate the differences in the TGF-ß1/TGF-ß2 proliferative and migratory responses noted in these cell types, or whether they just correlate with these differences in cellular responses has yet to be determined. Nonetheless, these data suggest that vascular cells derived from different vessel compartments (intima vs. media) exhibit differential sensitivities to various isoforms of TGF-ß. This differential sensitivity, and the broad range of effects elicited by TGF-ß1 and TGF-ß2, may have a significant role in modulating the regulation of inflammatory and healing responses observed throughout the vascular system.

CONCLUSIONS

In this chapter we have presented evidence supporting the concept that *in vitro* and *in vivo* specific extracellular matrix proteins elicit organization of integrin and non-integrin vascular cell surface matrix binding proteins and selected cytoskeletal elements while cytokines, such as PDGF and TGF-ß1, modulate the size of surface membrane-associated vascular cell integrin pools and influence cytoskeletal organization and matrix synthesis and protease-protease inhibitor systems.

Following balloon catheter denudation of rat carotid artery, the endothelial cells respond to the denudation injury initially by migration and proliferation. There is also an acute platelet response, plasma fibronectin deposition and synthesis by local vascular cells, and secretion and activation of TGF-ß1 by both endothelial and smooth muscle cells. Following these early responses, endothelial cell migration/proliferation is inhibited, creating a chronically de-endothelialized region of vessel wall. In addition, the deposition of plasma fibronectin (as well as other plasma components such as fibrinogen and fibrin split products), the development of a platelet releasate and the synthesis, secretion and activation of TGF-ß1 by both endothelial and smooth muscle cells stimulate medial smooth muscle cells to migrate to the area of injury, proliferate and synthesize, secrete and deposit matrix components (such as fibronectin). These events further enhance smooth muscle cell migration into the injured area. Thus, the complex interactions of vascular cell populations with surrounding and newly synthesized matrix components and a variety of cytokines elicit expression of "dysfunctional" phenotypes in local endothelial and smooth muscle cell populations which favor the development of arteriosclerosis (Figure 1).

Although still incomplete, our knowledge of how the extracellular matrix and soluble factors affect vascular cells is constantly growing. The mechanisms by which information is transduced across the plasma membrane are complex and most likely will involve a variety of matrix and soluble factor receptors, second messenger systems and dynamic cytoskeletal organizations[22,24].

ACKNOWLEDGEMENTS

I would like to thank my colleagues in the laboratory, June R. Merwin, Leonard Bell, Craig T. Basson, Olivier Kocher, Christian Prinz and Martin Marx for their continued interest and criticism.

REFERENCES

1.   Assoian, R.K. 1988. The role of growth factors in tissue repair IV: Type ß transforming growth factor and stimulation of fibrosis. In: Clark, R.F. and Henson, P., eds. The Molecular and Cellular Biology of Wound Repair. Plenum Press, New York, pp. 273-280.

2.   Basson, C.T., Knowles, W.J., Bell, L., Albelda, S.M., Castronovo, V., Liotta, L. and Madri, J.A. 1990. Spatiotemporal segregation of endothelial cell integrin and non-integrin extracellular matrix binding proteins during adhesion events. J. Cell Biol. 110: 789-802.

3.   Basson C.T., Asis, A., Reidy, M.A. and Madri, J.A. 1990. Extracellular matrix and soluble factors differentially modulate vascular cell integrins during bovine aortic cell adhesion in vitro and rat carotid atherogenesis in vivo.

4.   Bell, L. and Madri, J.A. 1989. Effect of platelet factors on migration of cultured bovine aortic endothelial and smooth muscle cells. Circ. Res. 65: 1057-1065.

5.   Cotran, R.S. and Pober, J.S. 1988. Endothelial activation: Its role in inflammatory and immune reactions. In: Simionescu, N. and Simionescu, M., eds. Endothelial Cell Biology in Health and Disease. Plenum Press, New York, pp. 335-348.

6.   Davis, B.H., Pratt, B.M. and Madri, J.A. 1987. Hepatic Ito cell culture: Modulation of collagen phenotype and cellular retinol binding protein by retinol and extracellular collagen matrix. J. Biol. Chem. 262: 10280-10286.

7.   Fishman, A.P. 1982. Endothelium. The New York Academy of Sciences, New York.

8.   Gimbrone, M.A. and Bevilacqua, M.P. 1988. Vascular endothelium: Functional modulation at the blood interface. In: Simionescu, N. and Simionescu, M., eds. Endothelial Cell Biology in Health and Disease. Plenum Press, New York, pp. 255-274.

9.   Hedin, U., Bottger, B.A., Forsberg, E., Johansson, S. and Tyberg, J. 1988. Diverse effects of fibronectin and laminin on phenotypic properties of cultured arterial smooth muscle cells. J. Cell Biol. 107: 307-320.

10.   Huang, J.S., Olsen, T.J. and Huang, S.H. 1988. The role of growth factors in tissue repair I: Platelet-derived growth factor. In: Clark, R.F. and Henson, P., eds. The Molecular and Cellular Biology of Wound Repair. Plenum Press, New York, pp. 243-252.

11.   Jennings, J.C., Mohan, S., Linkhart, T.A., Widstrom, R. and Baylink, D.J. 1988. Comparison of the biological actions of TGF ß-1 and TGF ß-2: Differential activity in endothelial cells. J. Cell Physiol. 137: 167-172.

12.   Kocher, O. and Madri, J.A. 1989. Modulation of actin mRNAs in cultured vascular cells by matrix components and TGF-ß1. In Vitro 25: 424-434.

13.   Leto, T.L., Pratt, B.M. and Madri, J.A. 1986. Mechanisms of cytoskeletal regulation: Modulation of aortic endothelial cell protein band 4.1 by the extracellular matrix. J. Cell Physiol. 127: 423-431.

14.   Madri, J.A., Dreyer, B., Pitlick, F. and Furthmayr, H. 1980. The collagenous components of subendothelium: Correlation of structure and function. Lab. Invest. 43: 303-315.

15.   Madri, J.A. 1982. The preparation of type V collagen. In: The Immunochemistry of the Extracellular Matrix, Vol. I, H. Furthmayr, ed. CRC Press, Boca Raton, Florida, pp. 75-90.

16.   Madri, J.A. and Williams, S.K. 1983. Capillary endothelial cell cultures: Phenotypic modulation by matrix components. J. Cell Biol. 97: 152-165.

17.   Madri, J.A. 1987. The extracellular matrix as a modulator of neovascularization. In: L. Gallo, ed. Cardiovascular Disease: Molecular and Cellular Mechanisms, Prevention, Treatment. Plenum Press, New York. pp. 177-184.

18.   Madri, J.A., Pratt, B.M. and Tucker, A.M. 1988. Phenotypic modulation of endothelial cells by transforming growth factor-ß depends upon the composition and organization of the extracellular matrix. J. Cell Biol. 106: 1375-1384.

19.   Madri, J.A., Pratt, B.M. and Yanniarello-Brown, J. 1988. Matrix-driven cell size changes modulate aortic endothelial cell proliferation and sheet migration. Amer. J. Pathol. 132: 18-27.

20.   Madri, J.A., Pratt, B.M. and Yannariello-Brown, J. 1988. Endothelial cell-extracellular matrix interactions: Matrix as a modulator of cell function. In: Simionescu, N. and Simionescu, M., eds. Endothelial Cell Biology in Health and Disease. Plenum Press, New York, pp. 167-190.

21.   Madri, J.A. and Pratt, B.M. 1988. Angiogenesis. In: Clark, R.F. and Henson, P., eds. The Molecular and Cellular Biology of Wound Repair. Plenum Press, New York. pp. 337-358.

22.   Madri, J.A., Kocher, O., Merwin, J.R., Bell, L. and Yannariello-Brown, J. 1989. The interactions of vascular cells with solid phase (matrix) and soluble factors. J. Cardiovasc. Pharmacol. 14: S70-S75.

23.   Madri, J.A., Reidy, M.A., Kocher, O. and Bell, L. 1989. Endothelial cell behavior following denudation injury is modulated by TGF-ß1 and fibronectin. Lab. Invest. 60: 755-765.

24.   Madri, J.A., Kocher, O., Merwin, J.R., Bell, L., Tucker, A. and Basson, C.T. 1989. Interactions of vascular cells with transforming growth factors ß. Ann. N.Y. Sci., in press.

25.   Merwin, J.R., Anderson, J., Kocher, O., van Itallie, C. and Madri, J.A. 1990. Transforming growth factor ß1 modulates extracellular matrix organization and cell-cell junctional complex formation during in vitro angiogenesis. J. Cell Physiol. 142: 117-128.

26.   Merwin, J.R., Newman, W., Beall, L.D., Tucker, A. and Madri, J.A. 1990. Vascular cells respond differentially to transforming growth factors-ß1 and ß2. In press.

27.   Munro, J.M. and Cotran, R.S.   1988.   The pathogenesis of atherosclerosis: Atherogenesis and inflammation. Lab. Invest. 58: 249-253.

28.   Pratt, B.M., Harris, A.S., Morrow, J.S. and Madri, J.A.   1984.   Mechanisms of cytoskeletal regulation: Modulation of aortic endothelial cell spectrin by the extracellular matrix. Amer. J. Pathol. 117: 337-342.

29.   Pratt, B.M., Form, D. and Madri, J.A.   1985. Endothelial cell-extracellular matrix interactions. In: Fleishmajer, R., Olsen, B. and Kuhn, K., eds. Biology, Chemistry and Pathology of Collagen. Ann. N.Y. Acad. Sci. 460: 274-288.

30.   Rizzino, A., Kazakoff, P., Ruff, E., Kuszynski, C. and Nebelsick, J.   1988. Regulatory effects of cell density on the binding of transforming growth factor ß, epidermal growth factor, platelet derived growth factor, and fibroblast growth factor. Cancer Res. 48: 4266-4271.

31.   Ross, R. 1986. Medical progress: The pathogenesis of atherosclerosis - an update. N. Engl. J. Med. 314: 488-500.

32.   Ross, R. 1988. Endothelial injury and atherosclerosis. In: Simionescu, N. and Simionescu, M., eds. Endothelial Cell Biology in Health and Disease. Plenum Press, New York. pp. 371-384.

33.   Sporn, M.B. 1990. The transforming growth factors-ß: Past, present and future. Ann. N.Y. Sci.. 593: 1-6.

34.   Terkeltaub, R.A. and Ginsberg, M.H. 1988. Platelets and response to injury. In: Clark, R.F. and Henson, P., eds. The Molecular and Cellular Biology of Wound Repair. Plenum Press, New York. pp. 35-56.

35.   Yannariello-Brown, J., Wewer, U., Liotta, L. and Madri, J.A. 1988. Distribution of a 69 kD laminin binding protein in aortic and microvascular endothelial cells: Modulation during cell attachment, spreading and migration. J. Cell Biol. 106: 1773-1786.

SECTION 4

AUTOCRINE AND PARACRINE CONTROL OF VASCULAR TISSUE

REGULATION OF PDGF EXPRESSION IN VASCULAR CELLS

Tucker Collins[1], Regina Young[2], Arturo E. Mendoza[1],
Jochen W.U. Fries[1], Amy J. Williams[1], Parvez Sultan[1]
and David T. Bonthron[3]

[1]Vascular Research Division, Department of Pathology
Brigham and Women's Hospital, 75 Francis Street
Boston, MA 02115
[2]Division of Hematology-Oncology, Children's Hospital
Department of Pediatrics, Harvard Medical School
Boston, MA 02115
[3]University of Edinburgh, Human Genetics Unit
Western General Hospital, Crewe Road
Edinburgh EH4 2XU, Scotland

INTRODUCTION

Recent experimental work on the cellular aspects of atherogenesis has been stimulated by the observation that, when blood platelets degranulate, they release protein growth factors that stimulate the migration and proliferation of vascular smooth muscle cells. One of the most well characterized of these mitogens was originally designated platelet-derived growth factor (PDGF). That PDGF has assumed major conceptual importance in understanding of atherosclerosis can be appreciated by examining a current model for atherosclerosis (reviewed in Ross, 1986; Ross *et al.*, 1986; 1990a). This model is summarized as follows: both nondenuding and denuding injury to endothelium leads to adherence of monocytes, which enter the vessel wall and secrete PDGF-like substances. Activated or injured endothelial cells can also secrete PDGF-like mitogens. Endothelial expression of PDGF may be important in signalling proliferation of adjacent perivascular cells. If injury is sufficiently extensive, platelets can aggregate, degranulate and release PDGF. The growth factor deposited in these sites is chemotactic and mitogenic for vascular smooth muscle cells. The smooth muscle cells may also be activated to secrete PDGF-like substances that may act in an autocrine fashion, inducing the intimal proliferative lesion characteristic of early atherosclerosis. The recruited smooth muscle cells elaborate the extracellular components of the atheromatous plaque (e.g., collagen and proteoglycans). If this model is correct, then understanding the structure of the PDGF-like material, as well as the regulatory mechanisms used by vascular cells to control production of the PDGF-like mitogens would be an important advance in understanding atherogenesis.

STRUCTURE OF THE PDGF GENES

Recent studies on the molecular biology of PDGF have determined the architecture of the PDGF genes and have shed new light on the potential diversity of control mechanisms.

Atherosclerosis, Edited by A. I. Gotlieb *et al.*
Plenum Press, New York, 1991

PDGF B-Chain Gene Structure

In 1983, a clear connection was established between PDGF, a naturally occurring mitogen, and the malignant transformation of cells. The acutely transforming simian sarcoma virus was isolated and the transforming gene (v-sis) sequenced (Devare *et al.*, 1983). Additionally, a normal human gene (c-sis) was noted to be highly homologous to the v-sis gene and was localized to chromosome 22 (Dalla-Favera *et al.*, 1982). Protein sequence analysis of purified PDGF revealed 92% homology between the large PDGF subunit, or the B chain, and the putative product of the v-sis transforming gene (Waterfield *et al.*, 1983; Doolittle, *et al.*, 1983). By inference, therefore, the normal cellular gene c-sis encoded the B chain of PDGF. Using the v-sis gene as a probe, the architecture of the homologous regions of the normal c-sis gene was established. Exons 2-6 and part of 7 contain the regions homologous to v-sis (Johnsson *et al.*, 1984; Josephs *et al.*, 1984). Study of the normal gene product completed the architecture of the gene (Fig. 1). The human c-sis gene is composed of seven exons spanning over 23 kb. Exon 1 contains the c-sis signal sequence and an unusually long 5'untranslated region; exon 7 contains the 3' untranslated region. Structural analysis of a PDGF B-chain cDNA from normal cultured endothelial cells (Collins *et al.*, 1985) revealed that the predicted protein was identical to that made in transformed cells (i.e., the B chain gene product made by a transformed cell did not contain point mutations). Additionally, analysis of the endothelial cDNA clone revealed that the 5' untranslated region of the B-chain was unusually long (about a kilobase) and contained multiple translational start sites upstream of the authentic open reading frame. This region of the B-chain was subsequently shown to inhibit PDGF B-chain translation in a cell free system (Ratner *et al.*, 1987) as well as in transfected cells (Rao *et al.*, 1988). The role of translational control in regulating PDGF expression is largely unexplored.

Transcriptional Regulation of PDGF B-Chain Expression

Regulation of PDGF B-chain expression in vascular cells may be important in the developing atherosclerotic lesion. Endothelial cells (Gajdusek *et al.*, 1980), smooth muscle cells (Seifert *et al.*, 1984; Walker *et al.*, 1986; Nilsson *et al.*, 1985) and macrophages (Shimokado *et al.*, 1985; Martinet *et al.*, 1986) can produce PDGF *in vitro*. All of these cell types are present in atherosclerotic plaques, but which cell type(s) synthesize PDGF in the vascular lesion is not resolved. Transcripts from the B-chain gene were present in surgically removed human carotid artery lesions at levels 5 fold greater than the low level of constitutive expression detected in the normal artery (Barrett and Benditt, 1987). Analysis of Northern blots of human carotid endarterectomy specimens revealed levels of PDGF B-chain mRNA which correlated with increased expression of c-fms (macrophage-specific) and to a lesser extent (about 10%) with von Willebrand factor (endothelial cell-specific) (Barrett and Benditt, 1988). Similarly, Northern analysis of RNA from advanced atherosclerotic lesions induced in nonhuman primates revealed increased PDGF B-chain expression (Ross *et al.*, 1990b). Additionally, *in situ* hybridization analysis of fibrotic atherosclerotic lesions suggested that the B-chain of PDGF was expressed by mesenchymal appearing intimal cells, whereas smooth muscle cells in the vascular lesions expressed predominantly PDGF-A chain (Wilcox *et al.*, 1988).

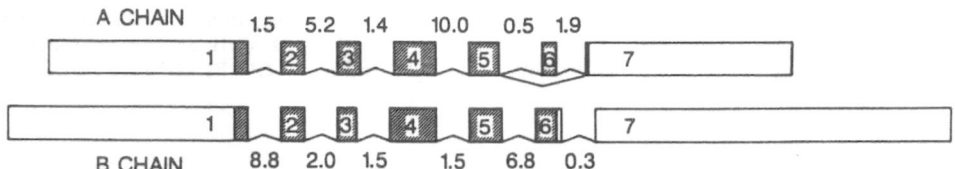

Figure 1. The structure of the A- and B-chain genes of PDGF. Exon-intron structure and splicing patterns of the PDGF A- and B-chain genes. Sizes of the exons are drawn to scale; lengths of the introns are indicated (kilobases). Modified from Bonthron *et al.*, 1988.

Recently, immunoreactive PDGF B-chain protein was found within macrophages in all stages of lesion development of both human and nonhuman primate atherosclerosis, suggesting that macrophages play an important role in providing PDGF to the intimal smooth muscle cells (Ross *et al.*, 1990b).

Because some of the increased B-chain transcript expression associated with atherosclerotic tissues may be transcriptionally mediated, we and others have initiated studies to characterize the promoter regions of the PDGF B-chain gene (Ratner *et al.*, 1987). Most eukaryotic genes appear to contain a promoter region spanning about 500 bp immediately upstream of the transcriptional initiation site. Within this region, there are several small DNA sequences (cis-acting elements) to which nuclear proteins (trans-acting factors) bind and regulate the rate of transcription initiation. In concert, these cis-acting elements and trans-acting factors mediate tissue specificity and they can have either a positive or negative effect on the rate of transcription. Outside of the promoter region, there may be other elements which can increase the levels of transcription in a position- and orientation-independent fashion (enhancers). By localizing the functional cis-acting elements in the PDGF A- and B-chain genes we hope to elucidate the normal transcriptional control of the genes, as well as mechanisms that may activate them in certain pathologic settings. Additionally, the identification of novel trans-acting factors interacting with the A and B chain genes may provide new insights into the general mechanism of transcriptional control.

The 5' flanking region of the PDGF B-chain gene was isolated and partially characterized to identify potential cis-acting transcriptional regulatory elements. The PDGF B-chain promoter was isolated by screening a human genomic library with a restriction fragment derived from the 5' end of the endothelial cDNA clone (B2-1). A recombinant phage clone was obtained which included the 5' portion of the human c-sis gene and exons 1-4 (Ratner *et al.*, 1987). Boundaries of the exons contained in this clone were delineated by comparison of the sequences of other genomic clones and the cDNA clones. The first exon is separated from the second exon by the largest intron in the c-sis gene (8.3 kb). The region of the genomic clone containing the putative promoter as well as the first exon was subjected to DNA analysis. The primary structure of this region of the gene has been reported by us (Ratner *et al.* 1987) and confirmed by others (e.g. Pech *et al.*, 1989). The B-chain RNA initiation site was determined by primer extension and S1 nuclease analysis and revealed a unique start site located downstream from a consensus TATAA box. No sequence consistent with a CCAAT consensus sequence was identified upstream of the B-chain RNA initiation site. In addition to the TATAA box site, analysis of the B-chain promoter sequence for consensus elements which may represent putative binding sites for trans-acting factors reveals four consensus recognition sequences for SP1 (GGGCGG), a CREB site ((T/G)(T/A)CGTCA) and a sequence GGAAGTG which is found in the adenovirus E4 gene, where it is recognized by a transcription factor designated E4TFI (reviewed in Mitchell and Tjian, 1989). Additionally, there are multiple sites with only a single base change from the AP1 consensus sequence (TGA(C/G)TCA).

To determine whether the 1.3 kb of PDGF B-chain upstream sequence was sufficient to direct B-chain transcription, the putative B-chain promoter was coupled to an indicator gene (CAT) and transfected into bovine aortic endothelial cells. The intact B-chain promoter directs synthesis of CAT in transiently transfected endothelial cells but not in dermal fibroblasts. The regions of the B-chain promoter which are responsible for efficient constitutive c-sis expression in bovine aortic endothelial cells were identified by deletion analysis. A series of unidirectional exonuclease III deletions were made in the PDGF B-chain promoter-CAT construct and the deletion end points defined by sequence analysis. These plasmids were transfected into cultured bovine aortic endothelial cells and CAT activity measured. The CAT activities in the transfections were normalized to total plasmid DNA content and use of the authentic transcription initiation site of the fusion gene was confirmed by primer extension analysis. Surprisingly, the constitutive activity of the B-chain promoter in endothelial cells is contained within a small (about 100 bp) region immediately upstream of the TATAA box. There are no consensus elements present in this sequence corresponding to binding sites of existing transcription factors (e.g., AP-1,2,3,4 or 5, SpI, NF-kB, CTF/NF-I, MLTF/USF,

C/EBP, CREB, OCT-1, E2F, EBP20, PEA2, E4EF2, or EF.C), as well as a variety of other interesting regulatory sequences (e.g., steroid regulatory elements, heat shock and metal response elements). Interestingly, this region of the B-chain promoter is strikingly conserved across species. Analysis of a similar PDGF B-chain promoter deletion series introduced into the megakaryocytic line K562 revealed a similar "basal" element required for constitutive B-chain expression (Pech *et al.*, 1989). Comparison of the human, feline and murine PDGF B-chain promoter sequences reveals almost complete conservation in this proximal area, suggesting the region's functional significance. In preliminary experiments, gel-shift analysis has been utilized to demonstrate that cultured endothelial cells constitutively synthesize a protein which binds to this region of the B-chain promoter (Bonthron *et al.*, manuscript in preparation). A major goal of this work is to further characterize this region of the B-chain promoter and determine the nature of this putatively novel constitutive trans-acting factor(s).

The definition of transcription factors regulating induced expression of the B chain of PDGF may produce valuable insights into the control of growth factor expression in the vessel wall. Expression of the B chain in cultured cells is induced by thrombin, transforming growth factor beta, epidermal growth factor and phorbol esters (Daniel *et al.*, 1986, 1987; Kavanaugh *et al.*, 1988; Silver *et al.*, 1989). Expression of the B chain is diminished by agents that increase cAMP accumulation (Daniel *et al.*, 1987; Kavanaugh, *et al.*, 1988). These effects appear to be mediated through effects upon rates of transcription (Daniel and Fen, 1988; Kavanaugh *et al.*, 1988). Some of these effects could be mediated through established transcription factors binding to previously defined elements. For example, cAMP regulatory domains usually consist of at least two classes of cis-acting elements: a CRE (TGACGTCA) or an AP-2 element (CCCCAGGC) (reviewed in Roesler *et al.*, 1988). However, sequences homologous to these consensus sequences are not found in the B-chain promoter sequence. Similarly there are no consensus AP1 binding sites in the B-chain promoter which could confer phorbol ester responsiveness. However, sequence analysis of the murine and human B-chain promoters reveals conserved regions as far as 850 bp upstream of the transcriptional start site which may contain elements capable of binding transcription factors regulating induced B-chain expression (Bonthron *et al.*, manuscript in preparation). Characterization of these B-chain regulatory elements, and identification of the factors involved in the regulation of B-chain expression in cultured cells, may define new regulatory genes whose altered expression is important in initiating the cellular events associated with the pathobiology of the vessel wall. Understanding these regulatory events may permit the development of new strategies for identifying patients at risk of forming atherosclerotic lesions.

PDGF A-Chain Gene Structure

The originally reported PDGF protein sequence information was interpreted to suggest that there were two protein chains of PDGF (Waterfield *et al.*, 1983; Doolittle, *et al.*, 1983). By screening a glioma cDNA library with synthetic oligonucleotides corresponding to the partial protein sequence, an A-chain cDNA was isolated and the gene mapped to chromosome 7 (Betsholtz *et al.*, 1986). Comparison of the predicted amino acid sequences of the PDGF A- and B-chains reveals extensive homology (56%). Interestingly, when PDGF A-chain cDNAs were obtained from endothelial cells, the structure was different from the reported A-chain sequence (Collins *et al.*, 1987; Tong *et al.*, 1987). The endothelial cDNA clones lacked an internal 69 base pair region near the C-terminus. The change in transcript structure shortens the endothelial transcript by 15 amino acids, removing a highly basic carboxyl-terminal region found in the glioma-derived precursor (Fig. 1). The functional implications of this change in C-terminal structure remain unclear.

To determine the basis for the two A-chain RNAs, the structure of the A-chain gene was determined (Bonthron *et al.*, 1988; Rorsman *et al.*, 1988). Examination of the gene revealed that the two different A-chain precursors, which differed by the presence or absence of a basic C-terminus, are generated as a result of alternative mRNA splicing events which include or exclude exon 6. Consistent with a common evolutionary origin, the intron-exon structures of the PDGF A- and B-chain genes are very similar (Fig. 1).

In regions of significant amino acid homology, for example, exons 1 and 2, the sequences and splice junctions can be precisely aligned. In contrast, exons 3 and 6 have few residues in common between A- and B-chains, although exon 6 in both genes does share the property of encoding basic C-terminal regions.

Three PDGF A-chain transcripts (2.8, 2.3 and 1.8 kb) are expressed by a variety of normal as well as transformed cells. Since the PDGF A-chain gene has a unique transcriptional start site (Bonthron et al., 1988), the three size classes of A-chain mRNAs probably arise by selection of alternative poly(A) sites in exon 7. However, the gene contains only one consensus polyadenylation signal (AATAAA). Report of a PDGF A-chain cDNA sequence from the human osteosarcoma cell line U-2OS (Hoppe et al., 1987) confirms that at least one of the transcripts contains an atypical polyadenylation signal (AATTAAA) located 470 bp downstream from the consensus signal. This transcript probably represents the largest PDGF A-chain mRNA (2.8 kb) observed on Northern blots. We speculate that the predominant A-chain transcript (2.3 kb) uses the consensus polyadenylation site and the remaining smaller species uses an upstream, as yet unidentified non-consensus polyadenylation site. The abundance of these three size classes of A-chain mRNA appears to be relatively invariant; in no instance has a relative increase of only one transcript size been substantiated. Thus, the functional significance of the alternative polyadenylation of the A-chain gene product is not known.

Transcriptional Regulation of PDGF A-Chain Gene Expression

Regulation of PDGF A-chain gene expression may be important in the development of vascular proliferative lesions. Endothelial cells (Collins et al., 1987b) and smooth muscle cells (Sejersen et al., 1986; Majesky et al., 1988) can express transcripts of the A-chain of PDGF in culture. Although smooth muscle cells cultured from normal adult rat carotid arteries make little PDGF, smooth muscle cells isolated from developing vessels secrete such a PDGF-like mitogen (Seifert et al., 1984). Smooth muscle cells taken from balloon injured rat carotid arteries produce a PDGF-like mitogen (Walker et al., 1986) and express PDGF A-chain transcripts (Majesky et al., 1990). This may imply that these proliferating myointimal cells have reassumed an immature, PDGF expressing phenotype. Interestingly, vascular cells from human carotid endarterectomy specimens secrete a PDGF-like mitogen and selectively express the A-chain of PDGF (Libby et al., 1988). Although the stimuli controlling increased PDGF A-chain expression in the vessel wall are not clearly defined, it is possible that the increased expression of PDGF seen in vascular proliferative lesions in adult vessels may represent an inappropriate reactivation of an autocrine mechanism operating during normal vascular development.

The pattern of expression of the A-chain gene suggests that the promoter may contain structural elements that are required for induced expression. Consistent with the independent pattern of regulation of the A- and B-chain genes, structural analysis of the PDGF A-chain promoter reveals that the architecture of this region of the gene is distinct from the B-chain gene (Fig. 2). Simple inspection of the DNA sequence reveals that the A-chain promoter contains consensus sequences for known DNA binding proteins. For example, the human PDGF A-chain promoter contains a consensus TATAA box located about 36 base pairs upstream from a unique transcriptional start site (Bonthron et al., 1988). The promoter has a high G+C content including multiple repeats of the sequence GGGCGG and its reverse complement (Fig. 2). This sequence is the core of the consensus binding site for a ubiquitous transcription factor, Sp1 (Kadonaga et al., 1987). The high G+C content, as well as the existence of potential Sp1 binding sites, are all common features of promoters which lack typical TATAA and CAAT box elements. Additionally, multiple putative AP2 binding sites (CCCCAGGC) are present in the A-chain promoter. Which of these elements are important in constitutive and induced A-chain expression is not known.

TWO TYPES OF PDGF RECEPTORS

PDGF exerts its mitogenic effects by means of high affinity receptors (reviewed in Williams, 1989). Analysis of the binding of the various PDGF isoforms to cultured

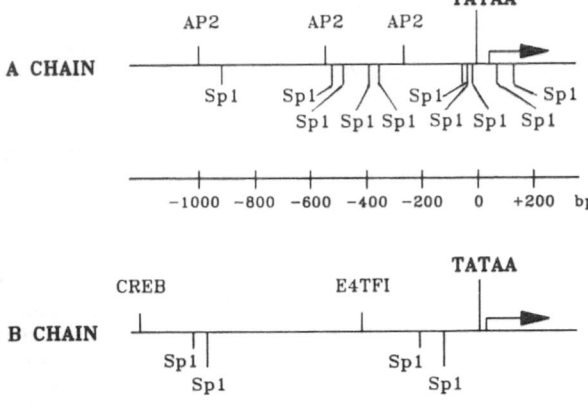

Figure 2. Comparison of the promoter regions of the PDGF A- and B-chain genes.

fibroblasts revealed two distinct receptor types, designated alpha and beta. The beta type PDGF receptor was first identified as a 180 kDa glycoprotein with ligand-activated protein tyrosine kinase activity. Analysis of the protein predicted by cDNAs revealed a signal sequence and a single membrane-spanning segment (Yarden *et al.*, 1986). The ligand-binding domain contained multiple potential acceptor sites for N-linked glycosylation and the spacing of the ten cysteine residues of this part of the receptor suggests that it consists of five immunoglobulin-like domains. The cytoplasmic domain contains a tyrosine kinase domain that is split into two parts by an intervening spacer region.

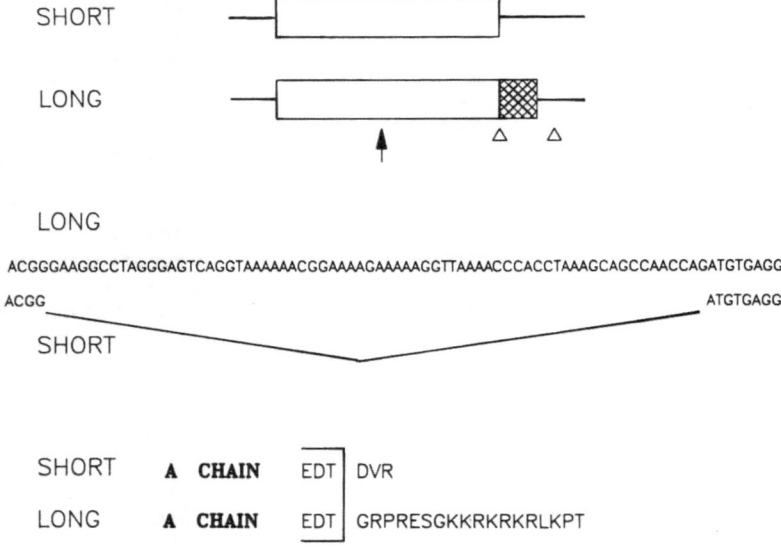

Figure 3. Comparison of the PDGF A-chain precursor sequences. Top, schematic representation of the short (endothelial) and the long (glioma) mRNAs. The shaded area bounded by triangles denotes the 69 base pair sequence in the long form which is absent from short form cDNA clones. Middle: alignment of the nucleotide sequences of the long and short forms of the A-chain in the region of the alternative splice. Bottom: alignment of the C-terminal amino-acid sequences of the A-chain precursors. (Modified from Collins *et al.*, 1987.)

Evidence now exists that a second PDGF receptor is present on many cells, and that its specificity for PDGF isoforms differs from the initially described PDGF receptor. A cDNA encoding a glycoprotein that efficiently binds PDGF A-chain homodimers but is structurally distinct from the PDGF-B type receptor was isolated (Matsui et al., 1989). Again, the predicted protein (designated alpha-type) was a transmembrane tyrosine kinase, which is organized in a manner similar to the beta-type receptor. The ligand binding domain is about 30% homologous to the beta-type receptor, and the cytoplasmic domain again contains a split tyrosine kinase domain, which is strikingly conserved.

The PDGF receptor has been proposed to exist as either a homo- or heterodimer on the target cells or to form a dimer on stimulation by PDGF (Heldin et al., 1989; Seifert et al., 1989). The high-affinity PDGF receptor consists of dimers of the two types of subunits: an alpha subunit, which could bind either an A- or a B-chain of a dimeric PDGF molecule, and a beta-subunit, which can bind only a B-chain of PDGF. The relative quantity of each receptor type expressed on the cell surface may dictate which dimeric receptor complex is generated on the responsive cell. Indeed, it is possible to differentially regulate the expression of PDGF-binding sites and the mitogenic responsiveness toward the three PDGF isoforms (Gronwald et al., 1989). Induction of PDGF receptors on smooth muscle cells may initiate vascular proliferative lesions (Rubin et al., 1988).

A-CHAIN ALTERNATIVE SPLICING IN VASCULAR CELLS

With a growth factor composed of two chains and at least two different cell surface receptors with tyrosine kinase activity, the potential complexity involved in the control of PDGF expression increases dramatically. Control of mitogen expression may occur at the level of PDGF A- and B-chain transcription, mRNA processing, translation, growth factor assembly, processing, or secretion. Additionally, the level of PDGF receptor expression, or the type of receptor subunits expressed, could be tightly coupled to growth-factor effects. How this receptor-ligand complex functions in the pathobiology of the vessel wall is not clearly understood, although there is evidence that multiple levels of regulation are employed. In the subsequent paragraphs, a focused analysis is presented of only one level of this control process: the regulation of PDGF A-chain alternative splicing.

Biological Function of the Basic Motif Encoded by the Sixth Exon of the A-Chain

The structural differences in the C-terminal regions of the short and long PDGF A-chain cDNAs suggests that there might be functional differences between the two forms of the mitogen. The conservation of the alternative splicing process during evolution (Mercola et al., 1988) further emphasizes the possibility that the functions of the two forms of the A-chain may be distinct. The basic region encoded by the sixth exon of the A-chain gene has been proposed to have several functions: first, this basic domain of the A- and B-chains might serve as a nuclear targeting sequence (Lee et al., 1987; Maher et al., 1989). The basic sequence present in the sixth exon of both the A- and B- chains is capable of targeting nonsecreted forms of the A-chain, as well as several cytoplasmic proteins, to the nucleus. Unfortunately, a role for PDGF in the nucleus remains to be determined. Similarly, it is difficult to imagine a normal mechanism for how the molecules would escape secretion. It is possible that the nuclear targeting function of this region of the A and B-chains of PDGF is a rather non-specific property of a basic domain in an appropriate context, as previously described for the SV40 large T antigen (Kalderon et al., 1984). A second possible function for the basic domain of the A-chain is regulation of local PDGF secretion. A peptide encoded by exon 6 induces the secretion of PDGF in cultured human endothelial cells, smooth muscle cells and fibroblasts (Raines et al., 1990). These observations are consistent with the possibility that release of the peptide encoded by exon 6 may regulate the release of PDGF from preformed storage pools. Thus, transcription and translation of A-chain containing exon 6 could further enhance the release of PDGF locally. Third, the basic motif common to the longer variant of PDGF A may be an "address peptide" for an extracellular target (Betsholtz *et al.*, 1990). Structural analysis of an endothelial growth factor designated vascular endothelial growth factor (VEGF) or vascular permeability factor (VPF) has revealed that this gene is distantly related to PDGF (Leung *et al.*, 1989; Keck *et al.*, 1989). The gene is alternatively spliced in a similar region as the A-chain of PDGF. The alignment of amino acid sequences in this region of the gene is shown in Table 1. The degree of sequence homology is greater in this region of the protein than in any other. These observations suggest that the basic motif common to the longer variants of PDGF-A chain and VEGF/VPF may serve a common function, possibly directing the growth factor to an interstitial structure or other extracellular target (Betsholtz *et al.*, 1990). Another approach to investigate the biological function of the alternatively spliced forms of the A-chain is to determine the normal spatial and temporal pattern of PDGF A-chain splicing and explore whether alterations in the A-chain splice pattern can be associated with pathophysiological responses.

Expression of the Long Form of PDGF A-Chain

To determine whether the PDGF long A-chain transcript is preferentially expressed in any normal tissue, the portion of the mouse A-chain gene encoding the alternatively spliced exon 6 was cloned and the distribution of A chain mRNAs analyzed in a variety of mouse tissues (Young *et al.*, 1990). Nuclease S1 and polymerase chain reaction analysis demonstrate that both the short and long forms of the A-chain transcripts are present in every adult tissue tested. The developmental specificity of PDGF A-chain expression was also examined. PDGF short and long chain transcripts were observed in embryonic stem cells, and 6, 10 and 13 day mouse embryos. The existence of the long A-chain protein was also demonstrated, by immunohistochemical techniques; analysis of murine tissues with a long A-chain specific antibody revealed scattered cytoplasmic staining which is present in skeletal muscle but absent in a variety of epithelia. An identical cellular distribution was observed for the short A-chain protein. These findings demonstrate that the A-chain splicing mechanism is conserved in the mouse and that expression of the two forms of the A-chain is not usually spatially or temporally distinct.

It is possible that the increased expression of PDGF A-chain in vascular cells during pathologic processes might coincide with an altered A-chain splicing pattern. To investigate this possibility, S1 nuclease mapping techniques were employed to examine the pattern of PDGF A-chain splicing in vascular cells. For S1 analysis, a 90 nucleotide probe spanning the exon 6-7 junction was synthesized and end-labelled. When this probe was hybridized to RNA isolated from cultured human umbilical vein endothelial cells or several human smooth muscle isolates, protected fragments of 70 and 80 nucleotides, which represent long and short A-chain RNA, respectively, were observed (Fig. 4). As we have noted previously using different techniques (Collins *et al.*, 1987a), the amount of long form PDGF A-chain in endothelial cells is very low.

Table 1. Homology between the long form of PDGF A and VEGF/VPF

hPDGF-A chain[1]	G	R	P	R	E	S	G	K	K	R	K	R	K	R	L	K	P	T
mPDGF-A chain[2]	G	R	R	R	E	S	G	K	K	R	K	R	K	R	L	K	P	T
xPDGF-A chain[3]	G	R	T	R	E	T	G	K	K	Q	K	R	K	K	L	K	P	T
hVEGF/VPF[4]	V	R	G	K	G	K	G	Q	K	R	K	R	K	K	S	R	Y	K

[1]Bonthron, D., *et al.*, Proc. Natl. Acad. Sci. 85: 1492, 1988;
[2]Young, R., *et al.*, Mol. Cell. Biol., manuscript in press;
[3]Mercola, M., *et al.*, Science 241: 1223, 1988;
[4]Leung, D., *et al.*, Science 246: 1306, 1989; and Keck, P., *et al.*, Science 246: 1309, 1989.

However, the endothelial cell does produce the long form of the A-chain, since it is possible to detect this transcript using appropriate primers and the polymerase chain reaction (Matoskova *et al.*, 1989; our unpublished data). Like the endothelial cells, either cultured human saphenous vein endothelial cells or human aortic smooth muscle cells express predominantly the short form of the A-chain. Interestingly, cells cultured from human carotid endarterectomy specimens-designated "atheroma cells", also express both forms of the A-chain. However, there is no significant alteration in the pattern of A-chain splicing in these cultured vascular cells.

Cellular hypertrophy is a pathophysiologic process in which expression of PDGF may play an important role. To explore the role of PDGF in a reversible adaptive process, expression of PDGF in tissue sections of human gestational myometrium was investigated by immunohistochemical techniques and confirmed by nuclease protection analysis (Mendoza *et al.*, 1990). Commensurate with an increase in immunoreactive PDGF expression in the myometrial smooth muscle cells, increased levels of PDGF A-chain mRNA but not PDGF B-chain or PDGF beta-type receptors, were seen in the gravid uterus relative to the non-gravid uterus. The amount of A-chain transcript increased during gestation and diminished during the puerperium. The increased PDGF A-chain expression was not associated with an alteration in the pattern of A-chain gene splicing. There was little of the A-chain transcript containing the sixth exon in gestational smooth muscle cells.

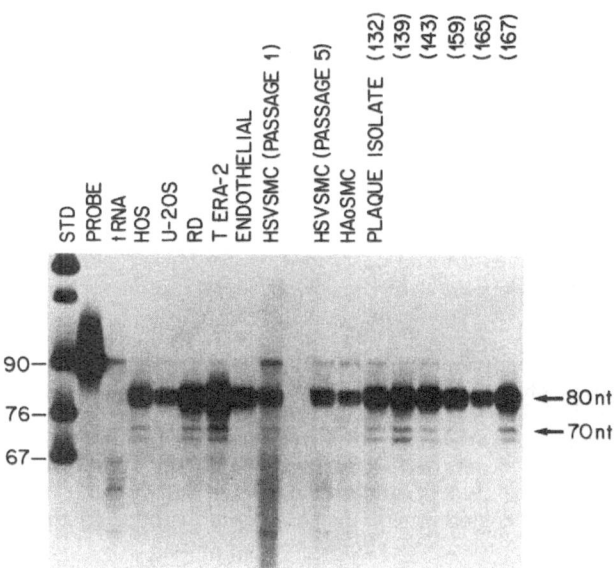

Figure 4. Expression of the short and long forms of the A-chain in vascular cells. RNA was isolated from various cell lines and primary cultures and hybridized to a 90 nucleotide (nt) 5'end-labelled oligonucleotide probe, treated with S1 nuclease, and the protected fragments sized on a 6% denaturing polyacry-lamide gel. RNA from several cell lines: 3 human osteosarcoma cell lines (HOS, U-2OS, and RD) and a human teratocarcinoma (T ERA-2) cell line, were compared with RNA from primary cultures of human umbilical vein endothelial cells (endothelial), human saphenous vein smooth muscle cells (HSVSMC), human aortic smooth muscle cells (HAoSMC), as well as RNA from human smooth muscle cells (Libby *et al.*, 1988) derived from human carotid endarterectomy specimens (plaque isolate). As described in detail in Young *et al.*, 1990, the 80 nt protected fragment corresponds to short form of the A-chain; the 70 nt protected species corresponds to the long form of the A-chain.

Table 2. Splice donor and acceptor sites flanking exons 5-7 of the mouse and human PDGF A-chain genes

Intron		5' Splice Donor Site	3' Splice Acceptor Site
		(C/A)AG/GURAGU	YnNCAG/G
5-6	HUMAN	C GG/GTGAGT	TTTTGTTAACAG/G
	MOUSE	C AG/GTGAGT	TTTCCTTGACAG/G
6-7	HUMAN	C AG/GTAGGA	TTCTCCCTGCAG/A
	MOUSE	C AG/GTAGGC	GCCTTCCCACAG/A

Thus, it appears that the 5-7 splice pattern (short form) is the predominant transcript species generated from the PDGF A-chain gene in many normal and pathologic settings. The mechanism(s) maintaining this relatively rigid splice pattern is not clear. The cis structural requirements involved in splice-site selection appear to be intact, surrounding exon 6 (Table 2). Sequence analysis of the human and murine PDGF-A chain genes encompassing exons 5 through 7 reveals 5' splice sites which are consistent with the consensus sequence (Bonthron *et al.*, 1988; Young *et al.*, 1990; Collins *et al.*, unpublished data). Similarly 3' splice sites and polypyrimidine tracts, as well as putative branch points, are present and conform to consensus. Thus, there is no *a priori* reason to suspect that 5-7 would represent the preferred splice pattern based upon the alignment of splice sites with consensus. Since those splice sites with a high match to consensus and a favorable adjacent exon environment are strongly competitive in cis-competition assays (reviewed in Smith *et al.*, 1989), we suspect that there are more subtle features of the sequence environment in this region of the A-chain gene which determine the splice-site compatibility. What these features consist of remain to be determined.

The recent advances in molecular biology have facilitated the structural analysis of both the A- and B- chain genes of PDGF, as well as the genes for the PDGF receptors. This has stimulated the generation of models illustrating how these components interact. Additionally, this structural information has provided insights into the regulation of the PDGF genes. With a detailed understanding of the structure and regulation of PDGF expression *in vitro*, models can be constructed for the function of PDGF *in vivo*. In the near future, these models will be tested and a detailed understanding of the biology of PDGF in the vessel wall should become clear.

ACKNOWLEDGEMENTS

Original research in the authors' laboratory was supported by NIH grants RO1 HL35716 and RO1 HL36028. T.C. is a fellow of the Pew Scholars Program. We wish to thank Dr. Peter Libby for providing the smooth muscle cell cultures used in these studies and Dr. Stuart Orkin for his enthusiastic support.

REFERENCES

Barrett, T.B. and Benditt, E.P. 1987. Sis (platelet-derived growth factor B-chain) gene transcript levels are elevated in human atherosclerotic lesion compared to normal artery. Proc. Natl. Acad. Sci. USA 84: 1099-1103.

Barrett, T.B. and Benditt, E.P. 1988. Platelet-derived growth factor gene expression in human atherosclerotic plaques and normal artery wall. Proc. Natl. Acad. Sci. USA 85: 2810-2814.

Bonthron, D.T., Morton, C.C., Orkin, S.H. and Collins, T. 1988. Platelet-derived growth factor A-chain: Gene structure, chromosomal location, and basis for alternative mR NA splicing. Proc. Natl. Acad. Sci. USA 85: 1492-1496.

Betsholtz, C., Johnsson, A., Heldin, C.-H., Westermark, B., Lind, P., Urdea, M.S., Eddy, R., Shows, T.B., Philpott, K., Mellor, A.L., Knott, T.J. and Scott, J. 1986. cDNA sequence and chromosomal location of human platelet-derived growth factor A-chain and its expression in tumour cell lines. Nature 320: 695-699.

Betsholtz, C., Rorsman, F., Westermark, B., Ostman, A. and Heldin, C.-H. 1990. Analogous alternative splicing. Nature 344: 299.

Collins, T., Ginsburg, D., Boss, J.M., Orkin, S.H. and Pober, J.S. 1985. Cultured human endothelial cells express platelet-derived growth factor chain 2: cDNA cloning and structural analysis. Nature 216: 748-750.

Collins, T., Pober, J.S., Gimbrone, M.A. Jr., Hammacher, A., Betsholtz, C., Westermark, B. and Heldin, C.-H. 1987a. Cultured human endothelial cells express platelet-derived growth factor A-chain. Am. J. Path. 126: 7-12.

Collins, T., Bonthron, D.T. and Orkin, S.H. 1987b. Alternative RNA splicing affects function of encoded platelet-derived growth factor A-chain. Nature 328: 621-624.

Dalla-Favera, R., Gallo, R.C., Giallongo, A., and Croce, C.M. 1982. Chromosomal localization of the human homolog of the (c-sis) gene of the simian sarcoma virus onc gene. Science 218: 686-688.

Daniel, T.O., Gibbs, V.C., Milfay, D.F., Garovoy, M.R. and Williams, L.T. 1986. Thrombin stimulates c-sis gene expression in microvascular endothelial cells. J. Biol. Chem. 261: 9579-9582.

Daniel, T.O., Gibbs, V.C., Milfay, D.F. and Williams, L.T. 1987. Agents that increase cAMP accumulation block endothelial c-sis induction by thrombin and transforming growth factor-beta. J. Biol. Chem. 262: 11893-11896.

Daniel, T.O. and Fen, Z. 1988. Distinct pathways mediate transcriptional regulation of platelet-derived growth factor B/c-sis expression. J. Biol. Chem. 263: 19815-19820.

Deck, P.J., Hauser, S.D., Drivi, G., Sanzo, K., Warren, T., Feder, J. and Connolly, D.T. 1989. Vascular permeability factor, an endothelial cell mitogen related to PDGF. Science 246: 1309-1312.

Devare, S.G., Reddy, E.P., Law, D.J., Robbins, K.C. and Aaronson, S.A. 1983. Nucleotide sequence of the simian sarcoma virus genome: demonstration that its acquired cellular sequences encode the transforming gene product p28sis. Proc. Natl. Acad. Sci. USA 30: 731-735.

Doolittle, R.F., Hunkapiller, M.W., Hood, L.E., Devare, S.G., Robbins, K.C., Aaronson, S.A. and Antoniades, H.N. 1983. Simian sarcoma virus onc gene, v-sis, is derived from the gene (or genes) encoding a platelet-derived growth factor. Science 221: 275-277.

Gajdusek, C., DiCorleto, P., Ross, R. and Schwartz, S.M. 1980. An endothelial cell-derived growth factor. J. Cell Biol. 85: 467-472.

Gronwald, R.G.K., Seifert, R.A. and Bowen-Pope, D.F. 1989. Differential regulation of expression of two platelet-derived growth factor receptor subunits by transforming growth factor beta. J. Biol. Chem. 264: 8120-8125.

Heldin, C.-H., Ernlund, A., Rorsman, C. and Ronnstrand, L. 1989. Dimerization of B-type platelet-derived growth factor receptors occurs after ligand binding and is closely associated with receptor kinase activation. J. Biol. Chem. 264: 8905-8912.

Hoppe, J., Schumacher, L., Eichner, W. and Weich, H.A. 1987. The long 3'untranslated regions of the PDGF-A and -B mRNAs are only distantly related. FEBS Lett. 223: 243-246.

Johnsson, A., Heldin, C.-H., Wasteson, A., Westermark, B., Deuel, T.F., Huang, J.S., Seeburg, P.H., Gray, A., Ullrich, A., Scrace, G., Stroobant, P. and Waterfield, M.D. 1984. The c-sis gene encodes a precursor of the B chain of platelet-derived growth factor. EMBO J. 3: 921-923.

Josephs, S.F., Guo, C., Ratner, L. and Wong-Staal, F. 1984. Human proto-oncogene nucleotide sequences corresponding to the transforming region of simian sarcoma virus. Science 223: 487-491.

Kadonaga, D., Richardson, W.D., Markham, A.F. and Smith, A.E. 1984. Sequence requirements for nuclear location of simian virus 40 large-T antigen. Nature 311: 499-509.

Kadonaga, J.T., Carner, K.R.,Masiarz, F.R. and Tjian, R. 1987. Isolation of cDNA encoding transcription factor Sp1 and functional analysis of the DNA binding domain. Cell 51: 1079-1090.

Kavanaugh, W.M., Harsh, G.R., Starksen, N.F., Rocco, C.M. and Williams, L.T. 1988. Transcriptional regulation of the A- and B-chain genes of platelet-derived growth factor in microvascular endothelial cells. J. Biol. Chem. 263: 8470-8472.

Lee, B.A., Maher, D.W., Hannink, M. and Donoghue, D.J. 1987. Identification of a signal for nuclear targeting in platelet-derived growth factor-related molecules. Mol Cell. Biol. 7: 3527-3537.

Leung, D.W., Cachianes, G., Kuang, W.-J., Goeddel, D.V. and Ferrara, N. 1989. Vascular endothelial growth factor is a secreted angiogenic mitogen. Science 246: 1306-1312.

Libby, P., Warner, S.J.C., Salomon, R.N. and Birinyi, L.K. 1988. Production of platelet-derived growth factor-like mitogen by smooth muscle cells from human atheroma. N. Engl. J. Med. 318: 1493-1497.

Maher, D.W., Lee, B.A. and Donoghue, D.J. 1989. The alternatively spliced exon of the platelet-derived growth factor A-chain encodes a nuclear targeting signal. Mol. Cell. Biol. 9: 2251-2253.

Majesky, M.W., Benditt, E.P. and Schwartz, S.M. 1988. Expression and developmental control of platelet-derived growth factor A-chain and B-chain/sis genes in rat aortic smooth muscle cells. Proc. Natl. Acad. Sci. USA 85: 1524-1528.

Majesky, M.W., Daemen, M.J.A.P. and Schwartz, S.M. 1990. Alpha-adrenergic stimulation of platelet-derived growth factor A-chain gene expression in rat aorta. J. Biol. Chem. 265: 1082-1088.

Martinet, Y., Bitterman, P.B., Mornex J., Grotendorst, G.R., Martin, G.R. and Crystal, R.G. 1986. Activated human monocytes express the c-sis proto-oncogene and release a mediator showing PDGF-like activity. Nature 319: 158-160.

Matoskova, B., Rorsman, F., Svensson, V. and Betsholtz, C. 1989. Alternative splicing of the platelet-derived growth factor A-chain transcript occurs in normal as well as tumor cells and is conserved among mammalian species. Mol. Cell. Biol. 9: 3148-3150.

Matsui, T., Heidaran, M., Miki, T., Popescu, N., La Rochelle, W., Kraus, M., Pierce, J. and Aaronson, S. 1989. Isolation of a novel receptor cDNA establishes the existence of two PDGF receptor genes. Science 243: 800-804.

Mendoza, A.E., Young, R., Orkin, S.H. and Collins, T. 1990. Increased platelet-derived growth factor A-chain expression in human uterine smooth muscle cells during the physiologic hypertrophy of pregnancy. Proc. Natl. Acad. Sci. USA 87: 2177-2181.

Mercola, M., Melton, D.A. and Stiles, C.D. 1988. Platelet-derived growth factor A-chain is maternally encoded in Xenopus embryos. Science 241: 1223-1225.

Mitchell, P.J., and Tjian, R. 1989. Transcriptional regulation in mammalian cells by sequence-specific DNA binding proteins. Science 245: 371-378.

Nilsson, J., Sjolund, M., Palmberg, L., Thyberg, J. and Heldin C.-H. 1985. Arterial smooth muscle cells in primary culture produce a platelet-derived growth factor-like protein. Proc. Natl. Acad. Sci. 82: 4418-4422.

Pech, M., Rao, C.D., Robbins, K.C. and Aaronson, S.A. 1989. Functional identification of regulatory elements within the promoter region of platelet-derived growth factor 2. Mol. Cell. Biol. 9: 396-405.

Raines, E.W. and Ross, R. 1990. A peptide encoded by the alternatively spliced exon 6 of PDGF A-chain induces the secretion of PDGF in cultured human endothelium, smooth muscle, and fibroblasts. FASEB J. 4: 480A.

Rao, C.D., Pech, M., Robbins, K.C. and Aaronson, S.A. 1988. The 5' untranslated region of the c-sis/platelet derived growth factor 2 transcript is a potent translational inhibitor. Mol. Cell. Biol. 8: 284-292.

Ratner, L., Thielan, B. and Collins, T. 1987. Sequences of the 5' portion of the human c-sis gene: characterization of the transcriptional promoter and regulation of expression of the protein product by 5' untranslated mRNA sequences. Nucl. Acids Res. 15: 6017-6036.

Roesler, W.J., Vandenbark, G.R. and Hanson, R.W. 1988. Cyclic AMP and the induction of eukaryotic gene transcription. J. Biol. Chem. 263: 9063-9066.

Rorsman, F., Bywater, M., Knott, T.J. Scott, J. and Betsholtz, C. 1988. Structural characterization of the human platelet-derived growth factor A-chain cDNA and gene: alternative exon usage predicts two different precursor proteins. Mol. Cell. Biol. 8: 571-577.

Ross, R. 1986. The pathogenesis of atherosclerosis - an update. N. Engl. J. Med. 314: 488-500.

Ross, R., Raines E.W. and Bowen-Pope, D.F. 1986. The biology of platelet-derived growth factor. Cell 46: 155-169.

Ross, R., Bowen-Pope, D.F. and Raines, E.W. 1990a. Platelet-derived growth factor and its role in health and disease. Phil. Trans. R. Soc. Lond. 327: 155-169.

Ross, R., Masuda, J., Raines, E.W., Gown, A.M., Katsuda, S., Sasahara, M., Maldern, L.T., Masuko, H. and Sato, H. 1990b. Localization of PDGF-B protein in macrophages in all phases of atherogenesis. Science 248: 1009-1012.

Rubin, K., Hansson, G.K., Ronnstrand, L., Claesson-Welsh, L., Fellstrom, B., Tingstrom, A., Larsson, E., Klareskog, L., Heldin, C.-H. and Terracio, L. 1988. Induction of B-type receptors for platelet-derived growth factor in vascular inflammossible implications for development of vascular proliferative lesions. Lancet i, 1353-1356.

Seifert, R.A., Hart, C.E., Phillips, P.E., Forstrom J.W., Ross, R., Murray, M.J. and Bowen-Pope, D.F. 1989. Two different subunits associate to create isoform-specific platelet-derived growth factor receptors. J. Biol. Chem. 264: 8771-8778.

Seifert, R.A., Schwartz, S.M. and Bowen-Pope, D.F. 1984. Developmentally regulated production of platelet-derived growth factor-like molecules. Nature 311: 669-671.

Sejersen, T., Betsholtz, C., Sjolung, M., Heldin, C.-H., Westermark, B. and Thyberg, J. 1986. Rat skeletal myoblasts and arterial smooth muscle cells express the gene for the A-chain but not the gene for the B-chain (c-sis) of platelet-derived growth factor (PDGF) and produce a PDGF-like protein. Proc. Natl. Acad. Sci. USA 83: 6844-6848.

Shimokado, K., Raines, E.W., Madtes, D.K., Barrett, T.B., Benditt, E.P. and Ross, R. 1985. A significant part of macrophage-derived growth factor consists of at least two forms of PDGF. Cell 43: 277-286.

Silver, B.J., Jaffer, F.E. and Abboud, H.E. 1989. Platelet-derived growth factor synthesis in mesangial cells: Induction by multiple peptide mitogens. Proc. Natl. Acad. Sci. 86: 1056-1060.

Smith, C.W.J., Patton, J.G. and Nadal-Ginard, B. 1989. Alternative splicing in the control of gene expression. Ann. Rev. Genet. 23: 527-577.

Tong, B.D., Auer, D.E., Jaye, M., Kaplow, J.M., Ricca, G., McConathy, E., Drohan, W. and Deuel, T.F. 1987. cDNA clones reveal differences between human glial and endothelial cell platelet-derived growth factor A-chains. Nature 328: 619-621.

Walker L.N., Bowen-Pope, D.F. and Reidy, M.A. 1986. Production of platelet-derived growth factor-like molecules by cultured arterial smooth muscle cells accompanies proliferation after arterial injury. Proc. Natl. Acad. Sci. USA 83: 7311-7315.

Waterfield, M.D., Scrace, G.T., Whittle, N., Stroobant, P., Johnsson, A., Wasteson, A., Westermark, B., Heldin, C.-H., Huang, J. S. and Deuel, T.F. 1983. Platelet-derived growth factor is structurally related to the putative transforming protein p28sis of simian sarcoma virus. Nature 304: 35-39.

Wilcox, J.N., Smith K.M., Williams, L.T., Schwartz, S.M. and Gordon, D. 1988. Platelet-derived growth factor mRNA detection in human atherosclerotic plaques by in situ hybridization. J. Clin. Invest. 82: 1134-1143.

Williams, L.T. 1989. Signal transduction by the platelet-derived growth factor receptor. Science 243: 1564-1570.

Yarden Y., Escobedo, J.A., Kuang, W-J., Yang-Feng, T.L., Daniel, T.O., Tremble, P.M., Chen, E.Y., Ando, M.E., Harkins, R.N., Francke, U., Fried, V.A., Ullrich, A. and Williams, L.T. 1986. Structure of the receptor for platelet-derived growth factor helps define a family of closely related growth factor receptors. Nature 323: 226-232.

Young, R., Mendoza, A., Collins, T. and Orkin, S.H. 1990. Alternatively spliced platelet-derived growth factor A-chain transcripts are not tumor-specific but rather encode normal cellular proteins. Mol. Cell. Biol., in press.

ENDOTHELIAL REGULATION OF VASOMOTOR TONE

IN ATHEROSCLEROSIS

David G. Harrison

Department of Internal Medicine
University of Iowa
Iowa City, Iowa 52242

INTRODUCTION

During the last ten years it has become clear that the vascular endothelium modulates blood flow and vascular resistance by elaborating a variety of potent vasoactive factors. These include the endothelium-derived relaxing factor[1-4], now recognized to be either nitric oxide or a related nitrosyl compound[5,6], a variety of vasoconstrictor and vasodilator prostaglandins[7,8,9], angiotensin II[10], and the peptide endothelin[11]. Of these, the most thoroughly studied has been the endothelium-derived relaxing factor. The existence of this substance was first discovered by Furchgott and co-workers who showed that the endothelium must be intact for acetylcholine to produce vascular relaxation[4]. It was subsequently realized that the endothelium released an extremely labile substance in response to a variety of stimuli which diffuses to the underlying vascular smooth muscle, activates guanylate cyclase[13,14,15], and subsequently produces potent vessel relaxation.

Shortly after Dr. Furchgott's original observation, several groups became interested in the possibility that this function of the endothelium might be impaired in a variety of diseases including atherosclerosis. The rationale underlying this hypothesis was that many of the neurohumoral agents that stimulate the release of EDRF also have direct constrictor effects on vascular smooth muscle. If the endothelium were dysfunctional, these substances might have a predominant unopposed vasoconstrictor effect which may predispose to vasospasm and hypertension often encountered in individuals with atherosclerosis. These considerations have led to an expanding area of research with implications regarding basic concepts of endothelial cell function and potential clinical relevance.

In this chapter, abnormalities of endothelial cell morphology and regulation of vasomotor tone in chronic hypercholesterolemia and atherosclerosis will be reviewed. Data regarding the effect of hypercholesterolemia on the microcirculation will be examined. Recent observations regarding the production of nitrogen oxides from normal and atherosclerotic vessels will be presented. Finally, we will speculate as to the mechanisms underlying abnormal endothelium-dependent vascular relaxation in atherosclerosis.

Alterations of Endothelial Cell Morphology Produced by Chronic Hypercholesterolemia

Several groups have examined the effect of diet induced hypercholesterolemia on endothelial cell morphology in experimental animals. Earlier studies suggested that endothelial cell denudation or desquamation was an integral and early result of hypercholesterolemia[16]. More recent studies using careful fixation techniques have conclusively shown that endothelial cell desquamation does not occur until very late in the atherosclerotic process. Taylor et al. recently examined the effect of cholesterol feeding on the morphology of endothelial cells in cynomolgus monkeys[17]. Endothelial cell desquamation was not apparent, even over sites of multilayered plaque. Over raised lesions, however, endothelial cell width was increased resulting in a loss of the normal orientation to the direction of flow. The endothelial cells conformed to the contours of the immediately underlying foam cells and often demonstrated marked thinning and attenuation of cell bodies over bulging foam cell profiles. On the abluminal surface of endothelial cells an abundance of 45-65 Å cytoplasmic elements insinuated between foam cells to maintain contact with the underlying elastic lamina. Thus, although endothelial cell loss is not a common finding in hypercholesterolemia, changes in morphology of the endothelium are striking.

Studies of Endothelium-Dependent Relaxation in Atherosclerosis

In 1986 and 1987 several publications clearly showed that endothelium-dependent vascular relaxations were impaired in animals with experimentally induced atherosclerosis. Habib and co-workers[18] studied aortic rings from cholesterol fed rabbits and found that relaxations to acetylcholine were markedly abnormal. Verbeuren and co-workers[19] studied rabbits fed cholesterol for a short term (8 weeks) and for longer periods (16 weeks). These investigators showed that relaxations to acetylcholine were dramatically impaired and contractions to serotonin selectively enhanced in both cholesterol fed groups. Relaxations to nitroglycerin were impaired to a slight extent only in the longer term cholesterol fed group. Jayakody et al.[20] also showed that cholesterol feeding of rabbits decreased aortic relaxations to acetylcholine. Studies by Freiman et al.[21] and Harrison et al.[22] showed that endothelium-dependent relaxations to acetylcholine, thrombin, and a calcium ionophore were impaired in iliac arteries from monkeys with diet induced atherosclerosis while relaxations to nitroglycerin were normal.

This defect in endothelial cell function seems to occur very soon after the onset of cholesterol feeding. Jayakody and co-workers have shown that endothelium-dependent vascular relaxation to acetylcholine is abnormal after only four weeks of cholesterol feeding[20]. Wines and co-workers[23] recently found that constriction to serotonin is enhanced and relaxation to methylcholine reduced in the aorta of Watanabe heritable hyperlipidemic rabbits at one month of age, before gross evidence of atherosclerosis exists.

Several groups have now shown that brief exposures to relatively high concentrations of purified low density lipoprotein can selectively inhibit endothelium-dependent relaxations of rabbit aorta[24]. There remains some controversy as to whether this is due to native LDL, oxidized LDL, or products of oxidized LDL. The latter have been shown to impair endothelium-dependent relaxations[25].

The Effect of Atherosclerosis on the Coronary Microcirculation

All of the earlier studies relevant to endothelial cell function in atherosclerosis were performed on large conduit vessels such as the aorta, the iliac artery, and the proximal coronary arteries. These vessels are not generally involved in the regulation of tissue perfusion. Until recently it has been unclear whether the resistance vasculature is equally affected as a result of hypercholesterolemia. This is particularly relevant because the resistance vasculature does not develop overt atherosclerosis. Two studies have examined the effect of hypercholesterolemia on true resistance vessels. Yamamoto showed that relaxations to acetylcholine were strikingly abnormal in 25 μm diameter arterioles of the cremaster muscle of cholesterol fed rabbits[26]. More recently Sellke et al. have studied coronary arterioles of cholesterol fed primates in vitro[29]. Relaxations to the endothelium-independent vasodilators nitroprusside and adenosine were not altered in these vessels. In contrast, endothelium-dependent vascular relaxations to bradykinin, acetylcholine and the calcium ionophore A23187 were strikingly abnormal in coronary arterioles from cholesterol fed animals. These studies of the microcirculation clearly show that hypercholesterolemia may affect endothelium-dependent vascular relaxation in vessels that do not develop overt atherosclerosis. These findings may have important implications regarding neurohumoral regulation of tissue perfusion in atherosclerosis.

Studies in Humans

Several groups have now shown that atherosclerosis also impairs endothelium-dependent vascular relaxations of human coronary arteries. This has been demonstrated by two groups using *in vitro* techniques[28,29] and by several groups in the cardiac catheterization laboratory. Ludmer and co-workers[30] showed that infusions of acetylcholine into the left anterior descending coronary artery produced minimal dilatation or no response in angiographically normal coronary arteries while producing constriction of coronary arteries which contained atherosclerosis. These investigators have subsequently shown that this propensity for vasodilatation versus vasoconstriction in response to intracoronary acetylcholine bears a strong relationship to the individual's number of risk factors for atherosclerosis[32]. Recently, it has been established that flow mediated vasodilatation, a phenomenon which is due to the release of EDRF, is abnormal in patients with even minimal atherosclerosis[32,33].

Mechanisms Underlying Impaired Endothelium-Dependent Relaxation in Atherosclerotic Blood Vessels

The predominant abnormality of endothelium-dependent vascular relaxation seems to be related to impaired synthesis or release of EDRF rather than altered responsiveness of vascular smooth muscle to EDRF. Atherosclerotic vessels relax normally to pharmacologic nitrovasodilators[18,19,21]. Bioassay studies have shown that EDRF released from atherosclerotic rabbit aorta in response to either acetylcholine or the calcium ionophore produces only about one-half as much relaxation of a normal, denuded segment of rabbit aorta[34]. In contrast, bioassay studies have also shown that aortic segments from cholesterol fed rabbits are supersensitive to EDRF and equally sensitive to nitric oxide compared to normal vessels[34]. Thus, the vascular smooth muscle of the aorta of cholesterol fed rabbits is capable of responding to EDRF.

It is conceivable that several other mechanisms may contribute to abnormal endothelium-dependent relaxation in atherosclerosis. Oxygen derived free radicals released from subintimal macrophages and other inflammatory cells present in the atherosclerotic lesion may inactivate EDRF[35] and intimal thickening may prevent the diffusion of EDRF to the underlying vascular smooth muscle.

The endothelium of atherosclerotic vessels may, under certain circumstances, release excess quantities of constrictor factors. Abnormal cell types present in the intima of atherosclerotic cell types may also release constrictor factors. Recently, Lopez and co-workers[36] have found that the tri-peptide f Met-Leu-Phe (fMLP) had essentially no effect when infused into the hindlimb of normal monkeys while producing pronounced constriction of large arteries in atherosclerotic monkeys. This peptide is capable of activating leukocytes and the authors suggested that the inflammatory cells within the intima of the atherosclerotic vessels were capable of releasing vasoconstrictor substances in response to fMLP. They further found that prostaglandin E2, which may be released from leukocytes, produced marked constriction of large arteries in atherosclerotic but not normal monkeys. Thus, not only do atherosclerotic vessels contain a variety of cell types which may release vasoactive agents, but these vessels exhibit increased responsiveness to at least one vasoactive substance released from leukocytes.

The Chemical Identity of the Endothelium-Derived Relaxing Factor: Alterations in Synthesis in Atherosclerosis

In 1987 Palmer and co-workers first showed that endothelial cells were able to release nitric oxide or a compound which would yield nitric oxide upon one electron reduction[5]. They further noted that this compound was released in sufficient quantities to account for all of the biological activity of the endothelium-derived relaxing factor. There remains however some debate regarding the precise identity of the EDRF. We have compared the release of nitric oxide (detected by chemiluminescence) and the biological activity of EDRF (determined by bioassay) from bovine aortic endothelial cells. The amount of nitric oxide present in the effluent of superfused aortic endothelial cells was only about one-fifth to one-eighth that necessary to account for the biological activity of the endothelium-derived relaxing factor[37]. Based on these observations, we concluded that there were either multiple EDRFs or that the endothelium-derived relaxing factor was not free nitric oxide but nitric oxide incorporated into a parent compound which was substantially more potent than nitric oxide. In support of this later hypothesis, we found that the potency of one such nitrosylated compound, S-nitroso-L-cysteine more closely resembled that of the endothelium-derived relaxing factor than did nitric oxide on a mole-to-mole basis[6]. Rubanyi et al. have shown that the endothelium-derived relaxing factor and S-nitrosocysteine in concentrations sufficient to produce vascular relaxation do not alter the electron paramagnetic resonance shift of hemoglobin while nitric oxide does[38]. Finally, Wei et al. have shown that pial vessels exposed to hydrogen peroxide do not relax to nitric oxide but do relax to either EDRF or S-nitroso-L-cysteine[39]. One effect of hydrogen peroxide is to oxidize sulfhydryl groups. Wei et al. have suggested that nitric oxide requires free sulfhydryl groups whereas the endothelium-derived relaxing factor and S-nitrosocysteine being nitrosothiols do not. Whether or not this interpretation of their data is accurate, their findings clearly show that EDRF and nitric oxide have very different characteristics and that S-nitroso-L-cysteine more closely mimics EDRF than does nitric oxide in pial vessels.

The concept that EDRF may be a nitrosothiol rather than authentic nitric oxide is also compatible with a number of other observations which have been made during the last several years. EDRF binds to anion exchange columns whereas nitric oxide does not, supporting the notion that it may be a polar compound[40]. At neutral pH's, S-nitroso-L-cysteine is a Zwitter ion and would therefore bind readily to either an anion or a cation exchange column. The endothelium-derived relaxing factor can be stabilized in an acidic environment[41]. In such an environment nitrosothiols are quite stable whereas they become very unstable at neutral pH's.

The incorporation of nitric oxide into a parent compound such as a nitrosothiol may have important implications regarding the effect of diseases on endothelium-dependent responses. Several diseases including atherosclerosis, diabetes[42], ischemia with reperfusion[43], and acute hypertension[44] are associated with abnormal endothelium-dependent vascular relaxations. One feature potentially common to all these processes is the excessive generation of oxygen free radicals. These may react with free sulfhydryl groups (or other EDRF substrates) making them unable to incorporate the nitroso moiety. This would allow the endothelial cell to produce nitric oxide while not producing an effective EDRF. This mechanism may underlie abnormal endothelium-dependent responses in many disease states.

This hypothesis has been supported by recent studies of the release of nitric oxide and related nitrogen oxides from the thoracic aorta of normal and cholesterol-fed rabbits[45]. In these studies, the quantity of nitric oxide recovered from cholesterol-fed rabbit aorta markedly exceeded that from normal rabbit aorta, although the vasorelaxant activity of EDRF from hypercholesterolemic animals was markedly reduced. This interesting finding of increased production of nitrogen oxides by the endothelium of atherosclerotic vessels may have important implications regarding abnormal endothelium-dependent vascular relaxation in atherosclerosis. They clearly show that there is not a defect in the enzymatic process leading to the production of nitric oxide in atherosclerotic vessels and are compatible with the concept that atherosclerosis may impair the incorporation of nitric oxide into a more potent nitrosylated compound.

Treatment or Prevention of Endothelium-Dependent Relaxation in Atherosclerotic Vessels

Several groups have attempted to prevent or correct abnormal endothelium-dependent responses in cholesterol fed animals. Habib and co-workers have shown that parenteral administration of the calcium antagonist PN200110 decreased the incorporation of cholesterol into the vessel wall and partially restored endothelium-dependent responses to normal in cholesterol fed rabbits[18]. In cynomolgus monkeys fed an atherogenic diet for 18 months, subsequent feeding of a normal diet for the ensuing 18 months returned endothelium-dependent relaxations to acetylcholine and thrombin to normal[46]. Shimokawa and Vanhoutte have shown that dietary cod liver oil improves endothelium-dependent responses in a pig model of focal coronary atherosclerosis produced by a combination of balloon denudation and cholesterol feeding[47]. Very recently, Vekshtein and co-workers[48] have presented preliminary data showing that six months of treatment with omega-3 fatty acids can reverse abnormal coronary vasoconstriction to acetylcholine in patients with early atherosclerosis.

SUMMARY

Abnormal modulation of vascular smooth muscle tone by the endothelium occurs both in experimental animals and in humans. The significance of this phenomenon, however, in terms of human disease has yet to be clearly defined. It has been demonstrated recently that the production of endothelium-derived constricting factors increases with age[49]. Since EDRF may play an important role in the maintenance of normal blood pressure, it is conceivable that the loss of EDRF with atherosclerosis together with the increased production of constrictor factors with age may predispose to "essential" hypertension which develops in the elderly.

Additionally, loss of EDRF may enhance or unmask constrictor effects of both humoral and locally produced substances (some released by intimal inflammatory cells). This may contribute to syndromes of altered vasomotion including unstable angina and variant angina encountered in the setting of atherosclerosis. The endothelium-derived relaxing factor also inhibits platelet aggregation[50,51]. It is possible that the loss of this property of the endothelium may enhance the propensity for local platelet deposition and the concomitant release of both vasoconstrictor factors from platelets and factors that may accelerate the development of atherosclerosis (such as the platelet-derived growth factor).

Finally, it is now apparent that hypercholesterolemia not only alters endothelial function in large vessels but also in the microcirculation. Thus, in addition to predisposing to altered vasomotion of large vessels, endothelial dysfunction in atherosclerosis may impair neurohumoral regulation of tissue perfusion at the microvascular level.

REFERENCES

1.   Carrier, G.O. and R.E. White. 1985. Enhancement of alpha-1 and alpha-2 adrenergic agonist-induced vasoconstriction by removal of endothelium in rat aorta. J. Pharm. Exp. Therap. 1985: 682-687.

2.   Cherry, P.D., R.F. Furchgott, J.V. Zawadski and D. Jothianandan. 1982. Role of endothelial cells in relaxation of isolated arteries by bradykinin. Proc. Natl. Acad. Sci. 79: 2106-2110.

3.   Cocks, T.M. and A.R. Angus. 1983. Endothelium dependent relaxation of coronary arteries by noradrenaline and serotonin. Nature 305: 627-630.

4.   Furchgott, R.F. and J.V. Zawadski. 1980. The obligatory role of endothelial cells in the relaxation of arterial smooth muscle by acetylcholine. Nature 228: 373-376.

5.   Palmer, R.M.J., A.G. Ferrige and S. Moncada. 1987. Nitric oxide release accounts for the biological activity of endothelium-derived relaxation factor. Nature (Lond.) 327: 524-526.

6.   Myers, P.R., R.L. Minor, R. Guerra Jr., J.N. Bates and D.G. Harrison. 1990. The vasorelaxant properties of the endothelium-derived relaxing factor more closely resemble S-nitrosocysteine than nitric oxide. Nature 345: 161-163.

7.   Needleman, P., G.R. Marshall and B.E. Sobel. 1975. Hormone interactions in the isolated rabbit heart; synthesis and coronary vasomotor effects of prostaglandins, angiotensin, and bradykinin. Circ. Res. 37: 802-898.

8.   Dusting, G.J. and J.R. Vane. 1980. Some cardiovascular properties of prostacyclin (PGI2) which are not shared by PGE2. Circ. Res. 44: 223-227.

9.   Luscher, T.F. and P.M. Vanhoutte. 1986. Endothelium-dependent contractions to acetylcholine in the aorta of the spontaneously hypertensive rat. Hypertension 8: 344-348.

10.   Dzau, V.J. 1986. Significance of the vascular renin-angiotensin pathway. Hypertension 8: 553-559.

11.   Yanagisawa, M., H. Kurihara, S. Kimura, Y. Tomobe, M. Kobayashi, Y. Mitsui, Y. Yazaki, K. Goto and T. Masaki. 1988. A novel potent vasoconstrictor peptide produced by vascular endothelial cells. Nature 332: 411-415.

12.   Gryglewski, R.J., S. Moncada and R.M.J. Palmer. 1986. Bioassay of prostacyclin and endothelium-derived relaxing factor (EDRF) from porcine aortic endothelial cells. Br. J. Pharmacol. 87: 685-694.

13.   Rapoport, R.M., M.B. Draznin and F. Murad. 1983. Endothelium-dependent relaxation in rat aorta may be mediated through cyclic GMP-dependent protein phosphorylation. Nature 306: 174-176.

14.   Förstermann, U., A. Mulsch, E. Bohme and R. Busse. 1986. Stimulation of soluble guanylate cyclase by an acetylcholine-induced endothelium-derived factor from rabbit and canine arteries. Circ. Res. 58: 531-538.

15.   Ignarro, L.J., R.E. Byrns and K.S. Wood. 1987. Endothelium-dependent modulation of cGMP levels and intrinsic smooth muscle tone in isolated bovine intrapulmonary artery and vein. Circ. Res. 60: 82-92.

16.   Faggiotto, A., R. Ross and L. Harker. 1984. Studies of hypercholesterolemia in the nonhuman primate. II. Fatty streak conversion in fibrous plaque. Arteriosclerosis 4: 341-356.

17.   Taylor, K.E., S. Glagov and C.K. Zarins. 1989. Preservation and structural adaptation of endothelium over experimental foam cell lesions. Quantitative ultrastructural study. Arteriosclerosis 9: 881-894.

18.   Habib, J.B., C. Bossaler, S. Wells, C. Williams, J.D. Morrisett and P.D. Henry. 1986. Preservation of endothelium-dependent vascular relaxation in cholesterol-fed rabbit by treatment with the calcium blocker PN 200110. Circ. Res. 58: 305-309.

19.   Verbeuren, T.J., F.H. Jordaens, L.L. Zonnekeyn, C.E. Van Hove, M.C. Coene and A.G. Herman. 1986. Effect of hypercholesterolemia on vascular reactivity in the rabbit. Circ. Res. 58: 552-564.

20.   Jayakody, L., M. Senaratne, A. Thomson and T. Kappagoda. 1987. Endothelium-dependent relaxation in experimental atherosclerosis in the rabbit. Circ. Res. 60: 251-264.

21. Freiman, P.C., G.C. Mitchell, D.D. Heistad, M.L. Armstrong and D.G. Harrison. 1986. Atherosclerosis impairs endothelium-dependent vascular relaxation to acetylcholine and thrombin in primates. Circ. Res. 58: 783-789.

22. Harrison, D.G., M.L. Armstrong, P.C. Freiman and D.D. Heistad. 1987. Restoration of endothelium-dependent relaxation by dietary treatment of atherosclerosis. J. Clin. Invest. 80: 1808-1811.

23. Wines, P.A., J.M. Schmitz, S.L. Pfister, F.J. Clubb Jr., L.M. Buja, J.T. Willerson and W.B. Campbell. 1989. Augmented vasoconstrictor responses to serotonin precede development of atherosclerosis in aorta of WHHL rabbit. Arteriosclerosis 9: 195-202.

24. Andrews, H.E., X.R. Bruckdorfer, R.C. Dunn and M. Jacobs. 1987. Low-density lipoproteins inhibit endothelium-dependent relaxation in rabbit aorta. Nature 327: 237-239.

25. Kugiyama, K., M. Bucay, J.D. Morrisett, R. Roberts and P.D. Henry. 1989. Oxidized LDL impairs endothelium-dependent arterial relaxation. Circulation 80: II-278.

26. Yamamoto, H., Bossaller, C., Cartwright, J.Jr. and Henry, P.D. 1988. Videomicroscopic demonstration of defective cholinergic arteriolar vasodilatation on atherosclerotic rabbit. J. Clin. Invest. 81: 1752-1758.

27. Sellke, F.W., M.L. Armstrong and D.G. Harrison. 1990. Endothelium-dependent vascular relaxation is abnormal in the coronary microcirculation of atherosclerotic primates. Circulation 81: 1586-1593.

28. Bossaller, C., G.B. Habib, H. Yamamoto, C. Williams, S. Wells and P.D. Henry. 1987. Impaired muscarinic endothelium-dependent relaxation and cyclic guanosine 5'-monophosphate formation in atherosclerotic human coronary artery and rabbit aorta. J. Clin. Invest. 79: 170-174.

29. Förstermann, U., A. Mügge, U. Alheid, A. Haverich and J.C. Frölich. 1988. Selective attenuation of endothelium-mediated vasodilation in atherosclerotic human coronary arteries. Circ. Res. 62: 185-190.

30. Ludmer, P.L., A.P. Selwyn, T.L. Shook, R.R. Wayne, G.H. Mudge, R.W. Alexander and P. Ganz. 1986. Paradoxial vasoconstriction induced by acetylcholine in atherosclerotic coronary arteries. N. Engl. J. Med. 315: 1046-1051.

31. Vita, J.A., C.B. Treasure, E.G. Nabel, R.D. Fish, J.M. McLenachan, A.C. Yeung, V.I. Vekshtein, A.P. Selwyn and P. Ganz. 1989. The coronary response to acetylcholine relates to coronary risk factors. Circulation 80: II-435.

32. Cox, D.A., J.A. Vita, C.B. Treasure, R.B. Fish, R.W. Alexander, P. Ganz and A.P. Selwyn. 1989. Atherosclerosis impairs flow-mediated dilation of coronary arteries in man. Circulation 80: 458-465.

33. Drexler, H., A.M. Zeiher, H. Wollschlager, T. Meinertz, J. Hanjorg and T. Bonzel. 1989. Demonstration of flow-dependent coronary dilatation in man. Circulation 80: 466-474.

34. Guerra, R. Jr., A.F.A. Brotherton, P.J. Goodwin, C.R. Clark, M.L. Armstrong and D.G. Harrison. 1990. Mechanisms of abnormal endothelium-dependent vascular relaxation in atherosclerosis. Blood Vessels (in press).

35. Gryglewski, R.J., Palmer, R.M.J. and Moncada, S. 1986. Superoxide anion is involved in breakdown of endothelium-derived vascular releasing factor. Nature 320: 454-456.

36. Lopez, J.A.G., M.L. Armstrong, D.G. Harrison, D.J. Piegors and D.D. Heistad. 1989. Vascular responses to leukocyte products in atherosclerotic primates. Circ. Res. 65: 1078-1086.

37. Myers, P.R., R. Guerra, Jr. and D.G. Harrison. 1989. Release of NO and EDRF from cultured bovine aortic endothelial cells. Am. J. Physiol. 256: H1030-H1037.

38. Rubanyi, G.M., A. Johns, D.G. Harrison and D. Wilcox. 1989. Evidence that EDRF may be identical with an S-nitrosothiol and not with free nitric oxide. Circulation 80 (Suppl II): II-281.

39. Wei, E.P. and H.A. Kontos. H_2O_2 and endothelium-dependent cerebral arteriolar dilation: Implications for the identity of EDRF from acetylcholine. Hypertension (in press).

40. Long, C.J., K. Shikano and B.A. Berkowitz. 1987. Anion exchange resins discriminate between nitric oxide and EDRF. Eur. J. Pharm. 142: 317-318.

41. Murray, J.J., I. Fridovich, R.G. Makhoul and P.O. Hagen. 1986. Stabilization and partial characterization of endothelium-derived relaxing factor from cultured bovine aortic endothelium cells. Biochem. Biophys. Res. Commun. 141: 689-696.

42. Mayhan, W.G. 1989. Impairment of endothelium-dependent dilatation of cerebral arterioles during diabetes mellitus. Am. J. Physiol. 256: H621-H625.

43. Ku, D. 1982. Coronary vascular reactivity after acute myocardial ischemia. Science 218: 576-578.

44. Wei, E.P., H.A. Kontos, C.W. Christman, D.S. DeWitt and J.T. Povlishock. 1985. Superoxide generation and reversal of acetylcholine-induced cerebral arteriolar dilation after acute hypertension. Circ. Res. 57: 781-787.

45. Minor, R.L., Jr., P.R. Myers, R. Guerra, Jr., J.N. Bate and D.G. Harrison. Diet induced atherosclerosis increases the release of nitrogen oxides from rabbit aorta. J. Clin. Inv. (in press).

46. Harrison, D.G., M.L. Armstrong, P.C. Freiman and D.D. Heistad. 1987. Restoration of endothelium-dependent relaxation by dietary treatment of atherosclerosis. J. Clin. Invest. 80: 1808-1811.

47. Shimokawa, H. and P.M. Vanhoutte. 1988. Dietary cod-liver oil improves endothelium-dependent responses in hypercholesterolemic and atherosclerotic porcine coronary arteries. Circulation 78: 1421-1430.

48. Vekshtein, V.I., A.C. Yeung, J.A. Vita, E.G. Nabel, R.D. Fish, J.A. Bittl, A.P. Selwyn and P. Ganz. 1989. Fish oil improves endothelium-dependent relaxation in patients with coronary artery disease. Circulation 80: II-434.

49. Koga, T., Y. Takata, K. Kobayashi, S. Takishita, Y. Yamashita and M. Fujishima. 1989. Age and hypertension promote endothelium-dependent contractions to acetylcholine in the aorta of the rat. Hypertension 14: 542-548.

50. Alheid, U., I. Reichwehr and U. Förstermann. 1989. Human endothelial cells inhibit platelet aggregation by separately stimulating platelet cyclic AMP and cyclic GMP. Eur. J. Pharm. 164: 103-110.

51. Busse, R., A. Luckhoff and E. Bassenge. 1987. Endothelium-derived relaxant factor inhibits platelets activation. Arch. Pharm. 336: 566-577.

PRODUCTION OF CYTOKINES BY VASCULAR WALL CELLS:

AN UPDATE AND IMPLICATIONS FOR ATHEROGENESIS

Peter Libby, Harald Loppnow, James C. Fleet,
Helen Palmer, Hong Mei Li, Stephen J.C. Warner,
Robert N. Salomon and Steven K. Clinton

Cardiovascular Research Laboratory
Departments of Medicine, Pathology, and
 Cellular and Molecular Physiology
New England Medical Center and
USDA Human Nutrition Research Center on Aging
Tufts University, 711 Washington Street
Boston, MA 02111

The cytokines are protein mediators of inflammation and immunity. Within the last decade, numerous cytokine activities originally defined by a variety of biological assays have become increasingly well defined; thanks to laborious protein purification and ultimately the application of recombinant DNA technology. The availability of well defined proteins expressed using molecular biologic techniques has provided material for precise definition of the activities and targets for these various mediators. Several general findings have emerged from this effort. One principle is that a single molecule or family of related molecules may exhibit multiple activities previously considered unrelated. A corollary is that cytokines tend to occur in families of closely related isoforms or structurally or functionally related proteins. Despite these simplifying generalizations, the list of recognized cytokines continues to expand. There are now some nine different interleukins (IL), many different colony stimulating factors (CSFs) and several distinct interferon classes. There is also an intriguing degree of overlap between the structure and function of molecules generally considered cytokines and certain peptide growth factors. For example IL1-ß has regions of amino acid sequence similarity with members of the fibroblast growth factor (FGF) superfamily[1].

The recognition that cytokines can alter key functions of vessel wall cells (e.g., compatibility with blood, adhesivity for leukocytes, immunologic functions, and growth state) piqued the interest of vascular biologists in these molecules. Vascular wall cells are not only targets for cytokine actions, but also can produce these mediators under some circumstances. The ability of vascular wall cells both to respond to and produce these multipotent mediators raises the possibility of complex autocrine and paracrine regulatory loops of potential significance in vascular homeostasis and pathophysiology. This chapter will review the current state of knowledge of cytokine production by vascular wall cells emphasizing new developments and the possible significance of vessel-derived cytokines in atherogenesis.

INTERLEUKIN 1 (IL1)

The IL1 family consists of two isoforms, denoted α and ß, very similar in their spectrum of functions despite significant divergence in their primary structure (amino

Atherosclerosis, Edited by A. I. Gotlieb *et al.*
Plenum Press, New York, 1991

acid sequence similarity of only 26%)[2,3,4]. Interleukin 1 possesses well known properties as an endogenous pyrogen and immunostimulator in addition to many less well known functions (e.g., neuroendocrine regulation, hematopoietic growth modulation)[5]. In addition, IL1 extensively alters functions of vascular endothelial and smooth muscle cells of importance in pathophysiology, as reviewed elsewhere[6,7]. The "classical" source of IL1 activity is the mononuclear phagocyte. However, human vascular endothelial cells can transcribe the genes that encode both isoforms of IL1[8,9,10]. Stimuli that elicit accumulation of IL1 mRNA in these cells include bacterial endotoxin (lipopolysaccharide), tumor necrosis factor (TNF), and IL1 itself.

Like the prototypical IL1 producing cell, the mononuclear phagocyte, human vascular endothelial cells express IL1 activity associated with their surfaces after exposure to inductive stimuli[11]. Again, as in the case of mononuclear phagocytes, this membrane-associated IL1 resembles the α isoform serologically. The classical biological activity of IL1 is thymocyte costimulation, the ability to augment the proliferation of mouse thymocytes incubated with a suboptimal concentration of mitogenic lectins such as phytohemagglutinin. When exposed to inducers, cultured human vascular endothelial cells elaborate into their incubation medium thymocyte costimulatory activity, a result generally interpreted as an indication of IL1 release by these cells[8,9,10]. However, more recent studies with target cell lines that exhibit great selectivity for IL1, and the use of monospecific antibody reagents raised against purified recombinant proteins, have failed to support the interpretation that these cells release substantial quantities of IL1 into the external medium following stimulation. Cytokines other than IL1 may contribute to the thymocyte costimulatory activity found in the conditioned medium of activated endothelial cells (Loppnow and Libby, unpublished observations).

Vascular smooth muscle cells also exhibit readily inducible accumulation of both species of IL1 mRNA[12,13]. The range of stimuli that evoke IL1 gene expression in smooth muscle cells resembles that previously documented in endothelial cells[13,14]. Our own experiments suggest that smooth muscle cells respond to inductive stimuli more rapidly and at lower concentrations than do endothelial cells also harvested from adult human veins. However, these *in vitro* observations may reflect differences in cell culture harvest and propagation techniques rather than *in vivo* differences. Like endothelial cells, smooth muscle cells express surface-associated IL1 activity after induction (Loppnow and Libby, unpublished observations). As in the case of endothelial cells, smooth muscle cells do not appear to elaborate large quantities of IL1 into the surrounding medium when assessed by current selective assay techniques. Recent experiments demonstrated that smooth muscle cells bearing surface IL1 can stimulate other cultured vascular cells when they come into contact (see below). This finding demonstrates the possibility of contact-mediated paracrine stimulation due to a cytokine produced *in situ* by vascular cells (Loppnow and Libby, unpublished observations).

Most of our knowledge of the capacity of vascular wall cells to express genes encoding IL1 was obtained *in vitro*. Recently we have investigated whether IL1 genes can be expressed in vascular tissue in intact animals. We chose the rabbit for these studies because of its widespread use in studies of inflammatory responses and because of the susceptibility of this species to diet-induced arterial abnormalities with features reminiscent of early stages of atherosclerotic lesions. We found that, in the basal state, extracts of rabbit aortae contained little or no mRNA encoding either species of IL1. However, intravenous administration of Gram-negative bacterial endotoxin elicited a transient increase in steady state message levels for IL1 α and ß. Extracts of aortic tissue from these rabbits exhibited increased IL1 biological activity. The peak level of IL1 mRNAs occurred one to two hours following the intravenous injection of endotoxin.

We also tested whether aortae of rabbits with diet-induced atheromatous changes expressed IL1 genes[15]. This is an important question as the state of activation of lipid-laden foam cells within the aortae of fat- and cholesterol-fed animals is uncertain. Using polymerase chain reaction analysis of cDNA prepared from RNA extracted from these experimental lesions, we found little or no elevation in the levels of mRNA encoding IL1 α or ß. However, cholesterol-fed rabbits had an accentuated response to an inductive stimulus since, after intravenous endotoxin administration, the levels of cDNA for both species of IL1 (as well as TNF) exceeded levels in aortae of animals fed control diets. The accentuated accumulation of mRNA encoding these cytokines was related to the concentration of cholesterol in the diet. We do not know what cell type within normal and atheromatous vessels transcribes the IL1 genes. Further studies using tissue fractionation or *in situ* hybridization techniques would be required to answer this question. Established atheroma in cholesterol-fed primates appear to contain endothelial and smooth muscle cells in addition to monocytes that express IL1 mRNA and protein[16]. Thus, vascular tissues *in vivo* may generate cytokines such as IL1 locally in ways that may be of importance in vascular pathophysiology.

INTERLEUKIN 6

Interleukin 6 is a relatively new name for a protein which is now known to cause a wide variety of biological activities. IL6 stimulates B cell proliferation, maturation into plasma cells and production of immunoglobulins. It also stimulates T cells and alters the biosynthetic profile of hepatocytes to favor the production of "acute phase" proteins in place of "housekeeping" proteins such as albumin[17,18,19,20,21]. Interleukin 6 is also an endogenous pyrogen. A number of groups recently defined the capacity of vascular endothelial cells to produce IL6 activity[22,23,24,25]. In contrast to IL1, IL6 contains a recognizable signal sequence in its primary structure. Instead of remaining associated with the cell layer as does IL1, IL6 appears to be secreted immediately into the surrounding medium, as shown by metabolic labelling and immunoprecipitation experiments[25]. The stimuli that evoke IL6 transcription and translation by human vascular endothelial cells resemble those that induce IL1 gene expression in these cells, and include bacterial endotoxin, tumor necrosis factor, and IL1. Smooth muscle cells appear to be an even more abundant source of IL6 than vascular endothelial cells. After exposure to IL1, nearly 4% of the newly-synthesized proteins released by these cells consist of IL6, as determined by quantitative metabolic labelling and immunoprecipitation studies[26].

The biological significance of this enormous output of a biologically active mediator by smooth muscle is unclear. IL6 seems to lack significant stimulatory or inhibitory efforts on the growth or other functions of vascular wall cells. Locally-elaborated IL6 could certainly participate in the modulation and function of B or T lymphocytes involved in regional immune and inflammatory responses[27,28]. In support of this concept, Duff and coworkers have localized substantial IL6 mRNA in the tunica media of vessels in inflamed rheumatoid synovia (personal communication). It is unknown whether vascular cells *in vivo* can express the IL6 gene. However, we recently found that short term organoid cultures of human vessels, like cultured cells, can secrete copious IL6 activity upon exposure to stimuli such as IL1[26].

TUMOR NECROSIS FACTOR (TNF-α)

Although vascular endothelial cells can transcribe both IL1 genes, we are unaware of convincing evidence that these cells can transcribe the TNF gene under any circumstance. We have sought TNF expression in these cells under a variety of inducing and "superinducing" conditions. Human vascular smooth muscle cells, although they appear not to contain TNF transcripts under ordinary culture conditions, can be coaxed to accumulate TNF mRNA when exposed to mediators such as IL1, particularly in the presence of inhibitors of protein synthesis such as cycloheximide or anisomycin ("superinduction" conditions)[29]. After superinduction of TNF mRNA and washing to remove the inhibitors of protein synthesis, we have demonstrated *de novo* synthesis of TNF protein and release of TNF biological activity from these cells. These studies establish the capacity of human vascular smooth muscle cells to transcribe and translate the TNF gene. In view of the rather extreme conditions required to elicit TNF gene expression for this important mediator in cultured smooth muscle cells, the biological significance of this observation remains unsettled. One group has localized immunoreactive TNF in human atherosclerotic plaques in association with smooth muscle cells, an indication that our *in vitro* observations may have relevance to human atherogenesis[30].

INTERLEUKIN 8 (IL8)

Interleukin 8 is a relatively recent designation for an activity also referred to as neutrophil activating factor[31]. This protein stimulates directed migration of both polymorphonuclear leukocytes and T lymphocytes, and promotes superoxide and lysosomal enzyme release from neutrophils. Interleukin 8 shares structural features with a family of protein mediators that includes ß-thromboglobulin, GRO, and other macrophage-derived mediators[32]. Several groups have demonstrated the capacity of human endothelial cells to express the IL8 gene, a finding of possible significance in acute inflammatory states that involve infiltration of polymorphonuclear leukocytes or lymphocytes. Recently Gimbrone's group has found that endothelial cells elaborate a possibly distinct form of IL8, a finding that emerged from their characterization of a leukocyte adhesion inhibitory activity[33]. This unexpected result illustrates the principle enunciated above that a single cytokine can exhibit seemingly unrelated activities (neutrophil chemotaxis and adhesion inhibition).

"INTERLEUKIN 9 (IL9)"

Interleukin 9 is a designation proposed to describe a macrophage chemoattractant and stimulatory activity, known variously as monocyte chemotactic and activating factor (MCAF) and monocyte chemoattractant protein-1 (MCP-1) among others. This protein is the human homolog of the mouse JE gene originally cloned from PDGF-stimulated fibroblasts[34]. This protein shares sequence similarity with IL8, making it a member of the same superfamily of mediators discussed above[31]. Endothelial cells express the gene for IL-9 and secrete the protein following stimulation with a variety of inflammatory mediators. Preliminary studies suggest a similar response by rat[35] and human (Clinton *et al.*, unpublished observations) vascular smooth muscle cells. The ability of vascular wall cells to express the gene for this monocyte chemoattractant protein has major implications for atherogenesis. Mononuclear phagocytes are the precursors of many of the foam cells that characterize fatty streaks in humans and arterial lesions associated with diets high in cholesterol and saturated fat. The local elaboration of monocyte chemoattractant and activating factors such as IL9 by activated vessel wall cells could play a crucial role in the initiation of these lesions.

COLONY STIMULATING FACTORS (CSFs)

The colony stimulating factors are hematopoietic growth factors that maintain the survival *in vitro* and stimulate proliferation of precursors of various cell lineages derived from bone marrow[36,37,38]. A prototypical CSF is granulocyte-macrophage colony stimulating factor (GM-CSF). Endothelial cells can produce GM-CSF upon exposure to stimuli such as IL1[39,40]. This finding implicates microvascular endothelial cells in regional paracrine regulation of the differentiation and growth of blood cell precursors within the bone marrow. In addition, endothelial cells express the GM-CSF receptor and the ligand induces endothelial cell migration *in vitro*[41].

The production of other CSFs by endothelial cells and the capacity of vascular smooth muscle cells to express CSF genes remain incompletely defined. Monocyte CSF (M-CSF) promotes the survival of monocytes and macrophages and accelerates the terminal differentiation of these cells in response to inflammatory mediators. We have found that both endothelial and smooth muscle cells express the gene for M-CSF in response to a variety of stimuli (IL-1, TNF, LPS) in culture (Clinton *et al.*, unpublished observations). We have also found M-CSF mRNA in the atheromatous aortas of rabbits fed diets high in cholesterol. This observation was stimulated by our finding of substantial extramedullary hematopoiesis and hyperplastic bone marrows in cholesterol- and fat-fed rabbits.

In relation to atherogenesis, oxidatively modified lipoproteins can elicit production of GM-CSF and M-CSF (monocyte-CSF) from endothelial cells[42]. Thus, endogenous vascular cells may elaborate factors that promote the expansion of populations of mononuclear phagocytes recruited by IL9/MCP-1/JE, in conjunction with the expression of monocyte-endothelial adhesion molecules whose expression is in turn subject to regulation by IL1 or TNF. These infiltrating mononuclear phagocytes give rise to many of the foam cells found in atheroma, and may themselves elaborate many cytokines, growth factors, and other mediators of substantial potential significance in atherogenesis[43].

INTERFERONS

The interferons are families of proteins that share the ability to induce an anti-viral state in target cells[44]. Various cell types can produce interferons. Leukocytes elaborate interferon α, fibroblasts are the classical source of interferon ß. Activated T lymphocytes and NK cells can produce interferon gamma, also known as immune interferon. In addition to their anti-viral effects, interferons exert anti-proliferative actions. We recently tested whether interferon gamma, a lymphokine probably present in human atheromata, inhibits the proliferation of human smooth muscle cells. We found that recombinant interferon gamma indeed slowed the proliferative response of human vascular smooth muscle cells exposed to growth stimulators such as platelet-derived growth factor (PDGF) and IL1[45]. Hansson and co-workers obtained similar results with rodent smooth muscle cells and obtained important *in vivo* data in rats supporting an antiproliferative effect of interferon gamma[46].

Because smooth muscle cells and fibroblasts share many characteristics in culture, we tested whether human smooth muscle cells might share with fibroblast the ability to express interferon ß. Our previous characterization of TNF-induced gene expression in smooth muscle cells also suggested that this cell type might produce an endogenous interferon because TNF elevated the levels of RNA encoding 2'-5'-oligoadenylate synthetase in these cells[29]. This enzyme is thought to be induced uniquely by interferons and is thus a candidate mediator of interferon's intracellular actions[47,48]. We found that, under classical interferon inducing conditions and in response to TNF, human smooth muscle cells accumulate interferon ß mRNA[49]. We have also found interferon-like antiviral activity in the supernatants of smooth muscle cell cultures under conditions in which the mRNA accumulates. These results are similar to those obtained in fibroblasts by Vilcek and co-workers[50]. We believe that endogenous production of interferon ß by human vascular smooth muscle cells may provide an autocrine inhibitory pathway as recombinant interferon ß, like immune interferon, inhibits the responsiveness of these cells to a number of mitogens.

THE SIGNIFICANCE OF CYTOKINE GENE EXPRESSION
BY VASCULAR WALL CELLS AND REMAINING QUESTIONS

The findings summarized in the foregoing discussion establish that intrinsic cells of the blood vessel wall, in addition to constituting targets for cytokine action, can also produce these multipotent mediators. At present, these functional capacities of vascular wall cells remain of uncertain biological significance. While cytokines likely play important roles in those systemic and local inflammatory reactions, the contribution of cytokines produced by vascular cells remains speculative. In the case of IL1, little of the biological activity produced by vascular endothelial and smooth muscle cells appears to be released. Thus, if IL1 derived from these vascular cells were to play a significant role in pathophysiology, it would likely involve direct contact between producing cells and their target cell type. On a quantitative basis, leukocytes would seem the likely sources of large amounts of cytokines required for systemic effects. In addition, many local inflammatory processes within blood vessels are associated with significant infiltration by leukocytes, cells that are "professional" producers of cytokines. Once within vascular lesions, activated leukocytes would seem a more plausible source for locally acting cytokines than the intrinsic vascular cells themselves. Interleukin 6, produced in astonishing quantities by smooth muscle cells, might prove an exception.

What then might be the pathophysiologic significance of the ability of vascular wall cells to elaborate cytokines? We speculate that, in the earliest phases of vascular injury or inflammation, cytokines generated by endothelium or smooth muscle cells provide key *initial* signals for autocrine or paracrine activation of the mechanisms that recruit and activate leukocytes, cells not normally present in the vessel wall, at specific foci. Thus we envisage both temporal and spatial importance for vascular cytokines. Considerable recent evidence supports roles for vascular endothelial and smooth muscle cells in local immune responses, including such functions as antigen presentation or stimulation of allogeneic reactions in the context of organ transplantation. In these situations, the vascular cells would subsume accessory cell functions traditionally ascribed to leukocytes. Cytokines secreted locally or expressed on the surfaces of vascular cells may also provide second signals at short distances that are critically important in the initiation and amplification of these immune responses. For these reasons, we propose that cytokines of vascular origin may contribute not only to the initial phases of atherogenesis but to transplant rejection, the vasculitides, and a variety of other pathologic processes that involve blood vessels. Definition of the capacities of vascular endothelial and smooth muscle cells to express various cytokine genes, as reviewed here, provides a firm foundation for the considerable experimental work that lies ahead to explore these possible functions.

REFERENCES

1. Thomas, K.A., Rios-Candelore, M., Gimenez-Gallego, G., DiSavlo, J., Bennett, C., Rodkey, J. and Fitzpatrick, S. 1985. Pure brain-derived acidic fibroblast growth factor is a potent angiogenic vascular endothelial cell mitogen with sequence homology to interleukin 1. Proc. Natl. Acad. Sci. USA 82: 6409-6413.

2. Auron, P.E., Webb, A.C., Rosenwasser, L.J., Mucci, S.F., Rich, A., Wolff, S.M. and Dinarello, C.A. 1984. Nucleotide sequence of human monocyte interleukin-1 precursor cDNA. Proc. Natl. Acad. Sci. USA 81: 7907-7911.

3. March, C.J., Mosley, B., Larsen, A., Cerretti, D.P., Braedt, G., Price, V., Gillis, S., Henney, C.S., Kronheim, S.R., Grabstein, K., Conlon, P.J., Hopp, T.P. and Cosman, D. 1985. Cloning, sequence and expression of two distinct human interleukin-1 complementary DNAs. Nature (Lond.) 315: 641-647.

4. Gubler, U., Chua, O.A., Stern, A.S., Hellman, C.P., Vitek, M.P., Dechiara, T.M., Benjamin, W.R., Collier, K.J., Dukovich, M., Familetti, P.C., Fiedler-Nagy, C., Jenson, J., Kaffka, K., Kilian, P.L., Stremlo, D., Wittreich, B.H., Woehle, D., Mizel, S.B. and Lomedico, P.T. 1986. Recombinant human interleukin 1α: Purification and biological characterization. J. Immunol. 136: 2492-2497.

5. Dinarello, C.A. 1989. Interleukin 1 and its biologically related cytokines. Adv. Immunol. 144: 153-205.

6. Bevilacqua, M.P., Pober, J.S., Wheeler, M.E., Cotran, R.S., Gimbrone, M.A.J. 1985. Interleukin-1 activation of vascular endothelium effects on procoagulant activity and leukocyte adhesion. Am. J. Path. 121: 393-403.

7. Pober, J.S. and Cotran, R.S. 1990. Cytokines and endothelial cell biology. Physiol. Rev. 70: 427-451.

8. Wagner, C.R., Vetto, R.M., Burger, D.R. 1985. Expression of I-region-associated antigen (Ia) and interleukin 1 by subcultured human endothelial cells. Cell. Immunol. 93: 91-104.

9. Libby, P., Ordovàs, J.M., Auger, K.R., Robbins, H., Birinyi, L.K. and Dinarello, C.A. 1986. Endotoxin and tumor necrosis factor induce interleukin-1 gene expression in adult human vascular endothelial cells. Am. J. Path. 124: 179-186.

10. Nawroth, P.P., Bank, I., Handley, D., Cassimeris, J., Chess, L. and Stern, D. 1986. Tumor necrosis factor/cachectin interacts with endothelial cell receptors to induce release of interleukin 1. J. Exp. Med. 163: 1363-1375.

11. Kurt-Jones, E.A., Fiers, W. and Pober, J.S. 1987. Membrane interleukin 1 induction on human endothelial cells and dermal fibroblasts. J. Immunol. 139: 2317-2324.

12. Moyer, C.F. and Reinisch, C.L. 1984. The role of vascular smooth muscle cells in experimental autoimmune vasculitis. I. The initiation of delayed type hypersensitivity angiitis. Am. J. Pathol. 117: 380-390.

13. Libby, P., Ordovàs, J.M., Birinyi, L.K., Auger, K.R. and Dinarello, C.A. 1986. Inducible interleukin-1 expression in human vascular smooth muscle cells. J. Clin. Invest. 78: 1432-1438.

14. Warner, S.J.C., Auger, K.R. and Libby, P. 1987. Human interleukin 1 induces interleukin 1 gene expression in human vascular smooth muscle cells. J. Exp. Med. 165: 1316-1331.

15. Fleet, J., Clinton, S., Salomon, R., Loppnow, H. and Libby, P. 1990. Atherogenic diets increase endotoxin-stimulated cytokine gene expression in rabbit aortae. FASEB J. 4: A1156.

16. Williams, K., Sajuthi, D., Tulli, H., Huggins, E. and Moyer, C. 1990. Vascular cells in atherosclerotic plaques of cynomolgous monkeys synthesize interleukin-1. FASEB J. 4: A1154.

17. Zilberstein, A., Ruggieri, R., Korn, J.H. and Revel, M. 1986. Structure and expression of cDNA and genes for human interferon-beta-2, a distinct species inducible by growth-stimulatory cytokines. EMBO J. 5: 2529-2537.

18. Brakenhoff, J.P., de Groot, E.R., Evers, R.F., Pannekoek, H. and Aarden, L.A. 1987. Molecular cloning and expression of hybridoma growth factor in Escherichia coli. J. Immunol. 139: 4116-4121.

19. Sehgal, P.B., May, L.T., Tamm, I. and Vilcek, J. 1987. Human ß^2 interferon and B-cell differentiation factor BSF-2 are identical. Science 235: 731-732.

20. Wong, G.G. and Clark, S.C. 1988. Multiple actions of IL6 within a cytokine network. Immunol. Today 9: 137-139.

21. Van Damme, J. 1989. Biochemical and biological properties of human HPGH/IL-6. Annals N.Y. Acad. Sci. 557: 104-112.

22. Norioka, K., Hara, M., Harigai, M., Kitani, A., Hirose, T., Suzuki, K., Kawakami, M., Tabata, H., Kawagoe, M. and Nakamura, H. 1988. Production of B cell stimulatory factor-2/interleukin-6 activity by human endothelial cells. Biochem. Biophys. Res. Comm. 153: 1045-1050.

23. Sironi, M., Breviario, F., Proserpio, P., Biondi, A., Vecchi, A., Van Damme, J., Dejana, E. and Mantovani A. 1989. IL1 stimulates IL6 production in endothelial cells. J. Immunol. 142: 549-553.

24. Jirik, F.R., Podor, T.J., Hirano, T., Kishimoto, T., Loskutoff, D.J., Carson, D.A. and Lotz, M. 1989. Bacterial lipopolysaccharides and inflammatory mediators augment IL6 secretion by human endothelial cells. J. Immunol. 142: 144-147.

25. Loppnow, H. and Libby, P. 1989. Adult human vascular endothelial cells express the IL6 gene differentially in response to LPS or IL1. Cell. Immunol. 122: 493-503.

26. Loppnow, H. and Libby, P. 1990. Proliferating or interleukin 1-activated human vascular smooth muscle cells secrete copious interleukin 6. J. Clin. Invest. 85: 731-738.

27. Le, J.M. and Vilcek, J. 1989. Interleukin 6: A multifunctional cytokine regulating immune reactions and the acute phase protein response. Lab Invest. 61: 588-602.

28. Le, J.M., Fredrickson, G., Pollack, M. and Vilcek, J. 1989. Activation of thymocytes and T cells by interleukin-6. Ann. N. Y. Acad. Sci. 557: 444-452.

29. Warner, S.J.C. and Libby, P. 1989. Human vascular smooth muscle cells: Target for and source of tumor necrosis factor. J. Immunol. 142: 100-109.

30. Barath, P., Fishbein, M.C., Cao, J., Berenson, J., Helfant, R.H. and Forrester, J.S. 1990. Detection and localization of tumor necrosis factor in human atheroma. Am. J. Cardiol. 65: 297-302.

31. Matsushima, K. and Oppenheim, J.J. 1989. Interleukin 8 and MCAF: Novel inflammatory cytokines inducible by IL1 and TNF. Cytokine 1: 2-13.

32. Yoshimura, T., Matsushima, K., Tanaka, S., Robinson, E.A., Appella, E., Oppenheim, J.J. and Leonard, E.J. 1987. Purification of a human monocyte-derived neutrophil chemotactic factor that has peptide sequence similarity to other host defense cytokines. Proc. Natl. Acad. Sci. USA 84: 9233-9237.

33. Gimbrone, M.A. Jr., Obin, M.S., Brock, A.F., Luis, E.A., Hass, P.E., Hébert, C.A., Yip, Y.K., Leung, D.W., Lowe, D.G., Kohr, W.J., Darbonne, W.C., Bechtol, K.B. and Baker, J.B. 1989. Endothelial interleukin-8: A novel inhibitor of leukocyte-endothelial interactions. Science 246: 1601-1603.

34. Rollins, B.J., Morrison, E.D. and Stiles, S.C. 1988. Cloning and expression of JE, a gene inducible by platelet-derived growth factor and whose product has cytokine-like properties. Proc. Natl. Acad. Sci. USA 85: 3738-3742.

35. Taubman, M.B., Rollins, B.J. and Nadal-Ginard, B. 1989. Expression of the JE gene in vascular smooth muscle cells: Differential effects of PDGF and angiotensin II. Circulation 80 (4): 11-451.

36. Metcalf, D. 1986. The molecular biology and functions of the granulocyte-macrophage colony-stimulating factors. Blood 67: 257-267.

37. Clark, S.C. and Kamen, R. 1987. The human hematopoietic colony-stimulating factors. Science 236: 1229-1237.

38. Groopman, J.E., Molina, J.M. and Scadden, D.T. 1989. Hematopoietic growth factors. Biology and clinical applications. N Engl J Med 321: 1449-1459.

39. Quesenberry, P.J. and Gimbrone, M.A. Jr. 1980. Vascular endothelium as a regulator of granulopoiesis: Production of colony-stimulating activity by cultured human endothelial cells. Blood 56: 1060-1067.

40. Bagby, G.C.J., Dinarello, C.A., Wallace, P., Wagner, C., Hefeneider, S. and McCall, E. 1986. Interleukin 1 stimulates granulocyte macrophage colony-stimulating activity release by vascular endothelial cells. J. Clin. Invest. 78: 1316-1323.

41. Bussolino, F., Wang, J.M., Defilippi, P., Turrini, F., Sanavio, F., S., E.C.-J., Aglietta, M., Arese, P. and Mantovani, A. 1989. Granulocyte- and granulocyte-macrophage-colony stimulating factors induce human endothelial cells to migrate and proliferate. Nature 337: 471-473.

42. Rajavashisth, T.B., Andalibi, A., Territo, M.C., Berliner, J.A., Navab, M., Fogelman, A.M. and Lusis, A.J. 1990. Induction of endothelial cell expression of granulocyte and

macrophage colony-stimulating factors by modified low-density lipoproteins. Nature 344: 254-257.

43. Ross, R. 1986. The pathogenesis of atherosclerosis - an update. New Engl. J. Med. 314: 488-500.

44. Pestka, S., Langer, J.A., Zoon, K.C. and Samuel, C.E. 1987. Interferons and their actions. Ann. Rev. Biochem. 56: 727-777.

45. Warner, S.J.C., Friedman, G.B. and Libby, P. 1989. Immune interferon inhibits proliferation and induces 2'-5'-oligoadenylate synthetase gene expression in human vascular smooth muscle cells. J. Clin. Invest. 83: 1174-1182.

46. Hansson, G.K., Jonasson, L., Holm, J., Clowes, M.K., Clowes, A. 1988. Gamma interferon regulates vascular smooth muscle proliferation and Ia expression *in vivo* and *in vitro*. Circ. Research 712-719.

47. Benech, P., Mory, Y., Revel, M. and Chebath, J. 1985. Structure of two forms of the interferon-induced (2'-5') oligo A synthetase of human cells based on cDNAs and gene sequences. EMBO J. 2249-2256.

48. Chebath, J., Benech, P., Mory, Y., Mallucci, L., Michalevicz, R. and Revel, M. 1986. IFN and (2'-5') oligo A synthetase in cell growth and in differentiation of hematopoietic cells. 351-363.

49. Palmer, H.J. and Libby, P. 1990. Interferon-beta: A potential autocrine regulator of human vascular smooth muscle cell (SMC) proliferation. FASEB J. 4: A906.

50. Reis, L.F., Ho, L.T. and Vilcek, J. 1989. Tumor necrosis factor acts synergistically with autocrine interferon-beta and increases interferon-beta mRNA levels in human fibroblasts. J. Biol. Chem. 264: 16351-16354.

SECTION 5

THROMBOSIS AND FIBRINOLYSIS

HYPOXIA AND ENDOTHELIAL CELL FUNCTION: ALTERATIONS IN

BARRIER AND COAGULANT PROPERTIES

Satoshi Ogawa, Masayasu Matsumoto*, Jerold Brett,
Matthias Clauss and David M. Stern

Department of Physiology and Cellular Biophysics
College of Physicians and Surgeons of Columbia University
630 West 168th Street, New York, NY 10032
*First Department of Medicine
Osaka University Medical School Hospital
Osaka 553, Japan

INTRODUCTION

Endothelial cells have a central role in regulation of vascular permeability and the coagulation mechanism[1,2]. Rather than passively controlling these functions, there are multiple active mechanisms through which endothelial cells contribute to the maintenance of barrier function and fluidity of blood. Control of vascular homeostasis by endothelial cells occurs in response to environmental stimuli. Recent work has focussed attention on cytokines, such as tumor necrosis factor/cachectin (TNF) and Interleukin 1, mediators of the host response which alter a range of endothelial properties and allow these cells to play a central role in the inflammatory response[1].

As the interface between blood and tissues, endothelium is exposed to alterations in the vascular milieu, including hypoxemia. Hypoxemia is a common denominator of a spectrum of vascular disorders including diminished blood flow and inadequate delivery of oxygenated blood to the microvasculature. Clinically, hypoxemia is associated with increased permeability of the vasculature and a prothrombotic tendency[3,4,5,6]. Furthermore, formation of thrombi in venous valve cusps correlates closely with hypoxemia in the venous valve pocket[5,6], leading Malone to hypothesize years ago that modulation of endothelial coagulant properties by hypoxia could contribute to thrombus formation[7]. These considerations led us to study the effect of hypoxia on the barrier and coagulant function of cultured endothelium. The results indicate that hypoxia shifts the balance of cell surface coagulant properties to favor activation of coagulation and increases endothelial monolayer permeability.

MATERIALS AND METHODS

Endothelial Cell Culture and Exposure to Hypoxia

Bovine aortic endothelial cells were grown from newborn calf aortas in minimal essential medium containing penicillin-streptomycin (50 U/ml, 5 µg/ml), glutamine and 10% fetal calf serum (Hyclone, Logan, Utah, USA) as described[8]. Bovine adrenal microvascular endothelial cells were isolated[9], and characterized based on the presence of von Willebrand factor and thrombomodulin[10,11]. Cells were subcultured using trypsin/EDTA, and cells from passage 3-12 were grown to confluence in different size culture dishes (Becton Dickson Labware, Lincoln Park, NJ). Permeability studies employed cultures grown, as described previously[12], on polycarbonate membranes placed on polystyrene inserts (Transwell, Coaster, Cambridge, MA). Cultures were characterized by determining cell density, determined by Coulter Counter (Model ZM, Coulter Electronics, Luton, England), time in culture, and labelling index using a kit from Amersham (Arlington Hts., Ill). After confluence was achieved in an ambient atmosphere, endothelial cells were transferred to the hypoxic environment for further study. Where indicated, capillary endothelial cultures were reoxygenated by placing them back into room air.

Confluent endothelial cultures were exposed to an atmosphere with lowered oxygen tensions by placing them in an incubator attached to a hypoxia chamber. This chamber controlled the oxygen concentration of the environment (this consisted of a humidified gas mixture containing 5% CO_2), and provided a work area in which experiments could be carried to completion in the controlled atmosphere (Coy Laboratory Products, Ann Arbor, MI). Throughout experiments, the oxygen concentration of culture medium bathing the cells was determined by analyzing dissolved gas with an ABL-2 apparatus (Radiometer, Sweden). In the figures, the partial pressures of oxygen in the medium are shown. During the course of these experiments, pH of the medium did not change significantly. In certain experiments, either cycloheximide (0.1 μg/ml for 48 hr; Sigma) or warfarin (1 μg/ml for 48 hr; Sigma) was added to the medium. Cell viability was assessed by trypan blue exclusion and release of LDH into the medium, which was quantitated using a commercially available kit (Sigma, St. Louis, MO).

Morphologic Studies

Monolayers of cultured endothelium were grown to confluence on coverslips in normoxia and placed in hypoxia for the indicated times. Cultures were then fixed in the hypoxic environment in phosphate-buffered saline (pH = 7.2), containing formalin (3.5%) and NP-40 (0.1%), and washed in phosphate-buffered saline. Rhodamine phalloidin was used to visualize F-actin (Molecular Probes, Junction City, OR). Cultures were examined in a Leitz Dialux 20 microscope using a 2.4 Ploempak filter block and water immersion fluorite objectives.

Assays of Endothelial Monolayer Permeability

Confluent capillary endothelial monolayers grown to confluence on Transwell inserts were placed in the hypoxic environment and after 96 hrs of hypoxia, a radiolabelled tracer was then added to the upper chamber, either ^{3}H-inulin (3 μg/ml; 24 Ci/g New England Nuclear), ^{3}H-sorbitol (38 ng/ml; 24 Ci/mmole, New England Nuclear) or ^{125}I-BSA (150 ng/ml; 5000 cpm/ng). Radiolabelled BSA was prepared by the lactoperoxidase method (David and Reisfeld, 1974) using Enzymobeads (Biorad, Sacramento, CA), gel-filtered with a Sephadex G25 column (Pharmacia, Piscataway, NJ), and dialyzed to remove non-covalently bound iodine. Transport of tracers from the inner to outer chamber, i.e., across the endothelial monolayer, was quantitated by dividing radioactivity emerging in the outer well by radioactivity remaining in the inner well after a 4 hour incubation period. This method has been described in detail previously[13].

Assays of Endothelial Thrombomodulin

Assays of thrombomodulin activity and antigen were assessed after exposing endothelial cultures to hypoxia. For functional assays[14], cultures were placed in an atmosphere with the indicated oxygen tension, washed three times with Hanke's balanced salt solution, and then incubated for 60 min at 37°C in HEPES (10 mM, pH = 7.45), NaCl (137 mM), glucose (11 mM), KCl (4 mM), $CaCl_2$ (2 mM) and BSA (1 mg/ml) containing protein C (100 μg/ml) and thrombin (0.1 U/ml). Antithrombin III (100 μg/ml) was added to terminate formation of activated protein C, and the amount of the enzyme formed was determined using a chromogenic assay (Spectrozyme, American Diagnostica, NY). Enzyme concentration was determined by comparison with a standard curve made in the presence of known amounts of activated protein C, as described previously[14].

A radioimmunoassay was used to assess total thrombomodulin antigen. Detergent lysates (1% NP-40) of endothelial cells were prepared in the presence of protease inhibitors, PMSF (2 mM; Sigma) and leupeptin (0.3 mM; Boehringer Mannheim, Houston, TX). The radioimmunoassay was performed according to a method described previously[15]. The sensitivity of this radioimmunoassay was 10 ng/ml, which corresponded to 80% binding on the standard curve.

The effect of hypoxia on overall protein biosynthesis was assessed by studying incorporation of ^{3}H-leucine (60 Ci/mmole; New England Nuclear) into trichloroacetic acid precipitable material, as described by Madri et al.[16]. In summary, ^{3}H-leucine (20 μCi) was added to endothelial cells 12 hr prior to harvesting cells. Then, the supernatant and cell-associated material were precipitated in ice cold trichloroacetic acid (10%). After 24 hr at 4°C, the precipitates were collected, washed, and counted in a RackBeta counter (LKB, Rockland, MD).

Northern blots to assess levels of thrombomodulin and fibronectin mRNA were performed by the procedure reported previously[17]. cDNA probes for thrombomodulin (generously provided by Dr. E. Sadler, Washington Univ., St. Louis, MO, USA)[18] and fibronectin (generously provided by Dr. R. Hynes, M.I.T., Cambridge, MA, USA)[19] were labelled using random labelling (Boehringer Mannheim random primer DNA labelling kit, Ind., IN, USA) and hybridization was performed at 42°C as described previously[20].

Assays of Endothelial-Dependent Factor X Activation

Endothelial procoagulant activity was initially studied by determining shortening of the clotting time of recalcified plasma in the presence of endothelial cells exposed to hypoxia: cultures were washed (either cells in monolayer or suspension were used, ≈ 10^6 cells/assay), incubated with a mixture containing equal volumes of citrated bovine plasma, veronal buffer, and $CaCl_2$ (20 mM), and formation of a fibrin clot was studied. Factor X activation by hypoxic endothelium was also studied by examining cleavage of ^{125}I-Factor X on SDS-PAGE. Factor X was radiolabelled by the lactoperoxidase method as described previously for albumin. ^{125}I-Factor X was activated with Russell's viper venom (American Diagnostica, Greenwich, USA). The Factor Xa clotting assay was performed by incubating purified Factor X (50 μg/ml) with endothelial cell suspensions or monolayers (derived from hypoxic or normoxic cultures) for the indicated times, removing an aliquot (60 μl) and adding it to factor VII/X deficient plasma (60 μl; Sigma) along with cephalin (60 μl) and $CaCl_2$ (20 mM; 60 μl). Enzyme concentration was determined by comparison with a standard curve made with known amounts of Factor Xa. In some experiments, endothelial cells were further incubated with either mercury chloride (0.1 mM) or PMSF (1 mM) for 30 min prior to carrying out the assay. Bovine Factors IX, X, Xa and prothrombin were purified as described previously[21,22]. Monospecific, neutralizing antibodies to bovine Factors IX, VIII and VII were provided by Dr. W. Kisiel (Univ. of New Mexico, Albuquerque, NM), and neutralizing monoclonal antibody to bovine tissue factor was provided by Dr. R. Bach (Mt. Sinai School of Medicine, New York). Purified recombinant tumor necrosis factor/cachectin was obtained from Hoffmann-LaRoche, Nutley, NJ.

Characterization of the Hypoxia-Induced Factor X Activator

Characterization of the hypoxia-induced Factor X activator of capillary endothelium present in veronal extracts of hypoxic endothelial cultures was carried out by isoelectric focusing and SDS-PAGE. In each case, the starting material was extracted by incubating cultures for 3 hr at 4°C with veronal buffer (20 mM, pH = 7.4) using ≈ 2×10^8 hypoxic endothelial cells, which had been exposed to a PO_2 of ≈ 14 mm Hg for 48 hr. For isoelectric focussing, 4 mg of protein were diluted in a solution containing 1.5% ampholyte (pH range 3-10, Biorad)/0.1% octyl-ß-glucoside, and isoelectric focussing was performed at 12 watt for 4 hr until 1200 volt was achieved. Fractions were dialyzed against 0.4 M Tris/HCl (pH = 7.5)/NaCl (0.2 M)/octyl-ß-glucoside (0.1%), concentrated using an Amicon (Lexington, MA) Centricon device to 0.2 ml, and then tested for their ability to activate Factor X using a Factor Xa coagulant assay. For SDS-PAGE, extracts of similar numbers of endothelial cells were run on non-reduced 10% gels by the method of Laemmli[23]. Proteins in the gel were either stained with Coomassie blue or subjected to elution in 0.2 M Tris/HCl (0.2 M; pH = 7.4)/NaCl (0.2 M)/octyl-ß-glucoside overnight at 4°C. After removing SDS from the samples[24], each fraction was tested for its ability to activate Factor X using the coagulant assay.

RESULTS

Underline:
General Properties and Barrier Function of Hypoxic Endothelial Cultures

When capillary endothelial cells are incubated under hypoxic conditions (PO_2's as low as 14 mm Hg for up to 5 days), cell viability was not altered based on trypan blue exclusion and release of LDH into the medium. This is in keeping with the results of previous studies using endothelium derived from large vessels[25,26], and supports the concept that endothelium has the facility to adapt to the hypoxic environment[13,25,27]. This adaptive process encompasses changes in a range of cellular properties, including alterations in endothelial morphology. In normoxia, confluent bovine capillary endothelial cells formed a continuous monolayer of elongated cells, each with a continuous narrow actin bundle at the extreme edge, and with variable numbers of central actin stress fibers in parallel axial arrays (Fig. 1A). After 48 hr in hypoxia ($PO_2 \approx 14$ mm Hg), most of the central stress fibers have disappeared, but the actin bundles which demarcate the periphery of the cell are intact (Fig. 1B). Occasional small elongate intercellular gaps (3-10 µm) were seen between the otherwise contiguous cells, interrupting the continuity of the monolayer (Fib. 1B). These changes in cell shape/cytoskeletal configuration became more apparent as the incubation time in hypoxia was longer. Monolayers subjected to hypoxia for 48 hr, and then allowed to recover in ambient air, are again continuous, without gaps and exhibit an increase in the number and development of their central axially-oriented stress fibers (Fig. 1C).

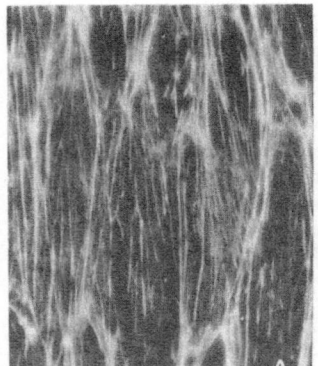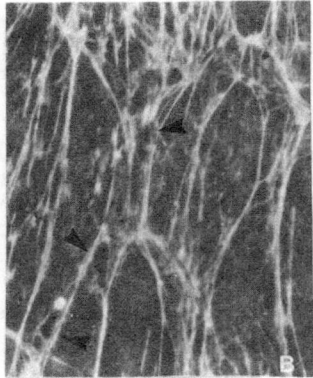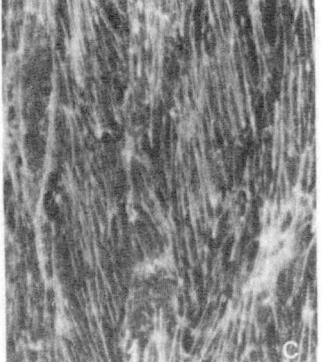

Figure 1. Effect of hypoxia on endothelial cell shape and the actin-based cytoskeleton. Endothelial monolayers were grown to confluence in normoxia and either maintained in normoxia (panel A) or transferred to hypoxia ($PO_2 \approx 14$ mm Hg) for 48 hr (panel B). Cultures exposed to hypoxia were then returned to the ambient atmosphere for 48 hr (panel C). The actin-based cytoskeleton was visualized by staining with rhodamine-conjugated phalloidin. Gaps in the monolayer are indicated by the arrows. Magnification: x 650.

Permeability of hypoxic endothelial cultures was studied using the monolayer-membrane system with Transwell plates[12,28,29,30]. Cultures exposed to hypoxia (PO$_2$ $\approx$ 14 mm Hg) for 96 hr showed increased diffusion of macromolecular (^{125}I-albumin, Mr $\approx$ 67,000 Da, and ^{3}H-inulin, Mr $\approx$ 5000 Da) and lower molecular weight (^{3}H-sorbitol, Mr $\approx$ 500 Da) tracers across the monolayer (Fig. 2). Hypoxia-mediated enhancement of monolayer permeability was dependent on the oxygen concentration (Fig. 2): at PO$_2$ = 14 mm Hg, there was approximately a three-fold increase in diffusion of the tracers across the monolayer. As the PO$_2$ increased, barrier function increased, approaching levels in normoxic controls by 21 mm Hg. Thus, at the PO$_2$'s present at the end of a capillary under normal resting conditions, $\approx$ 35-45 mm Hg, capillary endothelial monolayer permeability is comparable to that observed in normoxia. Restitution of cultures to an ambient air atmosphere reversed this defect in barrier function within 48 hrs (Fig. 2, bars marked R), in parallel with restoration of continuity of the monolayer (Fig. 1C) during the same time period.

Hypoxia-Mediated Suppression of Endothelial Thrombomodulin

To assess the effect of hypoxia on endothelial anticoagulant properties, we examined integrity of the protein C/protein S pathway[11], by studying alterations in thrombomodulin expression induced by an atmosphere with lowered oxygen tensions. Exposure of capillary endothelium to hypoxia resulted in a fall in cell surface thrombomodulin activity and antigen (Fig. 3A), beginning within 12 hr and reaching maximum by 48 hr, when cell surface thrombomodulin had declined about 80%. Hypoxia-induced attenuation of thrombomodulin also depended on the oxygen concentration, with a significant fall in thrombomodulin activity observed at lower oxygen concentrations starting at PO$_2$ $\approx$ 21 mm Hg (Fig. 3B). Suppression of thrombomodulin in hypoxia could be reversed, and when hypoxic cultures were placed again in ambient air, levels of cell surface thrombomodulin reached that of normoxic endothelium within 48 hr (data not shown). We next compared the effect of hypoxia on thrombomodulin with its effect on total protein synthesis, based on the incorporation of ^{3}H-leucine into material precipitable in trichloroacetic acid (Fig. 3A). Incorporation of ^{3}H-leucine fell with a time course similar to the decline in thrombomodulin, but general protein synthesis declined by only about 20% compared with normoxic controls (Fig. 3A). Consistent with this, another unrelated product of endothelium, fibronectin, showed a small increase during hypoxia (data not shown). Finally, Northern blots of endothelial RNA hybridized with cDNA probes for thrombomodulin and fibronectin showed a marked decrease in thrombomodulin, but not in fibronectin mRNA levels (Fig. 3C).

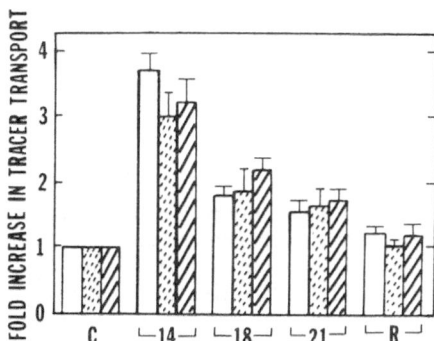

Figure 2. Effect of hypoxia on barrier function of endothelial monolayers. Monolayers on filters were incubated for 96 hr in an atmosphere with the indicated PO$_2$, and then barrier function was assessed by adding tracers (^{125}I-albumin, bars with lines; ^{3}H-inulin, bars with dashes; ^{3}H-sorbitol, open bars) to the compartment above the monolayer. Transfer of tracer to the compartment below the monolayer is shown as the ratio of transfer of tracer in hypoxic cultures to transfer of tracer in normoxic cultures (fold increase in hypoxia compared with normoxic controls). Transfer of tracer across normoxic cultures for each tracer was arbitrarily defined as 1 (mean ± SD is shown).

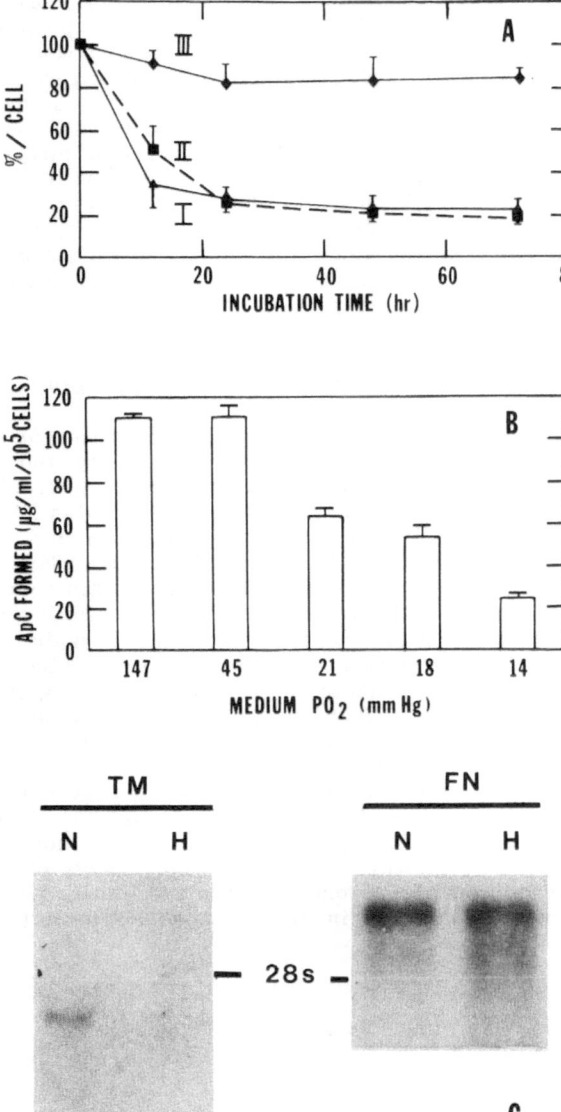

Figure 3. Effect of hypoxia on endothelial thrombomodulin. A. Time course. Confluent endothelial monolayers grown in normoxia were transferred to hypoxia ($PO_2 \approx 14$ mm Hg) and thrombomodulin activity (I), thrombin-mediated endothelial-dependent protein C activation, was assessed. For determination of thrombomodulin antigen (II), cultures were solubilized in detergent-containing buffer with protease inhibitors. Total protein synthesis (III) was estimated by determining the extent of incorporation of ^{3}H-leucine (added 12 hr before samples were harvested) into material precipitable in trichloroacetic acid. Data shown are a percent of the value in normoxic cultures per cell. B. Dose-response. Endothelial monolayers were incubated for 48 hr in an atmosphere with the indicated PO_2 and thrombin-mediated activated protein C (ApC) formation was assessed. Data shown are ApC formation/cell. The mean ± SD is shown. C. Northern blots of 24 hr hypoxic (H) and normoxic (N) endothelial cell RNA hybridized with cDNA probes to thrombomodulin (TM) and fibronectin (FN). (This blotting experiment was performed using aortic endothelial cells.)

Endothelial cultures exposed to hypoxia expressed clot-promoting activity, assessed by shortening of the recalcification time of plasma, which was not seen in normoxic cultures. Expression of cell surface procoagulant activity in hypoxia depended on the duration of exposure to the environment with low concentrations of oxygen (Fig. 4A), and on the oxygen concentration (data not shown). Previous observations, indicating that endothelial cells incubated with certain cytokines, such as Interleukin 1 and tumor necrosis factor/cachectin, express tissue factor[14,31], led us to examine the possible induction of tissue factor procoagulant activity by hypoxia. A blocking monoclonal antibody to bovine tissue factor, which inhibited the procoagulant activity of tumor necrosis factor-treated cells, failed to lengthen the clotting time of hypoxic endothelium (Figure 4B), suggesting that a tissue factor-independent procoagulant activity was being induced by hypoxia. Studies with ^{125}I-Factor X revealed direct cleavage of the coagulation factor with formation of the heavy chains of Factor Xaα and Xaβ (Fig. 5, lane A) (21) on reduced SDS-PAGE. In contrast, neither buffer controls (Fig. 5, lane A) nor normoxic cultures (Fig. 5, lane B) produced cleavage of Factor X. Control activation studies of the same iodinated Factor X by Russell's viper venom demonstrated activation of Factor X, mainly with formation of Factor Xaα heavy chain (Fig. 5 lane D), as previously reported[21]. Furthermore, Factor X activation did not require the presence of other coagulation proteins. This is similar to previous observations with large vessel native and cultured endothelium[26]. As capillary endothelial cultures were incubated in hypoxia (PO$_2$ ≈ 14 mm Hg), their ability to activate purified Factor X steadily increased up to 72 hr (Fig. 6A). Factor X activation by hypoxic endothelium was also dependent on the oxygen tension. It was expressed only at the lowest PO$_2$'s examined (Fig. 6B). Formation of Factor Xa by hypoxic endothelium was not affected by the presence of blocking antibodies to bovine tissue

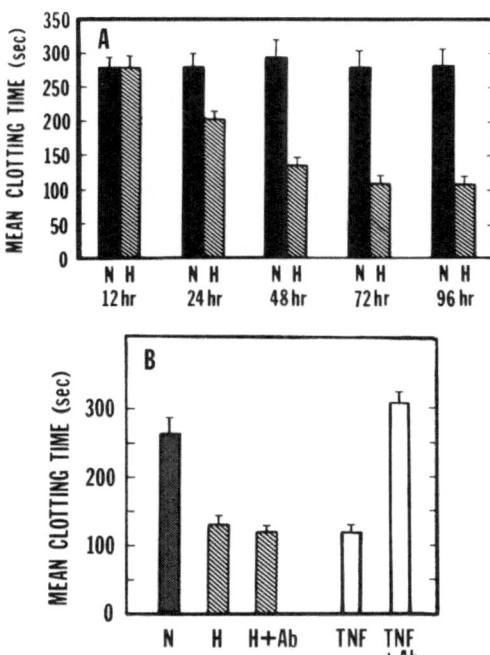

Figure 4. A. Time course of hypoxia-induced expression of procoagulant activity assessed by shortening of clotting time of recalcified plasma. N denotes normoxic controls. H denotes hypoxia. B. Effect of blocking tissue factor antibody on the procoagulant activity of hypoxic (H+Ab) and tumor necrosis factor (TNF) treated endothelial cultures (TNF+Ab). N, H, and TNF designate the procoagulant activity of endothelial normoxic, hypoxic, and TNF treated cultures, respectively. (This experiment was performed using aortic endothelial cells.)

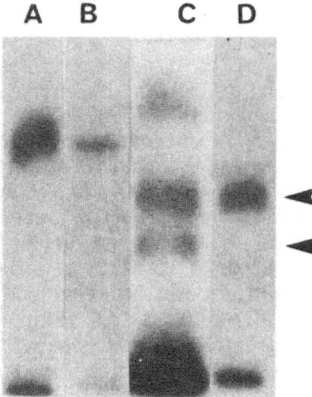

Figure 5. Activation of ^{125}I-Factor X by hypoxic endothelial cells (72 hr in hypoxia) seen on reduced SDS-PAGE. Each lane represents ^{125}I-Factor X incubated with buffer only (A), normoxic endothelial cells (B), hypoxic endothelial cells (C), or Russell's viper venom (D). Arrows correspond to the heavy chains of Factor Xaα and Xaß. (This experiment was performed using aortic endothelial cells.)

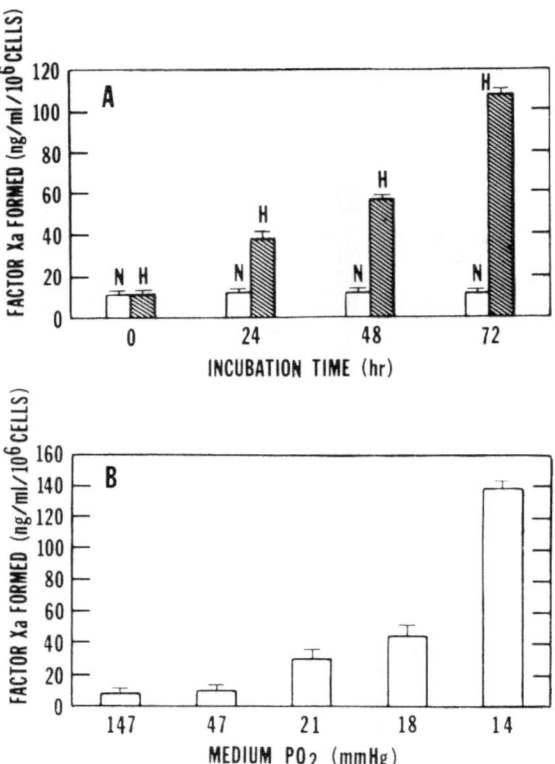

Figure 6. Factor X activation by hypoxic endothelial cells. A. Time course. Endothelial monolayers were exposed to hypoxia (PO_2 ≈ 14 mm Hg) for the indicated times, washed in calcium-free buffer, incubated with Factor X (1 μM) for 30 min at 37°C, and then Factor Xa formation was assessed. N = normoxic cultures; H = hypoxic cultures. B. Dose-response. Endothelial monolayers were exposed to the indicated PO_2 for 48 hr, washed, and Factor Xa formation was studied as described in A. The mean ± SD is shown.

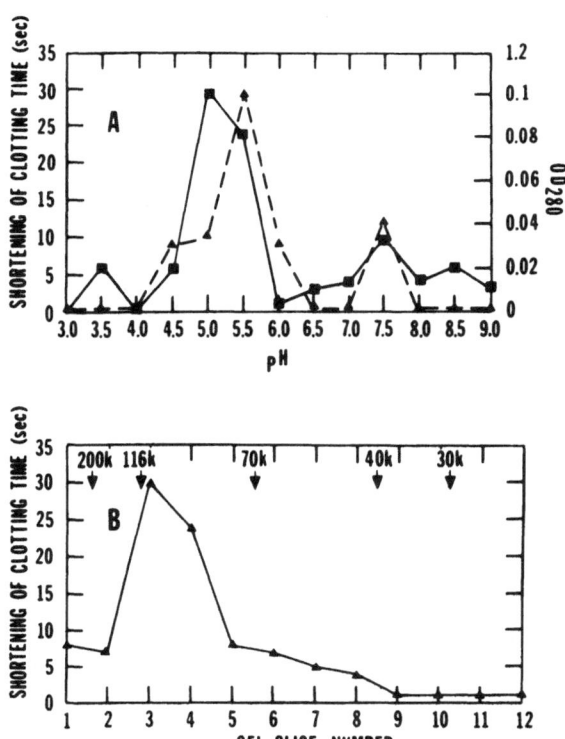

Figure 7. Characterization of the Factor X activator of hypoxic endothelium by
isoelectric focussing and SDS-PAGE. A. Isoelectric focussing. Endothelial cells
were exposed to hypoxia ($PO_2 \approx 14$ mm Hg) for 48 hr, extracted with veronal
buffer, and the extract was subjected to preparative isoelectric focussing.
Fraction number is plotted versus OD_{280} (triangles) and Factor X activating
ability (squares) of the samples, expressed as shortening of clotting time in Factors
VII/X deficient plasma. B. SDS-PAGE. Endothelial cells were exposed to
hypoxia ($PO_2 \approx 14$ mm Hg) for 48 hr, extracted with veronal buffer, and then
subjected to nonreduced SDS-PAGE (10%). Slices of the gel were then eluted,
dialyzed and tested for their ability to activate Factor X. Slice number of the gel
is plotted versus activation of Factor X, expressed as shortening of the clotting
time as above.   Apparent molecular weights were interpolated from semi-
logarithmic plots based on the migration of standard proteins run simultaneously.
K denotes kilodaltons.

factor, Factors IX, VIII or VII, demonstrating that neither the classical extrinsic or
intrinsic systems were involved (data not shown).   Furthermore, studies with
cycloheximide demonstrated that protein synthesis was required for hypoxic
endothelium to express the Factor X activator (data not shown). Once endothelial cells
had been incubated in hypoxia and expressed the Factor X activator, the latter activity
could be blocked by mercury chloride, an inhibitor of cysteine proteases, but not by
phenylmethylsulfonyl fluoride, an inhibitor of serine proteases.

These observations suggested that hypoxia induced the synthesis of an activator
of Factor X which seemed distinct from intrinsic and extrinsic factors/cofactors. To
further characterize this Factor X activator, isoelectric focussing and SDS-PAGE were
carried out on extracts of hypoxic capillary endothelial cells, and the capacity of the
fractionated material to activate Factor X was studied. Isoelectric focussing showed
a single peak of activity at pH $\approx$ 5.0 (Fig. 7A). Non-reduced SDS-PAGE of extracts
from hypoxic cultures showed a complex pattern when gels were stained for protein,

but gel elution indicated a single peak of Factor Xa forming activity at Mr $\approx$ 100 kDa (Fig. 7B).

DISCUSSION

The results reported here demonstrate that hypoxia can modulate two central properties of microvascular endothelium, barrier and coagulant function. The extent of increased permeability observed in response to hypoxia is at least comparable to that observed in previous studies with aortic endothelium and the cytokine, tumor necrosis factor/cachectin (TNF)[13,33]. In contrast to our previous studies with bovine aortic endothelium[26], where increased permeability of hypoxic monolayers was evident by 24 hrs, attenuated barrier function with capillary endothelial cells was first evident at 48 hrs and was not really striking until 96 hr (the time point used in Fig 2A). This could be due to a difference in the sensitivity of endothelium from different vascular beds to hypoxia-induced perturbation of monolayer permeability. Such variability in the endothelial response to hypoxia could be expected, since capillary, venous and arterial endothelium *in vivo* are exposed to different PO_2's in the blood. These studies indicate that alterations in endothelial monolayer barrier function could occur in hypoxia, thereby contributing to the clinically important syndrome of increased vascular leakage at high altitude ("high altitude pulmonary edema") or other situations where the PO_2 is rapidly lowered. However, our observations also demonstrate that marked shifts in oxygenation are required for endothelial permeability to be altered, consistent with the importance of normal barrier function in the maintenance of homeostasis.

The observations of Sevitt[5] and Hamer *et al.*[6], demonstrating that fibrin was deposited on the endothelial surface of venous valve cusps, an area subject to severe hypoxemia with stasis, led us to explore mechanisms which contribute to the apparent increased thrombogenicity of hypoxic endothelium. A previous study has demonstrated a small decline in net fibrinolytic activity of hypoxic endothelial cultures[34]. In the current study, hypoxia induced a coordinate decline in the anticoagulant cofactor thrombomodulin, and expression of an apparently novel procoagulant activity. This parallels our previous studies with hypoxic endothelial cells derived from the aorta[26]. In experiments examining modulation of endothelial procoagulant activity by cytokines, such as TNF, Interleukin 1, or endotoxin, induction of procoagulant activity has been shown to result from the synthesis and expression of tissue factor[14,31]. The situation is different in hypoxia: although protein synthesis was required for expression of the procoagulant activity (no similar activity was detected in normoxic cultures), a novel direct Factor X activator was induced. Furthermore, in the current study, the Factor X activator of hypoxic capillary endothelium had Mr $\approx$ 100 kDa, pI $\approx$ 5.0, and required a sulfhydryl group for its activity. Although the nature of the hypoxia-induced Factor X activator is not yet clear, based on its properties, there are similarities with the tumor procoagulant, a Factor X activator identified in malignant tissue which can be extracted with veronal buffer and is a cysteine protease[25]. However, the latter enzyme has been reported to have Mr $\approx$ 68 kDa. Further studies will be required to fully characterize the hypoxia-induced Factor X activator, and to assess its potential role in thrombosis observed in hypoxemic conditions.

These studies demonstrate that hypoxia exerts a complex effect on endothelial function. Although at the outset we considered that hypoxia could result in a general cessation of cellular function in endothelium, through a general slowing of metabolic and biosynthetic processes, the results of our studies show that hypoxia exerts more complex effects on endothelial function. A common denominator linking many hypoxia-induced changes in cellular function could relate to a redirection of protein biosynthesis with the suppression of certain proteins and induction of others (proteins whose expression is induced by hypoxia in other cell types have been termed *oxygen-regulated proteins*), as has been found in other cell types exposed to environments with low oxygen tensions[36,37,38,39]. For example, our data would suggest that hypoxia induces suppression of thrombomodulin synthesis and induction of the synthesis of a novel Factor X activator. The study of endothelial proteins whose expression is induced by hypoxia could provide insights into mechanisms operative in microvascular dysfunction in ischemia.

SUMMARY

In this study we demonstrated that hypoxia, with PO_2's as low as 12-14 mm Hg, was not lethal to either aortic or capillary endothelial cells, but it reversibly modulated central cellular functions essential for maintenance of homeostasis. Permeability of monolayers to solutes increased in a dose-dependent manner and cell surface coagulant properties were shifted to promote activation of coagulation. The anticoagulant cofactor thrombomodulin was suppressed by 85% and Northern blots demonstrated the complete suppression of thrombomodulin mRNA in hypoxic cultures. In addition, an apparently novel activator of Factor X, distinct from the classical extrinsic and intrinsic systems, was induced. The hypoxia-induced Factor X activator was cell surface-associated, had properties of a cysteine protease, had Mr corresponding to ≈ 100 kDa, based on SDS-PAGE, and pI ≈ 5.0. These findings that hypoxia dynamically modulates endothelial function provide insights into the contribution of macro- and microvascular endothelial dysfunction in the pathogenesis of vascular lesions.

REFERENCES

1.   Gimbrone, M.   1986.   Vascular Endothelium in Hemostasis and Thrombosis, Churchill and Livingstone, New York.

2.   Simionescu, N. and Simionescu, M.   1988.   Endothelial Cell Biology. Plenum Publishing Corp., New York.

3.   Kinasewitz, G., Groome, L., Marshall, R., Leslie, W. and Diana, H.   1986.   Effect of hypoxia on permeability of pulmonary endothelium of canine visceral pleura. J. Appl. Physiol. 61: 554-560.

4.   Olesen, S-P.   1986.   Rapid increase in blood-brain barrier permeability during severe hypoxia and metabolic inhibition. Brain Res. 368: 24-29.

5.   Sevitt, S.   1967.   The acutely swollen leg and deep vein thrombosis. Brit. J. Surg. 68: 166-170.

6.   Hamer, J., Malone, P. and Silver, I.   1981.   The PO_2 in venous valve pockets: its possible bearing on thrombogenesis. Brit. J. Surg. 68: 166-170.

7.   Malone, P.   1977.   A hypothesis concerning the aetiology of venous thrombosis. Med. Hypotheses 5: 189-201.

8.   Schwartz S.   1978. Selection and characterization of bovine aortic endothelial cells. In Vitro 14: 966-984.

9.   Furie, M., Cramer, E., Naprestek, B. and Silverstein, S.   1984.   Cultured endothelial monolayers that restrict the transendothelial passage of macromolecules and electrical current. J. Cell Biology 65: 1033-1042.

10.   Jaffe, E., Hoyer, L. and Nachman, R.   1973.   Synthesis of antihemophilic factor antigen by cultured human endothelial cells. J. Clin. Invest. 52: 2757-2765.

11.   Esmon, C.   1987.   The regulation of natural anticoagulant pathways. Science 235: 1348-1352.

12.   Brett, J., Gerlach, H., Nawroth, P., Steinberg, S., Godman, G. and Stern, D.   1989b. Tumor necrosis factor/cachectin increases permeability of endothelial monolayers by a mechanism involving regulatory G proteins. J. Exp. Med. 169: 1977-1991.

13.   Brett, J., Stern, D., Ogawa, S., Silverstein, S. and Loike, J.   1989a.   Energy metabolism in endothelial cells: central role of glycolysis in hypoxia. J. Cell Biol. 109: 1711 (abstract).

14.   Nawroth, P. and Stern, D.   1986.   Modulation of endothelial cell hemostatic properties by tumor necrosis factor. J. Exp. Med. 163: 740-745.

15.   Gerlach, H., Liebermann, H., Brett, J., Bach, R., Godman, G. and Stern, D.   1989. Growing/motile endothelium shows enhanced responsiveness to tumor necrosis factor/cachectin. J. Exp. Med. 170: 913-931.

16.   Madri, J., Pratt, B. and Tucker, A.   1988.   Phenotypic modulation of endothelial cells by transforming growth factor-ß depends upon the composition and organization of the extracellular matrix. J. Cell Biol. 106: 1375-1384.

17.   Chirgwin, J., Przbyla, R., MacDonald, R. and Rutter, W.   1979.   Isolation of biologically active ribonucleic acid from sucrose enriched ribonuclease. Biochem. 18: 5294-5299.

18.   Wen, D., Dittman, W., Ye, R., Deaven, L., Majerus, P. and Sadler J.   1987.   Human thrombomodulin: Complete cDNA sequence and chromosome localization of the gene. Biochem. 26: 4350-4357.

19.   Schwartzbauer, J., Tamkin, J., Lemischka, I. and Hynes R.   1983.   Three different fibronectin mRNAs arise by alternative splicing within the coding region. Cell 35: 421-423.

20.   Reth, M. and Alt, F.   1984.   Novel immunoglobulin heavy chains are produced from DJH gene segment rearrangements in lymphoid cells. Nature 321: 418-423.

21.   Fujikawa, K., Legaz, M. and Davie, E.   1972.   Bovine factor X_1 and X_2. Isolation and characterization. Biochemistry 11: 4882-4891.

22.   Fujikawa, K., Thompson, A., Legaz, M., Meyer, R. and Davie, E.   1973.   Isolation and characterization of bovine factor IX. Biochemistry 12: 4938-4944.

23.   Laemmli, U.   1970. Cleavage of structural proteins during the assembly of the head of bacteriophage T4. Nature 227: 680-685.

24.   Henderson, L. and Konigsberg, W.   1979.   A micromethod for complete removal of dodecyl sulfate from proteins by ion-pair extraction. Anal. Biochem. 93: 153-157.

25.   Lee, S-L. and Fanburg, B.   1987. Glycolytic activity and enhancement of serotonin uptake by endothelial cells exposed to hypoxia/anoxia. Circ. Res. 60: 653-688.

26.   Ogawa, S., Gerlach, H., Esposito C., Macaulay AP., Brett, J. and Stern, D.   1989. Hypoxia modulates the barrier and coagulant properties of cultured bovine endothelium: Increased monolayer permeability and induction of procoagulant properties. J. Clin. Invest. 85: 1090-1098.

27.   Cummiskey, J., Simon, L., Theodore, J., Ryan, U. and Robin, E.   1981.   Bioenergetic alterations in cultivated pulmonary artery and aortic endothelial cells exposed to normoxia and hypoxia. Exp. Lung Res. 2: 155-163.

28.   Albelda, S., Sampson, P., Haselton, F., McNiff, J., Mueller, S., Williams, S., Fishman, A. and Levine, E.   1988.   Permeability characteristics of cultured endothelial cell monolayers. J. Appl. Physiol. 64: 308-319.

29.   Del Vecchio, P., Siflinger-Birnboim, A., Shepard, J., Bizions, R., Cooper, J. and Malik, A.   1987.   Endothelial monolayer permeability to macromolecules. Fed. Proc. 46: 2511-2516.

30.   Shasby, D. and Roberts, R.   1987.   Transendothelial transfer of macromolecules in vitro. Fed. Proc. 46: 2506-2512.

31. Bevilacqua, M., Pober, J., Majeau, G., Cotran, R. and Gimbrone, M. 1986. Recombinant TNF induces procoagulant activity in endothelium. PNAS (USA) 83: 4533-4537.

32. Stern, D., Brett, J., Harris, K. and Nawroth, P. 1986. Participation of endothelial cells in the protein C-protein S anticoagulant pathway: The synthesis and release of protein S. J. Cell Biol. 102: 1971-1978.

33. Clark, M., Chen, M-J., Crooke, S. and Bomalaski, J. 1988. Tumor necrosis factor (cachectin) induces phospholipase A_2-activating protein in endothelial cells. Biochem. J. 250: 125-132.

34. Wojta, J., Jones, R., Binder, B. and Hoover, R. 1988. Reduction in PO_2 decreases the fibrinolytic potential of cultured bovine endothelial cells derived from pulmonary arteries and lung microvasculature. Blood 71: 1703-1706.

35. Falanga, A. and Gordon, S. 1985. Isolation and characterization of cancer procoagulant: A cysteine proteinase from malignant tissue. Biochemistry 24: 5558-5567.

36. Anderson, G., Stoler, D. and Scarcello, L. 1989. Normal fibroblasts responding to anoxia exhibit features of the malignant phenotype. J. Biol. Chem. 264: 14885-14892.

37. Sciandra, J., Subjeck, J. and Hughes, H. 1984. Induction of glucose-regulated proteins during anaerobic exposure and of heat-shock proteins after reoxygenation. PNAS (USA) 81: 4843-4847.

38. Subjeck, J. and Thung-Tai, S. 1986. Stress protein systems of mammalian cells. Am. J. Physiol. 250: C1-C17.

39. Wilson, R. and Sutherland, R. 1989. Enhanced synthesis of specific proteins, RNA and DNA caused by hypoxia and reoxygenation. J. Radiation Oncology Biol. Phys. 16: 957-961.

REGULATION OF TYPE ONE PLASMINOGEN ACTIVATOR INHIBITOR GENE

EXPRESSION IN CULTURED ENDOTHELIAL CELLS AND THE VESSEL WALL

M. Sawdey and D.J. Loskutoff

Research Institute of Scripps Clinic
La Jolla, CA 92037

Type one plasminogen activator inhibitor (PAI-1) is a rapid and specific inhibitor of both tissue- and urokinase-type plasminogen activators. Increases in PAI-1 activity may alter the normal hemostatic balance and promote intravascular fibrin deposition. Expression of the PAI-1 gene in cultured endothelial cells is strongly stimulated by agents implicated as mediators of inflammatory or repair processes. In the vessel wall, the action of these mediators may trigger synthesis and release of PAI-1, leading to local and systemic deficits in fibrinolytic activity and an increased risk of thrombotic disease.

In this chapter, we present an overview of the biology of PAI-1, and summarize current knowledge concerning the agents and mechanisms regulating its expression in cultured endothelial cells and the vessel wall. We discuss the implications of these results for the development and progression of atherosclerotic disease.

INTRODUCTION

Activation of the fibrinolytic system results in the generation of plasmin, a serine protease of broad substrate specificity. Plasmin functions in the blood to degrade vascular fibrin deposits, and in the tissues to mediate inflammatory and repair processes which require extracellular proteolysis. The expression of its activity depends upon a complex series of reactions involving proenzymes, cell surface receptors, solid-phase components, and specific enzyme inhibitors.

Plasminogen, the inactive precursor of plasmin, is abundant in the circulation and interstitial fluids. Cleavage of a single arginine-valine bond in plasminogen results in the formation of plasmin. This reaction is catalyzed by the plasminogen activators (PAs), of which there are two distinct molecular forms. Tissue-type PA, or t-PA, may be the relevant PA in physiologic fibrinolysis (1), as it possesses an enhanced affinity for plasminogen in the presence of fibrin (2). Urokinase, or u-PA, is secreted as a single-chain proenzyme (pro-u-PA) with limited PA activity (3-5). Pro-u-PA is cleaved by plasmin to form a two-chain disulfide-linked molecule with greatly enhanced catalytic efficiency (6). The existence of specific cellular receptors for u-PA (7,8) and plasminogen (9-11) suggests that u-PA may activate plasminogen on cell surfaces. Expression of u-PA activity is a hallmark of many cellular invasive or degradative processes in which extracellular proteolysis occurs, including inflammation and the migration of inflammatory cells (1,7,12,13), tissue remodeling (1,14,15) and wound healing (16,17).

Atherosclerosis, Edited by A. I. Gotlieb *et al.*
Plenum Press, New York, 1991

The elaboration of plasmin activity in these processes is subject to multiple levels of control. Sequestration within fibrin clots (18,19) or on cell surfaces (20,21) may otherwise protect newly-formed plasmin from its inhibitor, α-2 antiplasmin, and in effect localize its activity to these environments. Regulation of plasmin generation may additionally be achieved by alterations in the rate of PA synthesis (22,23), release (24,25), or by modulations in PA activity (1,2,6). However, the principal means by which plasmin generation is controlled is through the action of specific PA inhibitors (PAIs).

Several molecules possessing PAI activity have been described to date (26). Consideration of rate constants and detection of naturally occurring enzyme-inhibitor complexes suggest there are two physiologic PAIs (27). These are the endothelial or type-1 PAI (PAI-1; 28-30) and placental or type-2 PAI (PAI-2; 31,32). PAI-1 rapidly and efficiently inhibits the activity of both u-PA and t-PA (30,33-35). It appears to be the primary physiologic inhibitor of t-PA, since the majority of t-PA in plasma is found in complex with PAI-1 (36). PAI-2 is primarily a u-PA inhibitor which most likely functions to inhibit cellular u-PAs (31). Complexes between PAI-2 and t-PA have not been detected in plasma, even during pregnancy, when the concentration of PAI-2 exceeds that of PAI-1 (37).

PAI-1 inhibits both t-PA and u-PA by forming 1:1 stoichiometric complexes (33) that resist dissociation. Kinetic studies indicate that the dissociation constant for the interaction of human PAI-1 and t-PA is less than 1×10^{-12} M (38). The binding of PAI-1 to PAs is extremely rapid. The second-order rate constant for inhibition of t-PA is approximately $3.5 \times 10^{7} M^{-1}s^{-1}$ (38), and that for u-PA may be even higher (35). These values are at least two orders of magnitude greater than those for PAI-2 or other PA inhibitors (27).

Human PAI-1 has been extensively characterized. It is a single-chain glycoprotein of $M_r \approx 50,000$ that is synthesized and secreted by a variety of cell types in culture (for review, see 27), including endothelial cells (28,29,39-42). Its primary structure has been elucidated by molecular cloning studies (43-46). The pre-PAI-1 molecule is 402 amino acids in length, including a signal peptide of 23 amino acids. Removal of the signal peptide yields a mature secreted form of 379 amino acids, containing 3 potential N-linked glycosylation sites. The mature form is rich in methionine residues and lacks cysteine residues, consistent with many of its known biochemical properties (27; see below, "Biochemical Properties of PAI-1").

Homology searches revealed that PAI-1 is a member of the serine protease inhibitor or "serpin" superfamily. The serpins appear to have evolved from a common ancestral gene between 200 and 500 million years ago (47). Analysis of conserved amino acid residues between serpins suggests they share a common ternary structure, featuring an exposed loop in the COOH-terminal region of the molecule. This loop contains the reactive center of the inhibitor, which appears to act as a "bait" for the target protease (48). The serpin family includes other major plasma protease inhibitors such as α-2 antiplasmin and antithrombin III.

PAI-1 is normally detected in trace amounts in plasma, although its concentration may increase significantly in conditions such as gram-negative sepsis (49) or pregnancy (37; see below). The origin of plasma PAI-1 has yet to be determined. It is present in the α-granules of platelets (50-53), from which it can be released by physiologic concentrations of thrombin, collagen, or ADP (50-52). However, platelets are unlikely to constitute a major source of plasma PAI-1 under normal conditions (54). More likely sources include hepatocytes (55) and vascular endothelial cells. PAI-1 is synthesized and secreted by cultured endothelial cells of arterial (56), venous (28,29,39), and microvascular origin (57). It was initially purified from the conditioned media of cultured bovine aortic endothelial cells (BAEs) and is a major biosynthetic product of these cells, comprising from 2 to 12% of their secreted protein (42). It has also been detected in vascular smooth muscle cells (58) and a wide variety of cell types in culture (27). A recent immunohistochemical analysis of human tissues localized PAI-1 to venous endothelium and the muscularis layer of vein walls, umbilical artery endothelium and smooth muscle, and smooth muscle of numerous tissues. Quantitative analysis of tissue extracts revealed high levels of PAI-1 in the liver and spleen, possibly reflecting the clearance of PAI-1 from plasma or circulating platelets (59). However, cultured hepatocytes and hepatoma cells have also been shown to synthesize PAI-1 (55). Thus, hepatocytes also represent a potential source of plasma PAI-1.

BIOCHEMICAL PROPERTIES OF PAI-1

Initial experiments with the purified PAI-1 molecule suggested it was unusually stable, in that its activity could be detected following SDS-polyacrylamide gel electrophoresis, or prior treatment with ß-mercaptoethanol, or extremes of temperature or pH (42). The PAI-1 in conditioned media was subsequently shown to exist primarily in a "latent" form which lacked inhibitor activity, but could be converted to the active form by treatment with denaturants such as SDS or guanidine HCl (60). However, the PAI-1 in cell extracts is present primarily in the active form (61), and spontaneously decays into the latent form in solution (61-63). These results suggest that PAI-1 is synthesized in the active form, but decays into the latent form following its release from cells.

These and other observations indicate that the activity of PAI-1 may be regulated *in vivo*. Latent PAI-1 can be activated by micelles of negatively charged phospholipids (64), implying that membrane lipids may alter its activity. Despite this finding, the activation of latent PAI-1 by cell membranes has not been demonstrated, and no biologically relevant activators of latent PAI-1 have been identified. PAI-1 also circulates in plasma in complex with vitronectin (S protein; 65). The PAI-1 present in these complexes is functionally active (66), and is more stable in solution (65). However, in plasma, PAI-1 is rapidly cleared from the circulation (49), while its decay in solution requires several hours (67). The stabilization of plasma PAI-1 by vitronectin may therefore be insignificant. Of greater importance in this regard may be the interaction of PAI-1 with components of the extracellular matrix (ECM).

PAI-1 is present in the ECM of several cell types in culture (68-70), including endothelial cells (67,71) and vascular smooth muscle cells (72). Initial observations indicated that the ECM-bound form of PAI-1 was fully active, as the addition of exogenous u-PA or t-PA in solution resulted in complex formation and the dissociation of these complexes from the matrix. Interestingly, the PAI-1 present in ECM was biologically stable, with little loss of activity evident over a 24-hour period (67). More recent studies (73) have suggested that the binding and stabilization of PAI-1 by ECM is mediated in part by matrix-associated vitronectin. Thus, in contrast to the PAI-1 in conditioned media, which exists primarily in the latent form, the PAI-1 in ECM appears to be functionally active. The physiologic counterpart of the ECM *in vivo* (e.g., the subendothelium and interstitial matrices of the media) may therefore contain a reservoir of active PAI-1, which serves to limit plasmin-mediated proteolysis of matrix components by neutralizing PAs released from cells.

PAI-1 rapidly loses its activity in the presence of oxidants. The purified bovine (74) and human (27) PAI-1 molecules were found to be sensitive to concentrations of the oxidizing agent chloramine T which were ten-fold lower than those required to inactivate α-1-proteinase inhibitor (α-1-PI), an inhibitor of elastase known for its sensitivity to oxidants (75). Treatment of oxidatively inactivated PAI-1 with methionine sulfoxide reductase partially restored its activity, suggesting that methionine residues in PAI-1 may be important for its inhibitory function (74). *In vivo*, the liberation of oxidants at sites of inflammation may inactivate these inhibitors and allow plasmin- and elastase-mediated degradative processes to occur.

PAI-1 AND THROMBOTIC DISEASE

Clinical studies have demonstrated that PAI-1 activity is increased in a variety of conditions associated with thrombotic disease. Tenfold or greater elevations were observed in the plasma of patients with gram-negative sepsis (49), and during the second and third trimesters of pregnancy (37). Smaller but significant elevations were reported in patients with deep vein thrombosis (76,77). Increased PAI-1 activity was also found in young survivors of myocardial infarction. In these individuals, PAI-1 activity was correlated with elevated serum triglyceride levels, suggesting that PAI-1 may represent a risk factor in recurrent myocardial infarction (78). In addition, plasma PAI-1 activity rises rapidly in acute phase reactions following major surgery (79,80), trauma (80,81), and acute myocardial infarction (82).

Whether a causal relationship exists between these elevations in PAI-1 activity and thrombotic disease is unclear. For example, PAI-1 activity rises dramatically during pregnancy, yet thrombotic complications are generally not observed (37). One hypothesis to account for this fact is that PAI-1-induced deficits in fibrinolytic activity are inconsequential in the absence of thrombogenic stimuli (27). However, in conditions where such stimuli are present, fibrin formation would be independently initiated, due to the activation of the coagulation cascade. Where elevations in PAI-1 activity also exist, normal clearance would be inhibited, and the deposition and persistence of intravascular fibrin would result. In this way, increases in plasma PAI-1 activity, while clinically silent, would promote manifestation of the thrombotic state, and contribute to the development of thrombotic disease.

REGULATION OF PAI-1 BIOSYNTHESIS

In many of the clinical conditions where increased PAI-1 activity exists, comparable changes in antigen levels have also been detected, implying a stimulation of PAI-1 biosynthesis or release, as opposed to modulations in its activity. However, release from intracellular storage pools (e.g., platelets) is unlikely to account for prolonged systemic elevations in PAI-1, since it is rapidly cleared from the circulation (49). Furthermore, the magnitude of the increases in many cases would appear to be too large to reflect platelet release mechanisms alone. Finally, and perhaps most significantly, the biosynthesis of PAI-1 by cultured endothelial cells is regulated by a number of agents implicated as causative factors or mediators in these conditions (see Table 1).

One of the first such agents to be identified was endotoxin or lipopolysaccharide (LPS; 49), a component of the cell wall of gram-negative bacteria (83). Elevated t-PA inhibitor activity was observed both in the plasma of patients with gram-negative septicemia, and the plasma of rabbits following injection of low doses (i.e., 10 ng/kg body weight) of LPS. Treatment of cultured human endothelial cells with LPS increased the amount of PAI activity present in the conditioned media (49). This increase results from increases in PAI-1 (84,85). Other investigators have noted increased PAI activity in the plasma of rats injected with LPS (86), and in LPS-treated cultured endothelial cells of both human (86,87) and bovine (56,88) origin, an effect that requires ongoing RNA and protein synthesis (86). These data suggest that the alterations in plasma PAI-1 levels observed in gram-negative sepsis may result from the increased biosynthesis of PAI-1 by endothelial cells.

Serum also stimulates the PAI-1 activity of endothelial cells (Table 1). Trace amounts of LPS present as a contaminant in bovine serum may account in part for the serum-mediated suppression of the fibrinolytic activity of cultured bovine endothelial cells (89), and for serum-mediated increases in PAI-1 protein (85,86) and mRNA (90). However, it is likely that these serum effects are considerably more complex since serum also contains growth factors, cytokines, and proteases, all of which may stimulate PAI-1 biosynthesis.

A number of growth factors have been tested for their effects on PAI-1 synthesis. One of the most potent is transforming growth factor ß (TGF-ß), a polypeptide growth modulator present in platelets (for review, see 91) and released from platelet α-granules upon activation with thrombin (92). TGF-ß stimulates the synthesis of extracellular matrix components, including fibronectin and procollagen (93,94). In addition, picomolar concentrations of TGF-ß promote increased synthesis of PAI-1 in fibroblasts

Table 1. Agents stimulating the biosynthesis of PAI-1 *in vitro*

Stimulus	Cell type
LPS*	Human umbilical vein endothelial (49,84-87) Bovine aortic endothelial (85,124) Bovine pulmonary artery endothelial (88)
Serum*	Human umbilical vein endothelial (85,86) Bovine aortic endothelial (85,90)
TGF-ß*	Human lung fibroblasts (WI-38; 68,130,160) Human epidermoid carcinoma (A431; 95) Human lung carcinoma (A549; 160) Human fibrosarcoma (HT 1080; 129) Monkey kidney epithelial (BSC-1; 95) Bovine aortic endothelial (96,124) Bovine capillary endothelial (97) Bovine vascular smooth muscle (100) Mink lung epithelial (CCL 64; 95) Mouse embryo fibroblasts (AKR-2B; 95)
bFGF	Bovine capillary endothelial (97)
PDGF	Bovine vascular smooth muscle (100)
EGF*	Human hepatoma (Hep G2; 161)
Il-1*	Human umbilical vein endothelial (84,101,102,103) Human adult saphenous vein endothelial (84) Human fetal lung fibroblast (MRC-5; 162) Bovine aortic endothelial
TNF-α*	Human umbilical vein endothelial (103) Human fibrosarcoma (HT 1080; 128) Bovine aortic endothelial (124,163)
Thrombin*	Human umbilical vein endothelial (111,164) Human foreskin microvascular endothelial (112)
Glucocortico- steroids*	Human fibrosarcoma (HT 1080; 119,127) Human mammary carcinoma (MDA-MB-231; 117) Human foreskin fibroblasts (69,118) Rat hepatoma (HTC; 115,116)
Insulin*	Human hepatoma (Hep G2; 122) Human hepatocytes (123)

* Denotes agents that stimulate the accumulation of PAI-1 mRNA. References for mRNA are listed in boldface type.

191

(68), epithelial cells (95), and endothelial cells (97). Thus, the induction of PAI-1 appears to be one of the primary effects of TGF-ß in many cell types. Radiolabeling experiments with fibroblasts (68) and endothelial cells (96) have demonstrated that TGF-ß greatly increases deposition of PAI-1 into the ECM of these cells. *In vivo*, TGF-ß promotes the formation of granulation tissue and the growth of new blood vessels (94), and participates in wound healing (98). PAI-1 may protect newly-deposited tissue matrices from proteolysis in these processes.

Basic fibroblast growth factor (bFGF) is another growth factor that may modulate PAI-1 biosynthesis. It possesses angiogenic properties, and stimulates PAI-1 production by bovine capillary endothelial cells (97). However, bFGF also stimulates the production of u-PA, making interpretation of its effects on the overall fibrinolytic activity of these cells difficult. Neither epidermal growth factor (EGF), platelet-derived growth factor (PDGF), or endothelial cell growth factor (ECGF) stimulated PAI-1 synthesis in BAEs (96), although treatment with ECGF-heparin was reported to decrease PAI-1 production in umbilical vein endothelial cells (99). Interestingly, PDGF and TGF-ß have recently been shown to increase PAI-1 synthesis in smooth muscle cells (100). In the vessel wall, this may promote fibrin deposition at sites of endothelial injury.

Cytokines represent another group of molecules that regulate PAI-1 synthesis. For example, interleukin-1 (Il-1) stimulates the production of PAI-1 by cultured human umbilical vein endothelial cells (84,86,101-103). Purified monocyte Il-1 and the recombinant Il-1 species, Il-1-α or -ß, caused a 4- to 8-fold increase in active PAI-1 antigen in conditioned media (84) and cell extracts (103). Il-1 may also suppress t-PA production (84,103) suggesting that the antifibrinolytic effects of this monokine occur at the level of both PAI-1 and t-PA. *In vivo*, Il-1-induced alterations in endothelial PAI-1 synthesis may contribute to increases in circulating PAI-1 levels in acute phase reactions following major surgery, trauma, and myocardial infarction (27,104).

Tumor necrosis factor-α (TNF-α) is a cytokine that mediates septic shock and the hemorrhagic necrosis of certain tumors *in vivo* (105,106). It also stimulates PAI-1 synthesis by endothelial cells. The addition of 200 units/ml of TNF-α caused a 5-fold increase in the production of PAI-1 by human endothelial cells, again with concomitant decreases in t-PA production (103). TNF-α also induces tissue factor activity in endothelial cells (107,108). The decreased fibrinolytic activity and increased procoagulant activity of endothelial cells exposed to TNF-α may therefore promote both the formation and maintenance of fibrin. These responses may underlie the coagulopathy associated with gram-negative sepsis (109), and contribute to the necrotic effects of TNF-α on tumors by altering blood flow at tumor sites (110).

The induction of procoagulant activity by inflammatory mediators such as Il-1 and TNF-α (107,108), may lead to the increased generation of thrombin. The addition of thrombin (1.0 U/ml) to human umbilical vein endothelial cells caused a 6-fold increase in the accumulation of PAI-1 activity in the medium (111). This increase required active thrombin, and was abolished by inhibitors of RNA and protein synthesis. In contrast, human foreskin microvascular endothelial cells exhibited a twofold increase in PAI-1 activity at lower doses of thrombin (0.1 U/ml), but no increase in PAI-1 activity was found at doses of 1.0 U/ml or higher (112), although elevations in PAI-1 antigen were observed at this dose. These latter results may reflect the fact that while thrombin stimulates PAI-1 production, it also cleaves and inactivates PAI-1 when employed at high concentrations (113). Thrombin cleaves and inactivates PAI-1 present in the ECM (72), and stimulates the biosynthesis and release of t-PA from endothelial cells (114). The effects of thrombin on the fibrinolytic activity of these cells are therefore complex, and difficult to interpret.

Finally, a variety of observations indicate that hormones may regulate PAI-1 biosynthesis. For example, steroids stimulate PAI-1 synthesis in several systems, including rat hepatoma cells (115,116), human mammary carcinoma cells (117), and cultured human dermal fibroblasts (69,118). In HT 1080 fibrosarcoma cells, the level of PAI-1 protein was increased approximately tenfold upon the addition of 10^{-6} M dexamethasone (119). Despite these findings, no effect of steroids on PAI-1 activity has been demonstrated *in vivo*. Although plasma PAI-1 levels increased in parallel with cortisol following major surgery, experimental administration of ACTH that elevated plasma cortisol to similar levels failed to produce concomitant increases in PAI-1 activity (79).

PAI-1 and plasma insulin levels are also correlated (120,121). In studies with cultured cells, treatment with 10^{-8} M insulin produced a two-fold increase in secreted PAI-1 antigen and activity in human hepatoma cells, while no effect of insulin could be detected in human umbilical vein endothelial cells (122). Insulin also caused a two-fold increase in PAI-1 production by primary cultures of human hepatocytes (123). These observations suggest that hepatic synthesis may contribute to the elevated concentrations of PAI-1 found in the plasma of hyperinsulinemic subjects (120,121).

REGULATION OF PAI-1 GENE EXPRESSION

Many of the agents described above stimulate PAI-1 gene expression. These agents are indicated in Table 1. In many of these studies, the increases in mRNA and antigen levels were comparable. These and other studies (86,103,111) imply that mechanisms governing translation or secretion of the PAI-1 protein are unlikely to account for its induction in cells. The primary control mechanisms appear instead to exist at the level of gene transcription or mRNA processing.

In bovine aortic endothelial cells, the response of the PAI-1 gene to TGF-ß, LPS, and TNF-α was examined (124). Each agent increased steady-state mRNA levels within 1 hour, and a maximal response (30-100 fold increase) was observed at 6-18 hours (Figure 1). A return to near-basal levels was evident by 48 hours. Nuclear transcription run-on experiments indicated that both the magnitude and time course of mRNA induction was paralled by similar changes in the transcriptional activity of the gene (Figure 1). Additional experiments determined that the half-life of the mRNA was not increased by these agents. This finding is consistent with the conclusion that their effects were mediated by events at the level of transcription. Induction of the mRNA by each agent was independent of *de novo* protein synthesis, as determined by experiments employing the protein synthesis inhibitor cycloheximide. Interestingly, cycloheximide itself increased accumulation of the mRNA. This appeared to be due to an increase in message stability, as well as a modest induction of gene transcription. These results imply that labile protein factors may both repress transcription of the PAI-1 gene and stimulate turnover of its mRNA. Alternatively, cycloheximide, which "freezes" mRNA on ribosomes, may stabilize PAI-1 mRNA by inhibiting its translation, and thus excluding it from polysome-mediated degradative processes (125,126).

Nuclear run-on experiments have also demonstrated transcriptional control of the PAI-1 gene in other cell types. Changes in gene transcription appear to play a major role in the induction of PAI-1 in HT 1080 fibrosarcoma cells treated with glucocorticoids (127) or TNF-α (128). The response to TNF-α exhibits delayed kinetics in these cells relative to bovine aortic endothelial cells, suggesting an alternative or indirect mode of transcriptional activation. In human lung fibroblast (WI-38) cells, a modest increase in gene transcription was observed in response to TGF-ß (129). In previous experiments employing these cells, a 50-fold increase in steady-state mRNA levels in response to TGF-ß was reported (130). If consistent, these results suggest again that different mechanisms (e.g., mRNA stabilization) may distinguish responses to a given agonist in different cell types. However, further experiments are needed to directly assess whether changes in mRNA processing or stability are involved.

Experiments with human umbilical vein endothelial cells have also raised the possibility that cytokine induction of human PAI-1 mRNA may be due in part to mRNA stabilization. Initial analysis of total RNA isolated from these cells revealed two PAI-1 transcripts, approximately 3.2 and 2.3 kilobases (kb) in length. Sequencing of cDNAs (43-46) and analysis of the structure of the human PAI-1 gene (131) indicated that these mRNAs are colinear from their 5' termini, differing only in the length of

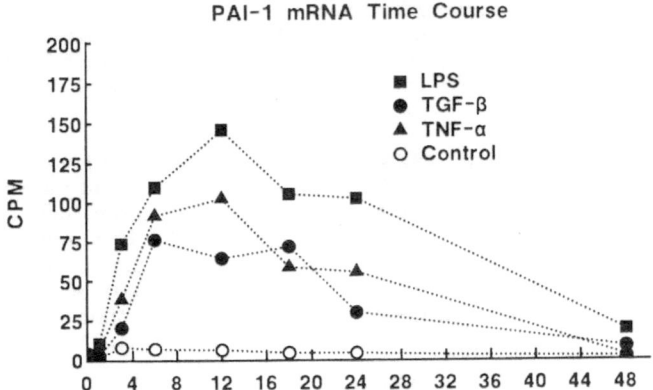

PAI-1 mRNA Time Course

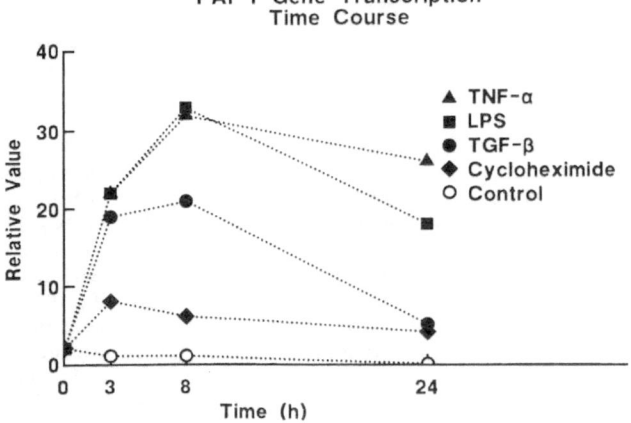

PAI-1 Gene Transcription
Time Course

Figure 1. Time course of induction of PAI-1 mRNA and gene transcription in bovine aortic endothelial cells in response to LPS, TGF-ß, and TNF-α. Confluent cultures of were pre-incubated for 24 hours in serum-free culture medium, and then refed with either serum-free medium alone (Control), or with serum-free medium containing 1 ng/ml TGF-ß, 10 ng/ml LPS, 2 ng/ml TNF-α, or 2 ug/ml cycloheximide. Total cytoplasmic RNA or cell nuclei were harvested at the indicated times. PAI-1 mRNA levels were determined by Northern blot hybridization and the results quantitated employing a radioisotopic scanning device. PAI-1 gene transcription was measured by nuclear transcription run-on assays and densitometric scanning of the autoradiograms. (Reprinted with permission from Sawdey et al., J. Biol. Chem. 1989; 264: 10396-10401).

their 3' untranslated regions. They appear to arise through the use of alternative polyadenylation signals (43,131). Treatment of umbilical vein cells with either Il-1 or TNF-α resulted in the increased accumulation of both PAI-1 mRNA species (103). However, a preferential induction of the 3.2-kb species was observed. This mRNA contains extensive AU-rich sequences not present in the 2.3-kb form. Similar sequences destabilize the mRNA encoding granulocyte-macrophage colony stimulating factor (132), and mediate mRNA induction in T lymphocytes (133). Taken together, these findings suggest that alterations in mRNA stability may account for the response of the 3.2-kb PAI-1 mRNA to cytokines. However, other mechanisms (e.g., selective processing of nuclear transcripts) may account for this result as well.

Little is known concerning the signal transduction pathways mediating activation of PAI-1 gene expression. The tumor promoter phorbol myristate acetate (PMA) stimulates accumulation of PAI-1 mRNA in human umbilical vein and bovine aortic endothelial cells (134,135), and in a rhabdomyosarcoma cell line (136). In nuclear run-on experiments, PMA stimulated PAI-1 gene transcription in both rhabdomyosarcoma cells (136) and U937 monocyte/macrophages (129). The known effects of PMA include activation of protein kinase C, which may transduce hormonal signals capable of altering gene transcription (137,138). In BAEs, down-regulation of protein kinase C levels by PMA pre-treatment partially abrogated the response of PAI-1 to LPS, but not to TGF-ß or TNF-α (139). These findings indicate that this pathway may in part mediate the response of the PAI-1 gene to LPS.

REGULATION OF PAI-1 PROMOTER ACTIVITY

There is now considerable evidence to suggest that gene transcription is regulated by alterations in the activity of specific nuclear factors which govern the rate of transcriptional initiation. Many of these factors interact with discrete DNA sequences in the promoter regions of genes. The sequence of the human PAI-1 promoter, as determined by van Zonneveld *et al.* (140), is shown in Figure 2. This sequence agrees well with that reported by Riccio *et al.* (129) and Bosma *et al.* (141), except for minor variations which may represent allelic variation or sequencing errors. The transcriptional initiation or "cap" site of human endothelial cell mRNA has been determined by nuclease protection experiments (140). The predominant cap site is located the conserved distance downstream from a typical "TATA" box (positions -28 to -23; see Figure 2). Comparison of the PAI-1 promoter sequence with consensus binding sites for known regulatory elements revealed the presence of several possible AP-1 sites (Figure 2). These sites represent the recognition sequence for a complex comprised of the transcription factors *jun* and *fos* (137,142), and have been implicated in the induction of gene transcription by phorbol esters. More recently, AP-1 sites have been shown to mediate the response of the procollagenase gene to TNF-α (143) and auto-induction of the TGF-ß gene (144). Their role in regulating PAI-1 gene transcription is unknown.

DNA transfection studies, utilizing fragments derived from the PAI-1 promoter and 5' flanking region linked to the firefly luciferase reporter gene, have begun to delineate the sequence elements governing the tissue-specificity and glucocorticoid-inducibility of the PAI-1 gene (140). Plasmids bearing either 1.5 kilobases (kb) or 187 base pairs (bp) of the sequence directly 5' to the cap site (p1.5KLuc and p187Luc, respectively) elaborated high levels of luciferase activity in mouse Ltk⁻ cells and bovine aortic endothelial cells, but were not active in HeLa cells. These observations are consistent with the pattern of endogenous PAI-1 gene expression in these cells, suggesting that tissue-specific expression of the PAI-1 gene may be conferred by elements within 187 bp of the cap site. In the rat hepatoma cell line FT02B, luciferase activity was increased by treatment of transfected cells with dexamethasone. A 27-fold induction was observed in the case of p1.5KLuc, while p187Luc was induced by approximately 8-fold. These results indicate that both proximal and distal promoter elements mediate the response to glucocorticoids in FT02B cells.

To further localize these elements, plasmids containing progressive 5' deletions of the promoter were employed (140). Constructs including 800 bp or more of 5' flanking sequence directed 30-fold or greater increases in luciferase activity. However, this response declined to approximately 10-fold if sequences between -800 and -549 of the cap site were deleted, indicating the presence of the distal element within this region. Similar experiments localized the proximal element to within 100 bp of the cap site. A 305 bp fragment containing this element conferred glucocorticoid responsiveness to a heterologous promoter in an orientation-independent fashion, suggesting it possessed enhancer-like properties.

These results implicate two distinct regions of the PAI-1 promoter in the response to glucocorticoids. The recent cloning of the rat (145) and mouse (146) PAI-1 promoters

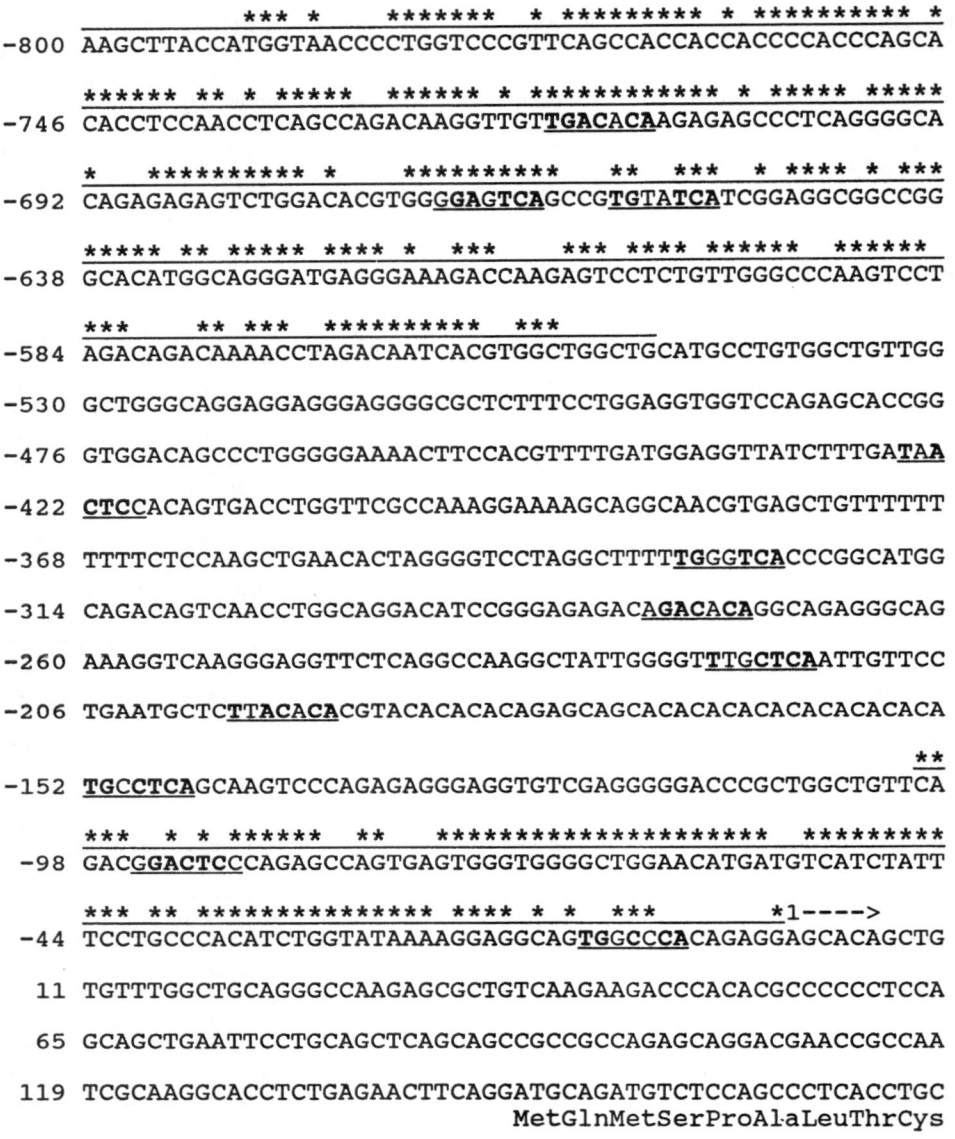

```
              *** *      *******   * ********* * ********** *
     -800  AAGCTTACCATGGTAACCCCTGGTCCCGTTCAGCCACCACCACCCCACCCAGCA

              ****** ** * *****     ****** * ************ * ***** *****
     -746  CACCTCCAACCTCAGCCAGACAAGGTTGTTGACACAAGAGAGCCCTCAGGGGCA

              *   ********** *      ********** *   **  ***  * **** * ***
     -692  CAGAGAGAGTCTGGACACGTGGGGGAGTCAGCCGTGTATCATCGGAGGCGGCCGG

              ***** ** ***** **** *   ***     *** **** ****** ******
     -638  GCACATGGCAGGGATGAGGGAAAGACCAAGAGTCCTCTGTTGGGCCCAAGTCCT

              ***     ** ***  **********  ***
     -584  AGACAGACAAAACCTAGACAATCACGTGGCTGGCTGCATGCCTGTGGCTGTTGG

     -530  GCTGGGCAGGAGGAGGGAGGGGCGCTCTTTCCTGGAGGTGGTCCAGAGCACCGG

     -476  GTGGACAGCCCTGGGGGAAAACTTCCACGTTTTGATGGAGGTTATCTTTGATAA

     -422  CTCCACAGTGACCTGGTTCGCCAAAGGAAAAGCAGGCAACGTGAGCTGTTTTTT

     -368  TTTTCTCCAAGCTGAACACTAGGGGTCCTAGGCTTTTTGGGTCACCCGGCATGG

     -314  CAGACAGTCAACCTGGCAGGACATCCGGGAGAGACAGACACAGGCAGAGGGCAG

     -260  AAAGGTCAAGGGAGGTTCTCAGGCCAAGGCTATTGGGGTTTGCTCAATTGTTCC

     -206  TGAATGCTCTTACACACGTACACACACAGAGCAGCACACACACACACACACACA

                                                               **
     -152  TGCCTCAGCAAGTCCCAGAGAGGGAGGTGTCGAGGGGGACCCGCTGGCTGTTCA

              ***   * * ****** ** ******************** *********
     -98   GACGGACTCCCAGAGCCAGTGAGTGGGTGGGGCTGGAACATGATGTCATCTATT

              *** **  ************** **** * *   ***        *1---->
     -44   TCCTGCCCACATCTGGTATAAAAGGAGGCAGTGGCCCACAGAGGAGCACAGCTG

      11   TGTTTGGCTGCAGGGCCAAGAGCGCTGTCAAGAAGACCCACACGCCCCCCTCCA

      65   GCAGCTGAATTCCTGCAGCTCAGCAGCCGCCGCCAGAGCAGGACGAACCGCCAA

     119   TCGCAAGGCACCTCTGAGAACTTCAGGATGCAGATGTCTCCAGCCCTCACCTGC
                                     MetGlnMetSerProAlaLeuThrCys
```

Figure 2. Nucleotide sequence of the human PAI-1 promoter and 5' flanking region. The sequence shown is from positions -800 to +173. The amino acid sequence of the coding region is shown under the DNA sequence. The major site of transcriptional initiation in endothelial cells (nucleotide 1) is indicated with an arrow. Regions of the promoter implicated in the response to dexamethasone (nucleotides -800 to -549 and -100 to -1) are overlined. Nucleotides conserved between the human (140), rat (145), and mouse (146) promoters within these regions are indicated with an *. Sequences containing homology to the binding site for the transcription factor AP-1 (137) are underlined. Bases homologous to the consensus sequence (TGACTCA) are shown in boldface type. (Adapted from van Zonneveld et al., Proc. Natl. Acad. Sci. USA 1988; 85: 5525-5529).

has provided further insight concerning these regions. Comparison of the promoter sequences of all three species revealed extensive homology both in the area proximal to the cap site, and in the region upstream, from nucleotides ≈ -800 to -549 (Figure 2). The high degree of conservation between species implies that these regions of the PAI-1 promoter may be of importance not only for transcriptional induction by steroids, but by other agents as well.

PAI-1 AND ATHEROSCLEROSIS

Atherosclerosis is probably a disease of multifactorial etiology, involving the altered metabolism and accumulation of blood lipids in the arterial wall, myointimal proliferation of smooth muscle cells, and ultimately, the stenosis of arteries leading to the formation of occlusive thrombi. The role of PAI-1 in these processes is largely unknown. However, PAI-1-induced deficits in fibrinolytic activity may potentially have an impact on both the development of atherosclerotic lesions and the incidence of thrombotic events in patients with advanced disease. For this reason, PAI-1 deserves careful consideration as a risk factor.

Deficient fibrinolysis due to increased plasma PAI-1 activity occurs in obese subjects (121), non-insulin dependent diabetics (147), hyperinsulinemia (120), and hyper-triglyceridemia (78,120). These conditions are recognized risk factors for the development of atherosclerosis. In addition, there is evidence for increased PAI-1 levels in patients with coronary artery disease (CAD), although conflicting results have been reported. For example, Paramo et al. found elevated PAI-1 levels in patients with angiographically documented CAD (148), while similar studies by Oseroff et al. (149) could demonstrate no such relationship. This discrepancy may in part reflect differences in the patient populations chosen for these studies and the exclusion of subjects with other risk factors (i.e., hyperlipidemia) codistributed with PAI-1 (149). In sum, these results imply that increases in PAI-1 activity, perhaps in association with disorders of lipid or carbohydrate metabolism, may predispose to the development of atherosclerosis. The incidence of increased PAI-1 levels in patients with thrombotic disease (see above, "PAI-1 and Thrombotic Disease") suggests further that systemic elevations in its activity may contribute to occlusive thrombosis in patients with advanced disease.

The mechanism whereby PAI-1-induced deficits in fibrinolytic activity may affect the progression of atherosclerotic disease is unknown, but reasonable speculation may be offered, based on current theories of atherogenesis. Ross (150) has proposed that atherosclerosis develops as a response to endothelial injury. In this view, perturbation of the endothelium results in the adherence of blood platelets and mononuclear phagocytes, which secrete mitogenic factors that induce smooth muscle cells to proliferate within the intima. Others, dating back to Rokitansky (151), have proposed that intimal deposition of blood components, including fibrin, might be involved. It is now clear from extensive histopathologic examination of human tissues that the organization and incorporation of mural thrombi plays a major role in the growth and development of atherosclerotic lesions (152). There is also considerable evidence to suggest that impairment of fibrinolytic activity, as proposed by Astrup (153), may contribute to the formation and persistence of such thrombi (154-157). It is therefore plausible that systemic elevations in PAI-1 activity, by stabilizing mural thrombi, may promote atherosclerotic disease.

Recent studies employing in situ hybridization have provided evidence that localized synthesis of PAI-1 in diseased vessels may also contribute to this process. Wilcox et al. employed this technique to demonstrate PAI-1 mRNA in cells in human atherosclerotic plaques (158). Endothelial cells expressing PAI-1 were a consistent feature of these lesions. In addition, mesenchymal-appearing cells were strongly positive for PAI-1 mRNA. The presence of both PAI-1 and tissue factor, which is extensively expressed within atheromatous plaques (159), may exacerbate the development of these lesions. Rupture of the plaque would expose the blood to tissue factor procoagulant activity and initiate clot formation and thrombosis. Where high local concentrations of PAI-1 exist, the otherwise normal clearance of these thrombi would be impaired. The recurrence of such episodes would promote the growth and development of the plaque, leading ultimately to severe stenosis and thrombotic occlusion of the vessel.

While the identity of the stimulus for PAI-1 synthesis within these lesions is unknown, it is likely that many of the agents described above (see above, "Regulation of PAI-1 Biosynthesis") are involved. For example, the release of PDGF from activated platelets, which may act as a mitogenic stimulus for smooth muscle cells (150), may also stimulate PAI-1 synthesis in these cells (100). Similarly, the release of TGF-ß from platelets, or TGF-ß, Il-1 or TNF-α from activated monocytes, may activate PAI-1 synthesis in endothelial cells.

In conclusion, systemic increases in PAI-1 activity, which may result from its increased biosynthesis in endothelial cells throughout the vasculature, or localized increases in PAI-1 activity, which may reflect the action of mediators of inflammatory or repair processes on specific cells of the vessel wall, may compromise normal fibrinolysis and contribute to the development of thrombotic disease. These responses may be effected by alterations in the activity of specific factors governing transcription of the PAI-1 gene or processing of its mRNA. Continued investigation into the nature of these factors will ultimately help to clarify the role of PAI-1 in vascular disease.

REFERENCES

1. Dano, K., Andreasen, P.A., Grondahl-Hansen, J., Kristensen, P., Nielsen, L.S. and Skriver, L. 1985. Plasminogen activators, tissue degradation and cancer. Adv. Cancer Res. 44: 139-226.

2. Hoylaerts, M., Rijken, D.C. and Collen, D. 1982. Kinetics of the activation of plasminogen by human tissue plasminogen activator. Role of fibrin. J. Biol. Chem. 257: 2912-2919.

3. Wun, T.-C., Schleuning, W.-D. and Reich, E. 1982. Isolation and characterization of urokinase from human plasma. J. Biol. Chem. 257: 3276-3283.

4. Verde, P., Stoppelli, M.P., Galeffi, P., Di Nocera, P. and Blasi, F. 1984. Identification and primary sequence of an unspliced human urokinase poly (A)+RNA. Proc. Natl. Acad. Sci. USA 81: 4727-4731.

5. Nielsen, L.S., Hansen, J.G., Skriver, L., Wilson, E.L., Kaltoft, K., Zeuthen, J. and Dano, K. 1982. Purification of zymogen to plasminogen activator from human glioblastoma cells by affinity chromatography with monoclonal antibody. Biochemistry 21: 6410-6415.

6. Stump, D.C., Lijnen, H.R. and Collen, D. 1986. Purification and characterization of a single-chain urokinase-type plasminogen activator from human cell cultures. J. Biol. Chem. 261: 1274-1278.

7. Vassalli, J.-D., Baccino, D. and Belin, D. 1985. A cellular binding site for the M_r 55,000 form of the human plasminogen activator, urokinase. J. Cell Biol. 100: 86-92.

8. Stopelli, M.P., Corti, M., Soffientini, A., Cassani, G., Blasi, F. and Assoian, R.K. 1985. Differentiation-enhanced binding of the amino-terminal fragment of human urokinase plasminogen activator to a specific receptor on U937 monocytes. Proc. Natl. Acad. Sci. USA 82: 4939-4943.

9. Miles, L.A. and Plow, E.F. 1985. Binding and activation of plasminogen on the platelet surface. J. Biol. Chem. 260: 4303-4311.

10. Hajjar, K.A., Harpel, P.C., Jaffe, E.A. and Nachman, R.L. 1986. Binding of plasminogen to cultured human endothelial cells. J. Biol. Chem. 261: 11656-11662.

11. Miles, L.A. and Plow, E.F. 1987. Receptor mediated binding of the fibrinolytic components, plasminogen and urokinase, to peripheral blood cells. Thromb. Haemost. 58: 936-942.

12. Granelli-Piperno, A. and Reich, E. 1978. A study of proteases and protease-inhibitor complexes in biological fluids. J. Exp. Med. 148: 223-234.

13. Reich, E. 1978. Activation of plasminogen: a general mechanism for producing localized extracellular proteolysis. In: Molecular Basis of Biological Degradative Processes. Berlin, R.D., Herrmann, H., Lepou, I.H. and Tanzer, J.M. (eds). Academic Press, New York, pp. 155-169.

14. Larsson, L.-I., Skriver, L., Nielsen, L.S., Grondahl-Hansen, J., Kristensen, P. and Dano, K. 1984. Distribution of urokinase-type plasminogen activator immunoreactivity in the mouse. J. Cell Biol. 98: 894-903.

15. Ossowski, L., Biegel, D. and Reich, E. 1979. Mammary plasminogen activator: correlation with involution, hormonal modulation, and comparison between normal and neoplastic tissues. Cell 16: 929-940.

16. Gross, J.L., Moscatelli, D., Jaffe, E.A. and Rifkin, D.B. 1982. Plasminogen activator and collagenase production by cultured capillary endothelial cells. J. Cell Biol. 95: 974-981.

17. Moscatelli, D., Presta, M. and Rifkin, D.B. 1986. Purification of a factor from human placenta that stimulates capillary endothelial cell protease production, DNA synthesis, and migration. Proc. Natl. Acad. Sci. USA 83: 2091-2095.

18. Aoki, N. and Harpel, P.C. 1984. Inhibitors of the fibrinolytic enzyme system. Sem. Thromb. Hemost. 10: 24-41.

19. Collen, D. 1980. On the regulation and control of fibrinolysis. Thromb. Haemost. 43: 77-89.

20. Plow, E.F., Freaney, D., Plescia, J. and Miles, L.A. 1986. The plasminogen system and cell surfaces: evidence for plasminogen and urokinase receptors on the same cell type. J. Cell. Biol. 103: 2411-2420.

21. Miles, L.A. and Plow, E.F. 1988. Plasminogen receptors: ubiquitous sites for cellular regulation of fibrinolysis. Fibrinolysis 2: 61-71.

22. Vassalli, J.-D., Hamilton, J. and Reich, E. 1976. Macrophage plasminogen activator: modulation of enzyme production by anti-inflammatory steroids, mitotic inhibitors, and cyclic nucleotides. Cell 8: 271-281.

23. Grant, P.J. and Medcalf, R.L. 1990. Hormonal regulation of haemostasis and the molecular biology of the fibrinolytic system. Clinical Science 78: 3-11.

24. Cash, J.C., Gader, A.M.A. and Da Costa, J. 1974. The release of plasminogen activator and factor VIII by LVP, AVP, DDAVP, ATIII and OT in man. Br. J. Haematol. 27: 363-364.

25. Mannucci, P.M., Aberg, M., Nilsson, I.M. and Robertson, B. 1975. Mechanism of plasminogen activator and factor VIII increase after vasoactive drugs. Br. J. Haematol. 30: 81-93.

26. Collen, D. 1986. Report of the meeting of the subcommittee on fibrinolysis. Jerusalem, Israel, June 2, 1986. Thromb. Haemost. 56: 415-416.

27. Loskutoff, D.J., Sawdey, M. and Mimuro, J. 1988. Type 1 plasminogen activator inhibitor. In: Progress in Hemostasis and Thrombosis. Coller, B. (ed). W.B. Saunders Co., Philadelphia, PA, 9: 87-115.

28. Emeis, J.J., van Hinsbergh, V.W.M., Verheijen, J.H. and Wijngaards, G. 1983. Inhibition of tissue-type plasminogen activator by conditioned medium from cultured human and porcine vascular endothelial cells. Biochem. Biophys. Res. Commun. 110: 392-398.

29. Philips, M., Juul, A.-G. and Thorsen, S. 1984. Human endothelial cells produce a plasminogen activator inhibitor and a tissue-type plasminogen activator-inhibitor complex. Biochim. Biophys. Acta 802: 99-110.

30. van Mourik, J.A., Lawrence, D.A. and Loskutoff, D.J. 1984. Purification of an inhibitor of plasminogen activators (antiactivator) synthesized by endothelial cells. J. Biol. Chem. 259: 14914-14921.

31. Astedt, B., Lecander, I. and Ny, T. 1987. The placental typeplasminogen activator inhibitor, PAI-2. Fibrinolysis 1: 203-208.

32. Kawano, T., Morimoto, K. and Uemura, Y. 1986. Urokinase inhibitor in human placenta. Nature 217: 253-254.

33. Wiman, B., Chmielewska, J. and Ranby, M. 1984. Inactivation of tissue plasminogen activator in plasma. J. Biol. Chem. 259: 3644-3647.

34. Coleman, P.L., Patel, P.D., Cwikel, B.J., Rafferty, U.M., Sznycer-Laszuk, R. and Gelehrter, T.D. 1986. Characterization of the dexamethasone-induced inhibitor of plasminogen activator in HTC hepatoma cells. J. Biol. Chem. 261: 4352-4435.

35. Hekman, C.M. and Loskutoff, D.J. 1988. Kinetic analysis of the interactions between plasminogen activator inhibitor 1 and both urokinase and tissue plasminogen activator. Arch. Biochem. Biophys. 262: 199-210.

36. Hanss, M. and Collen, D. 1987. Secretion of tissue-type plasminogen activator and plasminogen activator inhibitor by cultured human endothelial cells: modulation by thrombin,endotoxin, and histamine. J. Lab. Clin. Med. 109: 97-104.

37. Kruithof, E.K.O., Tran-Thang, C., Gudinchet, A., Hauert, J.,Nicoloso, G., Genton, C., Welti, H. and Bachmann, F.W. 1987. Fibrinolysis in pregnancy. A study of plasminogen activatorinhibitors. Blood 69: 460-466.

38. Colucci, M., Paramo, J.A. and Collen, D. 1986. Inhibition of one-chain and two-chain forms of human tissue-type plasminogen activator by the fast-acting inhibitor of plasminogen activator in vitro and in vivo. J. Lab. Clin. Med. 108: 53-59.

39. Dosne, A.M., Dupuy, E. and Bodevin, E. 1978. Production of a fibrinolytic inhibitor by cultured endothelial cells derived from human umbilical vein. Thromb. Res. 12: 377-378.

40. Loskutoff, D.J. and Edgington, T.S. 1977. Synthesis of a fibrinolytic activator and inhibitor by endothelial cells. Proc. Natl. Acad. Sci. USA 74: 3903-3907.

41. Loskutoff, D.J. and Edgington, T.S. 1981. An inhibitor of plasminogen activator in rabbit endothelial cells. J. Biol. Chem. 256: 4142-4145.

42. Loskutoff, D.J., van Mourik, J.A., Erickson, L.A. and Lawrence, D.A. 1983. Detection of an unusually stable fibrinolytic inhibitor produced by bovine endothelial cells. Proc. Natl. Acad. Sci. USA 80: 2956-2960.

43. Ny, T., Sawdey, M., Lawrence, D.A., Millan, J.L. and Loskutoff, D.J. 1986. Cloning and sequence of a cDNA coding for the human ß-migrating endothelial-cell-type plasminogen activator inhibitor. Proc. Natl. Acad. Sci. USA 83: 6776-6780.

44. Pannekoek, H., Veerman, H., Lambers, H., Diergaarde, P., Verweij, C.L, van Zonneveld, A.J. and van Mourik, J.A. 1986. Endothelial plasminogen activator inhibitor (PAI): a new member of Serpin gene family. EMBO J. 5: 2539-2544.

45. Ginsburg, D., Zeheb, R., Yang, A.V., Rafferty, U.M., Andreasen, P.A., Nielsen, L., Dano, K., Lebo, R.V. and Gelehrter, T.D. 1986. cDNA cloning of human plasminogen activator-inhibitor from endothelial cells. J. Clin. Invest. 78: 1673-1680.

46. Andreasen, P.A., Riccio, A., Welinder, K.G., Douglas, R., Sartorio, R., Nielsen, L.S., Oppenheimer, C., Blasi, F. and Dano, K. 1986. Plasminogen activator inhibitor type 1: reactive center and amino-terminal heterogeneity determined by protein and cDNA sequencing. FEBS Letters 209: 213-218.

47. Hunt, L.T. and Dayhoff, M.O. 1980. A surprising new protein superfamily containing ovalbumin, antithrombin-III, and α 1-proteinase inhibitor. Biochem. Biophys. Res. Commun.95: 864-871.

48. Carrell, R.W. and Travis, J. 1985. α-1-Antitrypsin and the serpins: Variation and countervariation. Trends Biochem. Sci. 10: 20-24.

49. Colucci, M., Paramo, J.A. and Collen, D. 1985. Generation in plasma of a fast-acting inhibitor of plasminogen activator in response to endotoxin stimulation. J. Clin. Invest. 75: 818-824.

50. Erickson, L.A., Ginsberg, M.H. and Loskutoff, D.J. 1984. Detection and partial characterization of an inhibitor of plasminogen activator in human platelets. J. Clin. Invest.74: 1465-1472.

51. Booth, N.A., Anderson, J.A. and Bennett, B. 1985. Platelet release protein which inhibits plasminogen activators. J. Clin. Pathol. 38: 825-830.

52. Kruithof, E.K.O., Tran-Thang, C. and Bachmann, F.W. 1986. Studies on the release of a plasminogen activator inhibitor by human platelets. Thromb. Haemost. 55: 201-205.

53. Sprengers, E.D., Akkerman, J.W.N. and Jansen, B.G. Blood platelet plasminogen activator inhibitor: two different pools of endothelial cell type. Thromb. Haemost. 55: 325-329.

54. Kruithof, E.K.O., Nicolosa, G. and Bachmann, F.W. 1987. Plasminogen activator inhibitor 1: Development of a radioimmunoassay and observations on its plasma concentration during venous occlusion and after platelet aggregation. Blood 70: 1645-1653.

55. Sprengers, E.D., Princen, H.M.G., Kooistra, T. and van Hinsbergh, V.W.M. 1985. Inhibition of plasminogen activators by conditioned medium of human hepatocytes and the hepatoma cell line Hep G2. J. Lab. Clin. Med. 105: 751-758.

56. Crutchley, D.J. and Conanan, L.B. 1986. Endotoxin induction of an inhibitor of plasminogen activator in bovine pulmonary artery endothelial cells. J. Biol. Chem. 261: 154-159.

57. Canfield, A.E., Schor, A.M., Loskutoff, D.J., Schor, S.L. and Grant, M.E. 1989. Plasminogen activator inhibitor-type 1 is a major biosynthetic product of retinal microvascular endothelial cells and pericytes in culture. Biochem. J. 259: 529-535.

58. Laug, W.E. 1985. Vascular smooth muscle cells inhibit plasminogen activators secreted by endothelial cells. Thromb. Haemost. 53: 165-169.

59. Simpson, A.J., Booth, N.A., Sewell, H., MacGregor, I.R., Bennett, B. 1989. Immunochemical localization of PAI-1 in human tissues. Thromb. Haemost. 62: 171.

60. Hekman, C.M. and Loskutoff, D.J. 1985. Endothelial cells produce a latent inhibitor of plasminogen activators that can be activated by denaturants. J. Biol. Chem. 260: 11581-11587.

61. Levin, E.G. and Santell, L. 1987. Conversion of active to latent plasminogen activator inhibitor from human endothelial cells. Blood 70: 1090-1098.

62. Kooistra, T., Sprengers, E.D. and van Hinsbergh, V.W.M. 1986. Rapid inactivation of plasminogen activator inhibitor upon secretion from cultured human endothelial cells. Biochem. J. 239: 497-503.

63. Hekman, C.M. and Loskutoff, D.J. 1988. Bovine plasminogen activator inhibitor 1: Specificity determinations and comparison of the active, latent and guanidine-activated forms. Biochemistry 27: 2911-2918.

64. Lambers, J.W.J., Cammenga, M., Konig, B., Pannekoek, H. and van Mourik, J.A. 1987. Activation of human endothelial type plasminogen activator inhibitor (PAI-1) by negatively charged phospholipids. J. Biol. Chem. 262: 17492-17496.

65. Declerck, P.J., De Mol, M., Alessi, M.-C., Baudner, S., Paques, E.-P., Preissner, K.T., Muller-Berghaus, G. and Collen, D. 1988. Purification and characterization of a plasminogen activator inhibitor 1 binding protein from human plasma. J. Biol. Chem. 263: 15454-15461.

66. Wiman, B., Lindahl, T. and Almqvist, A. 1988. Evidence for a discrete binding protein of plasminogen activator inhibitor in plasma. Thromb. Haemost. 59: 392-395.

67. Mimuro, J., Schleef, R.R. and Loskutoff, D.J. 1987. The extracellular matrix of cultured bovine aortic endothelial cells contains functionally active type 1 plasminogen activator inhibitor. Blood 70: 721-728.

68. Laiho, M., Saksela, O., Andreasen, P.A. and Keski-Oja, J. 1986. Enhanced production and extracellular deposition of the endothelial-type plasminogen activator inhibitor in cultured human lung fibroblasts by transforming growth factor-ß. J. Cell Biol. 103: 2403-2410.

69. Rheinwald, J.G., Jorgensen, J.L., Hahn, W.C., Terpstra, A.J., O'Connell, T.M. and Plummer, K.K. 1987. Mesosecrin: A secreted glycoprotein produced in abundance by human mesothelial, endothelial and kidney epithelial cells in culture. J. Cell Biol. 104: 263-275.

70. Pollanen, J., Saksela, O., Salonen, E.M., Andreasen, P.A., Nielsen, L., Dano, K. and Vaheri, A. 1987. Distinct localizations of urokinase-type plasminogen activator and its type 1 inhibitor under cultured human fibroblast and sarcoma cells. J. Cell Biol. 104: 1085-1096.

71. Levin, E.G. and Santell, L. 1987. Association of plasminogen activator inhibitor (PAI-1) with the growth substratum and membrane of human endothelial cells. J. Cell Biol. 105: 2543-2549.

72. Knudsen, B.S., Hapel, P.C. and Nachman, R.L. 1987. Plasminogen activator inhibitor is associated with the extracellular matrix of cultured bovine smooth muscle cells. J. Clin. Invest. 80: 1082-1089.

73. Seiffert, D., Wagner, N.N. and Loskutoff, D.J. 1990. Serum-derived vitronectin influences the pericellular distribution of type 1 plasminogen activator inhibitor. J. Cell Biol. (in press).

74. Lawrence, D.A. and Loskutoff, D.J. 1986. Inactivation of plasminogen activator inhibitor by oxidants. Biochemistry 25: 6351-6355.

75. Travis, J. and Salvesen, G.S. 1983. Human plasma protein inhibitors. Ann. Rev. Biochem. 52: 655-709.

76. Wiman, B. and Chmielewska, J. 1985. A novel fast inhibitor to tissue plasminogen activator in plasma, which may be of great pathophysiological significance. Scand. J. Clin. Lab. Invest. 177: 43-47.

77. Juhan-Vague, I., Valadier, J., Alessi, M.C., Aillaud, M.F., Ansaldi, J., Philips-Joet, C., Holvoet, P., Serradimigni, A. and Collen, D. 1987. Deficient t-PA release and elevated PA inhibitor levels in patients with spontaneous or recurrent deep venous thrombosis. Thromb. Haemost. 57: 67-72.

78. Hamsten, A., Wiman, B., deFaire, U. and Blomback, M. 1985. Increased plasma levels of a rapid inhibitor of tissue plasminogen activator in young survivors of myocardial infarction. N. Engl. J. Med. 313: 1557-1563.

79. Aillaud, M.F., Juhan-Vague, I., Alessi, M.C., Marecal, M., Vinson, M.F., Arnaud, C., Vague, P.H. and Collen, D. 1985. Increased PA-inhibitor levels in the postoperative period - no cause-effect relation with increased cortisol. Thromb. Haemost. 54: 466-468.

80. Kluft, C., Verheijen, J.H., Jie, A.F.H., Rijken, D.C., Preston, F.E., Sue-Ling, H.M., Jespersen, J. and Aasen, A.D. 1985. The postoperative fibrinolytic shutdown: A rapidly reverting acute phase pattern for the fast-acting inhibitor of tissue-type plasminogen activator after trauma. Scand. J. Clin. Lab. Invest. 45: 605-610.

81. Sprengers, E.D. and Kluft, C. 1987. Plasminogen activator inhibitors. Blood 69: 381-387.

82. Almer, L. and Ohlin, H. 1987. Elevated levels of the rapid inhibitor of plasminogen activator (t-PAI) in acute myocardialinfarction. Thromb. Res. 47: 335-339.

83. Morrison, D.C. and Ulevitch, R.J. 1978. The effects of bacterial endotoxins on host mediation systems. Am. J. Path. 93: 527-617.

84. Bevilacqua, M.P., Schleef, R.R., Gimbrone, M.A. Jr. and Loskutoff, D.J. 1986. Regulation of the fibrinolytic system of cultured human vascular endothelium by interleukin 1. J. Clin. Invest. 78: 587-591.

85. Podor, T.J., Curriden, S.A., Lawrence, D.A. and Loskutoff, D.J. 1986. Characterization of monoclonal antibodies to the ß-migrating PA inhibitor (ß-PAI) isolated from cultured bovine aortic endothelial cells (BAEs). Fibrinolysis 1: 35.

86. Emeis, J.J. and Kooistra, T. 1986. Interleukin 1 and lipopolysaccharide induce an inhibitor of tissue-type plasminogen activator in vivo and in cultured endothelial cells. J. Exp. Med. 163: 1260-1266.

87. Dubor, F., Dosne, A.M. and Chedid, L.A. 1986. Effect of polymyxin B and colimycin on induction of plasminogen antiactivator by lipopolysaccharide in human endothelial cell culture. Infect. Immun. 52: 725-729.

88. Podor, T., Sawdey, M., Mathison, J., Tobias, P., Ulevitch, R. and Loskutoff, D.J. 1988. Serum-derived lipopolysaccharide (LPS) binding factors enhance the LPS-mediated induction of type 1 plasminogen activator inhibitor in endothelial cells. Fibrinolysis 2(supp 1): 149.

89. Levin, E.G. and Loskutoff, D.J. 1980. Serum-mediated suppression of cell-associated plasminogen activator activity in cultured endothelial cells. Cell 22: 701-707.

90. Sawdey, M., Ny, T. and Loskutoff, D.J. 1986. Messenger RNA for plasminogen activator inhibitor. Thromb. Res. 41: 151-160.

91. Sporn, M.B., Roberts, A.B., Wakefield, L.M. and de Crombrugghe, B. 1987. Some recent advances in the chemistry and biology of transforming growth factor-ß. J. Cell Biol. 105: 1039-1045.

92. Assoian, R.K. and Sporn, M.B. 1986. Type-ß transforming growth factor in human platelets: Release during platelet degranulation and action on vascular smooth muscle cells. J. Cell Biol. 102: 1217-1223.

93. Ignotz, R. and Massague, J. 1986. Transforming growth factor-ß stimulates the expression of fibronectin and collagen and their incorporation into the extracellular matrix. J. Biol. Chem. 261: 4337-4345.

94. Roberts, A.B., Sporn, M.B., Assoian, R.K., Smith, J.M., Roche, N.S., Wakefield, L.M., Heine, U.I., Liotta, L.A., Falanga, V., Kehrl, J.H. and Fauci, A.S. 1986. Transforming growth factor type-ß: Rapid induction of fibrosis and angiogenesis in vivo and stimulation of collagen formation in vitro. Proc. Natl. Acad. Sci. USA 83: 4167-4171.

95. Thalacker, F.W. and Nilsen-Hamilton, M. 1987. Specific induction of secreted proteins by transforming growth factor-ß and 12-0-tetradecanoylphorbol-13-acetate. J. Biol. Chem. 262: 2283-2290.

96. Mimuro, J. and Loskutoff, D.J. 1987. Effect of transforming growth factor-ß (TGFß) on the fibrinolytic system of cultured bovine aortic endothelial cells (BAEs). Thromb. Haemost. 58: 1647.

97. Saksela, O., Moscatelli, D. and Rifkin, D.B. 1987. The opposing effects of basic fibroblast growth factor and transforming growth factor ß on the regulation of plasminogen activator activity in capillary endothelial cells. J. Cell Biol. 105: 957-963.

98. Pierce, G.F., Mustoe, T.A. and Deuel, T.F. 1988. Transforming growth factor ß induces increased directed cellular migration and tissue repair in rats. Prog. Clin. Biol. Res. 266: 93-102.

99. Konkle, B.A. and Ginsburg, D. 1988. The addition of endothelial cell growth factor and heparin to human umbilical vein endothelial cell cultures decreases plasminogen activator inhibitor-1 expression. J. Clin. Invest. 82: 579-585.

100. McFall, B.C. and Reilly, C.F. 1990. TGFß and PDGF induce plasminogen activator inhibitor release from vascular smooth muscle cells. FASEB J. 4: A892.

101. Gramse, M., Breviario, F., Pintucci, G., Millet, I., Dejana, E., Van Damme, J., Donati, M.B. and Mussoni, L.M. 1986. Enhancement by interleukin-1 (IL-1) of plasminogen activator inhibitor (PAI) activity in cultured human endothelial cells. Biochem. Biophys. Res. Commun. 139: 720-727.

102. Nachman, R.L., Hajjar, K.A., Silverstein, R.L. and Dinarello, C.A. 1986. Interleukin 1 induces endothelial cell synthesis of plasminogen activator inhibitor. J. Exp. Med. 163: 1595-1600.

103. Schleef, R.R., Bevilacqua, M.P., Sawdey, M., Gimbrone, M.A. Jr. and Loskutoff, D.J. 1988. Cytokine activation of vascular endothelium: Effects on tissue-type plasminogen activator and type 1 plasminogen activator inhibitor. J. Biol. Chem. 263: 5797-5803.

104. Dinarello, C. 1984. Interleukin 1 and the pathogenesis of the acute-phase response. New Engl. J. Med. 311: 1413-1418.

105. Beutler, B. and Cerami, A. 1987. Cachectin: More than a tumor necrosis factor. N. Engl. J. Med. 316: 379-385.

106. Old, L.J. 1985. Tumor necrosis factor (TNF). Science 230: 630-632.

107. Bevilacqua, M.P., Pober, J.S., Majeau, G.R., Fiers, W., Cotran, R.S. and Gimbrone, M.A. Jr. 1986. Recombinant tumor necrosis factor induces procoagulant activity in cultured human vascular endothelium: Characterization and comparison with the actions of interleukin 1. Proc. Natl. Acad. Sci. USA 83: 4533-4537.

108. Nawroth, P.P. and Stern, D.M. 1986. Modulation of endothelial cell hemostatic properties by tumor necrosis factor. J. Exp. Med. 163: 740-745.

109. Mathison, J., Wolfson, E. and Ulevitch, R. 1988. Participation of tumor necrosis factor in the mediation of gram negative bacterial lipopolysaccharide-induced injury in rabbits. J. Clin. Invest. 81: 1925-1937.

110. Le, J. and Vilcek, J. 1987. Tumor necrosis factor and interleukin 1: Cytokines with multiple overlapping biologicalactivities. Lab. Invest. 56: 234-248.

111. Gelehrter, T.D. and Sznycer-Laszuk, R. 1986. Thrombin induction of plasminogen activator inhibitor in cultured human endothelial cells. J. Clin. Invest. 77: 165-169.

112. van Hinsbergh, V.W.M., Sprengers, E.D. and Kooistra, T. 1987. Effect of thrombin on the production of plasminogen activators and PA inhibitor-1 by human foreskin microvascular endothelial cells. Thromb. Haemost. 57: 148-153.

113. de Fouw, N.J., van Hinsbergh, V.W.M., de Jong, Y.F., Haverkate, F. and Bertina, R.M. 1987. The interaction of activated protein C and thrombin with the plasminogen activator inhibitor released from human endothelial cells. Thromb. Haemost. 57: 176-182.

114. Levin, E.G., Marzec, U., Anderson, J. and Harker, L.A. 1984. Thrombin stimulates tissue plasminogen activator release from cultured human endothelial cells. J. Clin. Invest. 74: 1988-1995.

115. Loskutoff, D.J., Roegner, K., Erickson, L.A., Schleef, R.R.,Huttenlocher, A., Coleman, P.L. and Gelehrter, T.D. 1986. The dexamethasone-induced inhibitor of plasminogen activator in hepatoma cells is antigenically related to an inhibitor produced by bovine aortic endothelial cells. Thromb. Haemost. 55: 8-11.

116. Coleman, P.L., Barouski-Miller, P.A. and Gelehrter, T.D. 1982. The dexamethasone-induced inhibitor of fibrinolytic activity in hepatoma cells - a cellular product which specifically inhibits plasminogen activation. J. Biol. Chem. 257: 4260-4264.

117. Busso, N., Belin, D., Failly-Crepin, C. and Vassalli, J.-D. 1987. Glucocorticoid modulation of plasminogen activators and of one of their inhibitors in the human mammary carcinoma cell line MDA-MB-231. Cancer Res. 47: 364-370.

118. Crutchley, D.J., Conanan, L.B. and Maynard, J.R. 1981. Human fibroblasts produce an inhibitor directed against plasminogen activator when treated with glucocorticoids. Ann. N.Y. Acad. Sci. 370: 609-616.

119. Andreasen, P.A., Pyke, C., Riccio, A., Kristensen, P., Nielsen, L.S., Lund, L.R., Blasi, F. and Dano, K. 1987. Plasminogen activator inhibitor type 1 biosynthesis and mRNA level are increased by dexamethasone in human fibrosarcoma cells. Mol. Cell Biol. 7: 3021-3025.

120. Juhan-Vague, I., Vague, P., Alessi, M.C., Badier, C., Valadier, J., Aillaud, M.F. and Atlan, C. 1987. Relationship between plasma insulin, triglyceride, body mass index, and plasminogen activator inhibitor 1. Diabetes Metab. 13: 331-336.

121. Vague, P., Juhan-Vague, I., Aillaud, M.F., Badier, C., Viard, R., Alessi, M.C. and Collen, D. 1986. Correlation between blood fibrinolytic activity, plasminogen activator

inhibitor level, plasma insulin level and relative body weight in normal and obese subjects. Metabolism 35: 250-253.

122. Alessi, M.C., Juhan-Vague, I., Kooistra, T., Declerck, P.J. and Collen, D. 1988. Insulin stimulates the synthesis of plasminogen activator inhibitor 1 by the human hepatocellular cell line Hep G2. Thromb. Haemost. 60: 491-494.

123. Kooistra, T., Bosma, P.J., Tons, H.A.M., van den Berg, A.P., Meyer, P. and Princen, H.M.G. 1989. Plasminogen activator inhibitor 1: Biosynthesis and mRNA level are increased by insulin in cultured human hepatocytes. Thromb. Haemost. 62: 723-728.

124. Sawdey, M., Podor, T.J. and Loskutoff, D.J. 1989. Regulation of type 1 plasminogen activator inhibitor gene expression in cultured bovine aortic endothelial cells: Induction by transforming growth factor-ß, lipopolysaccharide, and tumor necrosis factor-α. J. Biol. Chem. 264: 10396-10401.

125. Graves, R.A., Pandey, N.B, Chodchoy, N. and Marzluff, W.F. 1987. Translation is required for regulation of histone mRNA degradation. Cell 48: 615-626.

126. Ross, J. and Kobs, G. 1986. H4 histone messenger RNA decay in cell-free extracts initiates at or near the 3' terminus and proceeds 3' to 5'. J. Mol. Biol. 188: 579-593.

127. Medcalf, R.L., Van den Berg, E. and Schleuning, W.-D. 1988. Glucocorticoid-modulated gene expression of tissue- and urinary-type plasminogen activator and plasminogen activator inhibitor 1 and 2. J. Cell Biol. 106: 971-978.

128. Medcalf, R.L., Kruithof, E.K.O. and Schleuning, W.-D. 1988. Plasminogen activator inhibitor 1 and 2 are tumor necrosis factor/cachetin responsive genes. J. Exp. Med. 168: 751-759.

129. Riccio, A., Lund, L.R., Sartorio, R., Lania, A., Andreasen, P.A., Dano, K. and Blasi, F. 1988. The regulatory region of the human plasminogen activator inhibitor type-1 (PAI-1) gene. Nucleic Acids Res. 16: 2805-2823.

130. Lund, L.R., Riccio, A., Andreasen, P.A., Nielsen, L.S., Kristensen, P., Laiho, M., Saksela, O., Blasi, F. and Dano, K. 1987. Transforming growth factor-ß is a strong and fast acting positive regulator of the level of type-1 plasminogen activator inhibitor mRNA in WI-38 human lung fibroblasts. EMBO J. 6: 1281-1286.

131. Loskutoff, D.J., Linders, M., Keijer, J., Veerman, H., vanHeerikhuizen, H. and Pannekoek, H. 1987. The structure of the human plasminogen activator inhibitor 1 gene: Non-random distribution of introns. Biochemistry 26: 3763-3768.

132. Shaw, G. and Kamen, R. 1986. A conserved AU sequence from the 3' untranslated region of GM-CSF mRNA mediates selective mRNA degradation. Cell 46: 659-667.

133. Lindsten, T., June, C.H., Ledbetter, J.A., Stella, G., Thompson, C.B. 1989. Regulation of lymphokine messenger RNA stability by a surface-mediated T cell activation pathway. Science 244: 339-343.

134. Levin, E.G. and Santell, L. 1988. Regulation of tPA and PAI-1 expression in human endothelial cells by protein kinase C and cAMP. Fibrinolysis 2(supp 1): 19.

135. Sawdey, M. and Loskutoff, D.J. Unpublished results.

136. Mayer, M., Lund, L.R., Riccio, A., Skouv, J., Nielsen, L.S., Stacey, S.N., Dano, K. and Andreasen, P.A. 1988. Plasminogen activator inhibitor type-1 protein, mRNA and gene transcription are increased by phorbol esters in human rhabdomyosarcoma cells. J. Biol. Chem. 263: 15688-15693.

137. Lee, W., Mitchell, P. and Tjian, R. 1987. Purified transcription factor AP-1 interacts with TPA-inducible enhancer elements. Cell 49: 741-752.

138. Imagawa, M., Chiu, R. and Karin, M. 1987. Transcription factor AP-2 mediates induction by two different signal transduction pathways: Protein kinase C and cAMP. Cell 51: 251-260.

139. Slivka, S., Podor, T.J. and Loskutoff, D.J. 1989. Evaluation of second messengers and protein kinases involved in the regulation of type 1 plasminogen activator inhibitor synthesis. Fibrinolysis 3(supp 1): 40.

140. van Zonneveld, A.-J., Curriden, S.A. and Loskutoff, D.J. 1988. Type 1 plasminogen activator inhibitor gene: Functional analysis and glucocorticoid regulation of its promoter. Proc. Natl. Acad. Sci. USA 85: 5525-5529.

141. Bosma, P.J., van den Berg, E.A., Kooistra, T., Siemieniak, D.R. and Slightom, J.L. 1988. Human plasminogen activator inhibitor-1 gene. J. Biol. Chem. 263: 9129-9141.

142. Rauscher, F.J., Cohen, D.R., Curran, T., Bos, T.J., Vogt, P.K., Bohmann, D., Tjian, R., Franza, B.R. Jr. 1988. Fos-associated protein p39 is the product of the jun proto-oncogene. Science 240: 1010-1016.

143. Brenner, D.A., O'Hara, M., Angel, P., Chojkier, M. and Karin, M. 1989. Prolonged activation of jun and collagenase genes by tumour necrosis factor-α. Nature 337: 661-663.

144. Kim, S.-J., Angel, P., Lafyatis, R., Hattori, K., Kim, K.Y., Sporn, M.B., Karin, M. and Roberts, A.B. 1990. Autoinduction of transforming growth factor ß1 is mediated by the AP-1 complex. Mol. Cell Biol. 10: 1492-1497.

145. Bruzdzinski, C.J., Riordan-Johnson, M., Nordby, E.C., Suter, S.M. and Gelehrter, T.D. 1990. Isolation and characterization of the rat plasminogen activator inhibitor-1 gene. J. Biol. Chem. 265: 2078-2085.

146. Prendergast, G.C., Diamond, L.E., Dahl, D. and Cole, M.D. 1990. The c-myc-regulated gene mr1 encodes plasminogen activator inhibitor 1. Mol. Cell Biol. 10: 1265-1269.

147. Auwerx, J., Bouillon, R., Collen, D. and Geboers, J. 1988. Tissue-type plasminogen activator inhibitor activity in diabetes mellitus. Arteriosclerosis 8: 68-72.

148. Paramo, J.A., Colucci, M. and Collen, D. 1985. Plasminogen activator inhibitor in the blood of patients with coronary artery disease. Br. Med. J. 291: 573-574.

149. Oseroff, A., Krishnamurti, C., Hassett, A., Tang, D. and Alving, B. 1989. Plasminogen activator and plasminogen activator inhibitor activities in men with coronary artery disease. J. Lab. Clin. Med. 113: 88-93.

150. Ross, R. 1986. The pathogenesis of artherosclerosis: An update. N. Engl. J. Med. 314: 488-500.

151. von Rokitansky, C. 1852. A manual of pathological anatomy. Sydenham Society, London.

152. Schwartz, C., Valente, A.J., Kelley, J.L., Sprague, E.A., Edwards, E.H. 1988. Thrombosis and the development of atherosclerosis: Rokitansky revisited. Sem. Thromb. Hemost. 14: 189-195.

153. Astrup, T. 1956. Biological significance of fibrinolysis. Lancet 2: 565-570.

154. Ardlie, N.G. and Schwartz, C.J. 1968. A comparison of the organization and fate of autologous pulmonary emboli and of artificial plasma thrombi in the anterior chamber of the eye in normocholesterolemic rabbits. J. Pathol. Bacteriol. 95: 1-18.

155. Naimi, S., Goldenstein, R. and Proger, S. 1963. Studies of coagulation and fibrinolysis of arterial and venous blood in normal subjects and patients with atherosclerosis. Circulation 27: 904-918.

156. Chacrabarti, R., Hocking, E.O. and Fearnley, G.R. 1968. Fibrinolytic activity and coronary-artery disease. Lancet 1: 987-992.

157. Kwaan, H.C. 1979. Physiologic and pharmacologic implications of fibrinolysis. Artery 5: 285-291.

158. Gordon, D., Augustine, A.J., Smith, K.M., Schwartz, S.M. and Wilcox, J.N. 1989. Localization of cells expressing tPA, PAI1, and urokinase by *in situ* hybridization in human atherosclerotic plaques and in the normal rhesus monkey. Thromb. Haemost. 62: 131.

159. Wilcox, J.N., Smith, K.M., Schwartz, S.M. and Gordon, D. 1989. Localization of tissue factor in the normal vessel wall and in the atherosclerotic plaque. Proc. Natl. Acad. Sci. USA 86: 2839-2843.

160. Keski-Oja, J., Raghow, R., Sawdey, M., Loskutoff, D.J. Postlethwaite, A.E., Kang, A.H. and Moses, H.L. 1988. Regulation of mRNAs for type-1 plasminogen activator inhibitor, fibronectin, and type I procollagen by transforming growth factor-ß. J. Biol. Chem. 263: 3111-3115.

161. Lucore, C.L., Fugii, S., Wun, T.C., Sobel, B.E. and Billadello, J.J. 1988. Regulation of the expression of type 1 plasminogen activator inhibitor in Hep G2 cells by epidermal growth factor. J. Biol. Chem. 263: 15845-15848.

162. Michel, J.B. and Quertermous, T. 1989. Modulation of mRNA levels for urinary- and tissue-type plasminogen activator and plasminogen activator inhibitors 1 and 2 in human fibroblasts by interleukin 1. J. Immunol. 143: 890-895.

163. Medina, R., Socher, S.H., Han, J.H. and Friedman, P.A. 1989. Interleukin-1, endotoxin or tumor necrosis factor/cachectin enhance the level of plasminogen activator inhibitor messenger RNA in bovine aortic endothelial cells. Thromb. Res. 54: 41-52.

164. Dichek, D. and Quertermous, T. 1989. Thrombin regulation of mRNA levels of tissue plasminogen activator and plasminogen activator inhibitor-1 in cultured human umbilical vein endothelial cells. Blood 74: 222-228.

PLATELET REACTIONS IN THROMBOSIS

Marian A. Packham

Department of Biochemistry
University of Toronto
Toronto, Canada M5S 1A8

INTRODUCTION

Platelets contribute to atherosclerosis in a number of ways. At sites of injury of arterial walls, they can promote smooth muscle cell proliferation by releasing growth factors[1,2]. Platelet-fibrin thrombi that form at injury sites on vessel walls or on advanced atherosclerotic lesions can be organized and contribute to vessel wall thickening[3-6], and arterial thrombi composed of platelets and fibrin are responsible for clinical complications of atherosclerosis: thromboembolism and occlusive thrombosis[7,8]. Activated platelets in a thrombus can also release materials that cause vasospasm[9].

Several recent reviews of specific aspects of platelet reactions have been published[8,10-13]. This review will be focused on the role of platelets in the formation of arterial thrombi and thromboemboli.

THROMBUS FORMATION

Platelets circulate in a disc shape and do not adhere to the normal, undamaged endothelial surface that lines the blood vessels. However, platelets respond rapidly to any abnormal or unnatural surface that becomes exposed to circulating blood[14-16]. Such surfaces include damaged or altered endothelial cells, the subendothelium, ruptured atherosclerotic plaques, persistent non-occlusive thrombi, and artificial surfaces. When a platelet contacts one of these surfaces, it adheres, puts out pseudopods, and spreads.

If the subendothelium is exposed, a layer of platelets is rapidly deposited from the circulation. A number of internal reactions occur that result in the release of materials that may affect other platelets and other cells in the vicinity[11,13]. The platelets release the contents of their dense granules (ATP, ADP, Ca^{2+}, serotonin) and the contents of their alpha granules which include many proteins: fibrinogen, von Willebrand factor, fibronectin, thrombospondin, platelet factor 4, beta-thromboglobulin, factor V of the intrinsic coagulation pathway, growth factors (PDGF, TGFß), plasminogen activator inhibitor (PAI-1), alpha$_2$-antiplasmin, and many others. Lysosomal granule contents are released only upon strong stimulation. Adherence also stimulates the platelets to mobilize arachidonic acid from membrane phospholipids and form thromboxane A_2. Thromboxane A_2 (TXA_2) and the released ADP and serotonin are all aggregating agents. If blood flow is rapid and laminar as it is in a normal blood vessel, ADP, serotonin and TXA_2 are swept away, but if flow is disturbed, perhaps because of atherosclerotic lesions, these aggregating agents can accumulate and stimulate platelets in the vicinity to aggregate on the platelets that are adherent at the injury site.

Atherosclerosis, Edited by A. I. Gotlieb *et al.*
Plenum Press, New York, 1991

In addition, phospholipids become available on the membrane of activated platelets that have secreted granule contents, and two main steps of the intrinsic pathway of coagulation are greatly accelerated, leading to the formation of thrombin[17]. Thrombin may also be formed through the extrinsic coagulation pathway as a result of tissue factor exposed at the injury site[18]. Thrombin has two important roles in the formation of arterial thrombi. It is a strong aggregating agent for platelets that can act when TXA_2 formation is blocked and ADP is removed, and thrombin converts fibrinogen to fibrin which forms in and around the mass of aggregating platelets and stabilizes it.

Thus, an arterial thrombus is largely made up of aggregated platelets and fibrin, although there may be a few red and white cells in it. Whether the thrombus persists or embolizes depends on lysis of fibrin, deaggregation of the platelets, and the forces of blood flow. In considering these reactions of platelets in thrombus formation in more detail, this review will be focused on adhesion reactions, aggregation, the release of granule contents, and deaggregation. Adhesion and aggregation involve reactions at the platelet surface with adhesive proteins, many of which are released from the platelet alpha granules, in addition to being present in plasma and in the subendothelium.

MEMBRANE GLYCOPROTEINS

All the interactions of platelets begin at their surface, and many involve glycoproteins that span the membrane. Congenital abnormalities have made it possible to determine the function of some of these glycoproteins[19]. Several of the glycoproteins have been identified as receptors for proteins that take part in thrombus formation, and several of these are members of the integrin family (Table 1). Glycoprotein IIb/IIIa (GPIIb/IIIa) is responsible for aggregation and binds fibrinogen, von Willebrand factor and fibronectin[20]. GPIa/IIa is the collagen receptor[21,22], GPIb/IX is the von Willebrand factor receptor[23] and a receptor for thrombin, although it does not appear to be the major thrombin receptor[10,24]. Glycoprotein IV is the receptor for thrombospondin[25,26] which may be involved in stabilizing platelet aggregates by binding to fibrinogen[27]. A number of receptors that we know must exist on the surface of platelets are not yet identified: notably the main receptor for thrombin and the ADP receptor. Also not listed in Table 1 are the receptors for inhibitors of platelet function such as the receptor for prostacyclin that increases cyclic AMP in platelets. When the receptors are occupied they transmit signals to intracellular signalling systems to activate platelet reactions.

Table 1. Membrane Glycoproteins of Platelets that Act as Receptors

Receptor	Ligand
Integrins	
GPIIb/IIIa	Fibrinogen, von Willebrand factor, fibronectin, vitronectin
GPIa/IIa	Collagen
GPIc/IIa	Fibronectin
GPIc'/IIa	Laminin
VnR	Vitronectin
Others	
GPIb/IX	von Willebrand factor (thrombin)
GPV	? (hydrolysed by thrombin)
GPIV (GPIIIb)	Thrombospondin (collagen)

Normally, the non-thrombogenic endothelium covers the adhesive proteins (von Willebrand factor, fibronectin, collagen) that are in the subendothelium and thus the endothelium prevents platelet adhesion to blood vessel walls. However, when the subendothelium is exposed, platelets adhere to its surface. In many studies with experimental animals, the endothelium has been removed by passage of a balloon catheter through the aorta or a large artery. Interactions of platelets with the subendothelium exposed in this way have received a great deal of study, although one must be cautious and remember that they may not be entirely analogous to the reactions that result in thrombus formation on ruptured atherosclerotic plaques.

One device that has been used extensively is the Baumgartner flow chamber in which segments of a de-endothelialized vessel are inverted on a rod in the chamber and blood is allowed to flow past the injured surface[14]. The number of platelets in contact with the surface, the number of platelets that are spread out on the surface, and the size of thrombi are quantitated morphometrically. Alternatively, platelets labelled with [51]chromium or [111]indium have been used for quantitation of adhesion *in vitro* and *in vivo*[28-30], but these radioisotopic techniques measure platelet accumulation and do not distinguish the adhesion of the initial layer of platelets from the formation of platelet-rich thrombi.

It has been recognized for some time that GPIb on the platelet surface is involved in adhesion since patients with a deficiency of this glycoprotein (Bernard-Soulier syndrome) have bleeding problems[19] and fewer platelets adhere in flow chambers[31]. GPIb is a receptor for von Willebrand factor[23] which is required for platelet adhesion under conditions of high shear[14,32]. Most investigators have agreed that if shear forces are low, von Willebrand factor is not involved in platelet adhesion, although Fuster's group[33] has presented evidence indicating that von Willebrand factor may also be involved under low shear conditions. von Willebrand factor is present in the platelet alpha granules, in plasma, and in the subendothelial matrix. It is secreted by stimulated endothelial cells in which it is stored in the Weibel-Palade bodies[34,35]. Work by Sixma's group has shown that although von Willebrand factor in the matrix supports platelet adhesion, it is insufficient for optimal platelet adhesion, and plasma von Willebrand factor is also required for normal adhesion[32,36].

The role of fibronectin, also present in plasma, platelets and the subendothelium, has not been widely studied, but it is clear that it affects platelet adhesion to the subendothelium or the extracellular matrix of cultured endothelial cells[37,38]. Sixma's group has shown that F(ab')$_2$ fragments of an antibody to fibronectin partially inhibit platelet adhesion to the subendothelium[39]. Although fibronectin binding to platelets through GPIIb/IIIa and GPIc/IIa can be inhibited with peptides containing the arginyl-glycyl-aspartyl (RGD) sequence[40,41], such peptides do not inhibit fibronectin-dependent adhesion under flow conditions, an observation that has led Nievelstein and Sixma[42] to conclude that there may be another binding system for the interaction of platelets with fibronectin that may only appear when fibronectin is on a surface.

Fibrinogen, however, does not appear to play a part in the adhesion of the initial layer of platelets to a damaged vessel surface, and adhesion has been reported to be normal with blood from patients with afibrinogenemia[43,44]. In contrast, fibrinogen-coated artificial surfaces do promote platelet adhesion[16].

The fact that platelets adhere to some types of collagen has been recognized for decades[15]. However, it is only recently that the platelet glycoprotein, GPIa/IIa, that is involved in this reaction was identified when patients with a deficiency of this glycoprotein were found because of bleeding problems[45,46]. Platelets are not strongly stimulated by collagen that is in the subendothelium (types IV and V), but they are strongly stimulated by collagen that is deeper in the vessel walls (types I and III) and by the collagen in atherosclerotic plaques, type III. GPIV (CD36[47]) is a primary receptor for platelet-collagen adhesion, but other, GPIV-independent mechanisms may be responsible for subsequent anchorage of the adherent platelets[48]. Adherence to collagen activates the signalling reactions within platelets that result in the formation of TXA_2 and the release of their granule contents[49].

If blood flow is laminar, materials released from adherent platelets cannot accumulate, and only a thin layer of platelets covers the vessel wall, as shown by electron micrographs of vessels from which the endothelium has been removed with a balloon catheter[14,50]. If blood flow is disturbed, as it is at vessel bifurcations and branches, stenoses, or atherosclerotic plaques, a thrombus may form on the adherent platelets. Thus in angioplasty, even if the vessel is badly damaged and presents a surface to which platelets adhere, restoration of laminar flow will reduce the likelihood of rethrombosis[12,51,52].

PLATELET AGGREGATION

Platelet aggregation is dependent on the normal functioning of the GPIIb/IIIa complex, a heterodimer on the platelet surface[19,53]. This complex belongs to the integrin family of receptors that have been identified on the surface of many cells[54]. When this complex is lacking or defective as in thrombasthenia, platelets do not aggregate in response to any aggregating agent, no matter how strong[19]. This complex is not available on unactivated platelets in a conformation that will bind adhesive proteins, so unactivated platelets in the circulation do not aggregate. All the aggregating agents (ADP, TXA_2, thrombin, platelet activating factor) make the GPIIb/IIIa complex available to bind adhesive proteins such as fibrinogen and von Willebrand factor. Aggregation involves bifunctional adhesive proteins that bind to the GPIIb/IIIa complexes on adjacent platelets and form bridges between the activated platelets[20,55].

The first protein that was shown to bind to activated platelets was fibrinogen. In the late 1970's, we[56] and then others[57] used ^{125}I-labelled fibrinogen to show that within a few seconds of the addition of ADP, fibrinogen is bound to platelets as they aggregate and then dissociates during deaggregation. Other investigators have studied equilibrium binding at 30 minutes and determined that there are 38,000 ± 10,000 receptors for fibrinogen on each activated platelet[58,59]. With platelets from patients with thrombasthenia, lacking the GPIIb/IIIa complex, the labelled fibrinogen does not bind although the platelets change shape in response to ADP, indicating that they are activated[57,60].

Fibrinogen is a glycoprotein that occurs in plasma and in the alpha granules of platelets. It is made up of three pairs of nonidentical subunits, Aα, Bβ and gamma, covalently linked by disulfide bridges. Peptides containing two different amino acid sequences in fibrinogen can bind to GPIIb/IIIa; these are the RGD sequences in the Aα chains and a dodecapeptide near the carboxyl terminus (400-411) of each gamma chain[55]. The RGD sequences in the Aα chains are in RGDF at amino acid sequences Aα95-98 and in RGDS at Aα572-575. However, the latter RGD sequence may not be necessary for fibrinogen binding to platelets[61]. Recent evidence from experiments with antibodies raised against RGDF and RGDS indicate that the RGDF sequence at Aα95-98 and the gamma-chain 400-411 sequence are two recognition sites that interact with the same site, or mutually exclusive sites, on GPIIb/IIIa[62]. Recently, the amino acid sequences of the parts of GPIIb and GPIIIa to which these peptides bind have been identified[63,64]. RGD binds to residues 109-171 of GPIIIa and the dodecapeptide binds to residues 294-314 on GPIIb. One current concept is that fibrinogen binds in a prone, rather than an upright, position so that multiple attachments to the platelet surface are possible[55].

Without fibrinogen in the medium, platelets aggregate poorly in response to aggregating agents such as ADP that do not cause the release of granule contents[65,66], but aggregation in response to release-inducing agonists such as thrombin or collagen does not require exogenous fibrinogen. It is thought that fibrinogen released from the alpha granules takes part in aggregation, although some of it may remain bound at the site of discharge[67-69].

Other molecules that contain the RGD sequence also bind to the GPIIb/IIIa complex on activated platelets; these include von Willebrand factor, fibronectin and vitronectin. Of these, von Willebrand factor has received considerable attention because of reports that under some conditions it can substitute for fibrinogen in binding platelets together[70,71]. Most aggregation studies have shown that at the concentrations present in normal plasma, fibrinogen competes successfully for the receptor[72-74]. In addition, we found that von Willebrand factor, even at a concentration three-fold greater than that in plasma, was ineffective in supporting ADP-induced aggregation when the concentration of Ca^{2+} in the suspending medium was in the physiological range of 1 to 2 mM[65]. However, there is some evidence that, under conditions of very high shear such as might occur at a stenosis in an atherosclerotic vessel, von Willebrand factor may be the adhesive protein that binds platelets together through GPIIb/IIIa[43,44].

Various small peptides containing the RGD sequence block fibrinogen (or von Willebrand factor) binding to GPIIb/IIIa[75-77]. Consequently, these peptides inhibit aggregation, particularly in response to low concentrations of aggregating agents. Some naturally occurring peptides such as trigramin from a snake venom are inhibitory because they contain this RGD sequence, and they are attracting attention as possible short-acting antithrombotic agents[78]. The dodecapeptide of the gamma chain of fibrinogen also inhibits hemostatic plug formation in small branches of the rabbit mesenteric artery[79.] Antibodies to the GPIIb/IIIa complex such as those produced by Coller[80-82] have had limited tests in experimental animals and in man as inhibitors of thrombosis.

Although current dogma is that fibrinogen is an absolute requirement for platelet aggregation, there are several pieces of evidence that fibrinogen may not be required for platelet aggregation in response to strong agonists:

(i) Several groups of investigators have shown that platelets from afibrinogenemic patients, who have practically no fibrinogen in their plasma or platelets will aggregate well in response to thrombin, even in artificial media[76,83-85]. They also release a large proportion of their granule contents[83]. The amount of von Willebrand factor that these platelets could release is insufficient to support aggregation unless one postulates that enough of it remains bound to the platelets at the point of discharge to give a high local concentration.

(ii) Direct visualization of fibrinogen using an immunogold labelling technique also shows many points where aggregated platelets are in contact with each other without any fibrinogen visible between them[69]. In these experiments, fibrinogen was evident only within the alpha granules and at sites of discharge of granule contents.

(iii) Our studies, and those of others, with platelets degranulated by treatment with thrombin also indicate that little or no fibrinogen may be required for aggregation in response to strong agonists[86,87]. These platelets released over 90% of their dense granule contents and a similar percentage of their alpha granule contents. Although fibrinogen has to be added to support the aggregation of these degranulated platelets in response to weaker agonists such as platelet activating factor, arachidonic acid, or the TXA_2 mimetic U46619, these platelets will aggregate without added fibrinogen in response to a high concentration of thrombin or the calcium ionophore A23187[87].

All of these experiments seem to indicate that there may be ways in which platelets are linked to each other that do not involve fibrinogen binding, and possibly not von Willebrand factor binding either. What this linkage may be, we do not know. It is evident, however, that GPIIb/IIIa is required.

Another adhesive protein, thrombospondin, released from platelet alpha granules, has been implicated in stabilization of the binding of fibrinogen to platelets[27]. Thrombospondin is a lectin-like molecule whose receptor on the platelet surface has recently been identified as GPIV[25], also known as GPIIIb[26].

A granule membrane protein with a molecular weight of 140 (GMP-140) described by McEver and his colleagues[88], which is the PADGEM (platelet activation-dependent granule-external membrane) protein found by Furie's group[89], has been identified by monoclonal antibodies on the surface of platelets that have undergone a release reaction, and also on the surface of stimulated endothelial cells[90,91] (thus the GMP-140 name is more appropriate because the molecule is not restricted to platelets). GMP-140 is on the inside of the membrane of the alpha granules of platelets and on the inside of the Weibel-Palade bodies of endothelial cells[91]. (The Weibel-Palade bodies contain von Willebrand factor.) When the contents of the granules are released, GMP-140 becomes part of the external surface of the cells. Labelled antibodies to GMP-140 have been used to detect platelets in the circulation that have undergone a release reaction[92] or been incorporated into thrombi[93]. Fluorescent antibodies to GMP-140 are being used with flow cytometry to identify and quantify platelets that have undergone a release reaction[94,95].

Leukocytes and macrophages have receptors for GMP-140 and it is implicated in the adhesion of these cells to stimulated endothelium[96]. Stimulated platelets also bind to leukocytes through GMP-140[97,98], which may be responsible for the recruitment of leukocytes into some thrombi. GMP-140 is homologous with ELAM-1 and Mel 14 which are other members of the selectin family of receptors that mediate the interaction of leukocytes with other cells of the vascular system[99].

STIMULUS-RESPONSE COUPLING

Up to this point, only the reactions that occur at the platelet surface and the release of the contents of platelet granules have been outlined. What goes on inside platelets when they are stimulated to cause them to change shape; form pseudopods; centralize their granules and then release the granule contents; mobilize arachidonate from membrane phospholipids and form TXA_2; make GPIIb/IIIa available as a receptor for adhesive proteins; change the phospholipids that are exposed on their surface to ones that form part of the tenase and prothrombinase complexes of the intrinsic coagulation pathway; and eventually take part in clot retraction and the consolidation of the thrombus?

Stimulus-response coupling in platelets has been most extensively studied in experiments in which thrombin has been used as the agonist[10], and extrapolation of the results to the effects of other aggregating agents may not always be justified. Thrombin is the strongest aggregating and release-inducing agent to which platelets are likely to be exposed in vivo, and it is being increasingly recognized as important in arterial thrombosis under some circumstances. Internal changes in platelets upon stimulation with thrombin appear to be similar to the reactions to other stimuli that have been observed in many other types of cells.

The major receptor for the extracellular signal, thrombin, has not been identified[10,24]. Although thrombin interacts with GPIb, if this glycoprotein is absent or defective, as in the Bernard-Soulier syndrome, platelets do aggregate in response to thrombin, although their responsiveness and their ability to bind thrombin is somewhat decreased[100]. Thrombin hydrolyses GPV[101], but the rate is too slow to account for the rapid responses of platelets to thrombin. Nevertheless, the proteolytic activity of thrombin may be important because thrombin whose proteolytic activity has been blocked binds to platelets in the same amounts as native thrombin, but does not induce aggregation or the release of granule contents[102]. Thus, the binding of thrombin is not sufficient to initiate the reactions it causes. Whatever the main receptor may be, it appears that a guanine nucleotide binding G protein (G_p) couples the receptor to a phospholipase C which is specific for phosphoinositides. Phosphatidylinositol bisphosphate (PIP_2) is hydrolysed to two second messengers, inositol 1,4,5-trisphosphate (IP_3) and diacylglycerol. These messenger molecules stimulate different reactions that contribute synergistically to platelet responses. IP_3 causes an increase in cytoplasmic free Ca^{2+} and diacylglycerol activates protein kinase C, both of which are responsible for platelet responses[10].

The concentration of Ca^{2+} in the cytoplasm is raised from approximately 100 nM to micromolar concentrations as a result of the IP_3-mediated mobilization of Ca^{2+} from a sequestered pool in the dense tubular system, and also as a result of translocation of Ca^{2+} from the external medium by a mechanism that probably involves a receptor-operated channel and may be mediated by inositol 1,3,4,5-tetrakisphosphate (IP_4)[10,103]. The increase in cytoplasmic Ca^{2+} can be measured by preloading the platelets with a fluorophore such as quin 2, fura 2 or indo 1, whose fluorescence intensities increase when the cytosolic Ca^{2+} increases in response to stimulation. The concentration of Ca^{2+} remains elevated while active thrombin is present[104].

Increased cytoplasmic Ca^{2+} has a number of effects[10]. It stimulates a Ca^{2+}/calmodulin-dependent protein kinase which phosphorylates the 20 kDa myosin light chain[105]. It activates Ca^{2+}-dependent proteases (notably calpain), leading to cytoskeletal reorganization. It activates phospholipase A_2, thus freeing arachidonic acid which is converted to TXA_2, an aggregating agent. Cytosolic Ca^{2+} also has roles in activating phospholipase C and protein kinase C. The activation of protein kinase C by diacylglycerol occurs at the inner aspect of the platelet membrane in the presence of Ca^{2+} and phosphatidylserine. Activated protein kinase C phosphorylates a 47 kDa protein which may modulate actin polymerization and contribute to the cytoskeletal reorganization needed for secretion of granule contents[105]. Activated protein kinase C may also enhance the activity of phospholipase A_2 by phosphorylating and thus inhibiting lipocortin which, in its unphosphorylated state, antagonizes phospholipase A_2[10].

It should be emphasized that platelet aggregation and the release of granule contents in response to thrombin are not dependent on the formation of TXA_2 or on ADP released from the dense granules. Extensive aggregation and release of granule contents occur in the presence of aspirin, which block thromboxane A_2 formation, plus creatine phosphate/creatine phosphokinase (CP/CPK) which rapidly converts released ADP to ATP[106]. Thus, thrombin can act independently of thromboxane A_2 and released ADP, although they do make some contribution with very low concentrations of thrombin. Consequently, when thrombin has a major role in thrombus formation, one would predict that blocking thromboxane A_2 formation with aspirin will have very little inhibitory effect.

Responses to ADP, a weak agonist, are somewhat different than the responses to thrombin. The observations reviewed here are those that occur when the concentration of Ca^{2+} in the suspending medium is in the physiological range (1 to 2 mM). Under this condition, only the primary aggregation response occurs; it is readily reversible and does not involve the formation of TXA_2 or the release of granule contents[107]. It should be pointed out, however, that at micromolar concentrations of external Ca^{2+}, such as those in citrated platelet-rich plasma, more extensive reactions occur[107], but results obtained under these unphysiological conditions will not be discussed here.

The receptor for ADP is unidentified. Since granule contents are not released from platelets stimulated with ADP, the presence of external fibrinogen is required for extensive aggregation when platelets have been freed from plasma. In keeping with the observation that TXA_2 is not produced during the primary aggregation response to ADP, is the finding that detectable IP_3 is not formed when platelets are stimulated with ADP[108,109]. The kinetics of the rise in cytosolic Ca^{2+} is quite different from that observed with thrombin. The concentration of Ca^{2+} in the cytosol rises within a fraction of a second[104,110] and this increase is delayed by about 200 ms in the absence of external Ca^{2+}[110]. It occurs without detectable aggregation. In contrast to the increase in Ca^{2+} in response to thrombin, the increase in Ca^{2+} in platelets stimulated with ADP reaches its peak in 2 seconds and then declines[110]. It is puzzling that the increase in cytoplasmic Ca^{2+} induced by ADP does not result in the activation of phospholipase A_2, as evidenced by the lack of formation of thromboxane A_2, or the release of granule contents. Possibly, a more sustained rise in cytosolic Ca^{2+} is required to initiate these reactions.

Aggregation in response to collagen is mediated by the synergistic effects of TXA_2 and released ADP[106,111]. Inhibition of TXA_2 formation with aspirin is partially inhibitory; removal of released ADP with CP/CPK is also partially inhibitory; but together these inhibitors essentially abolish the response to collagen. These results indicate that TXA_2 has a major role in collagen-induced aggregation. Although both thromboxane mimetics and collagen activate phospholipase C[49,112], it appears that the other consequences of phospholipase C activation are relatively unimportant in response to agonists that act through the thromboxane receptor.

Contrary to the reports in the literature, epinephrine does not function as an aggregating agent if the concentration of Ca^{2+} in the medium is in the physiological range[113,114]. The major role of epinephrine is its strong potentiation of the effects of other aggregating agents[115].

STABILIZATION OF PLATELET AGGREGATES

Some time ago, our group began investigating ways of deaggregating platelets to try to determine what contributes to the stability of platelet aggregates[116]. To break up an arterial thrombus, in addition to lysing fibrin with fibrinolytic agents, the platelet-to-platelet bonds also have to be broken. When platelets are aggregated without the release reaction occurring, for example by ADP, deaggregation can be readily induced by the addition of prostaglandin E_1 or prostacyclin to raise cyclic AMP levels and cause sequestration of internal Ca^{2+}, by the addition of EDTA to chelate external Ca^{2+} and Mg^{2+} and thus prevent the binding of adhesive proteins by the GPIIb/IIIa complex, or by the addition of enzyme systems that remove ADP, such as CP/CPK or apyrase[117]. However, when platelets are aggregated by agents such as thrombin that induce the release of the contents of platelet granules, deaggregation is not readily induced by any of these inhibitors. The ease with which deaggregation can be achieved depends on the concentration of thrombin, the length of time it is allowed to act, and the length of time that elapses before the addition of inhibitors that disrupt the aggregates. These observations are relevant to the recognized need to administer plasminogen activators and inhibitors of thrombin and of platelet function as soon as possible after a thrombus has formed to be successful in disrupting the thrombus.

In the early studies with human platelets[117], it was found that to deaggregate the platelets after extensive thrombin-induced release of granule contents had occurred, it was necessary to add a combination of inhibitors to the artificial medium in which the platelets were suspended. This combination was composed of a high concentration of heparin (added before thrombin), CP/CPK or prostaglandin E_1 (to inhibit the effect of ADP), and a proteolytic enzyme such as chymotrypsin or plasmin. There appeared to be a heparin-sensitive reaction that may involve the binding of released proteins with which heparin interacts, such as platelet factor 4, thrombospondin, or fibronectin.

Further experiments on the factors that contribute to the stability of platelet aggregates have been reported recently[118]. The studies were made possible by the availability of patients with platelets with delta storage pool deficiency which do not have releasable dense granule contents (ADP, serotonin) and a unique patient, designated VR, whose platelets have a severe selective impairment of platelet aggregation and fibrinogen binding induced by ADP although they respond normally to other aggregating agents. In these experiments, a combination of hirudin, PGE_1 and chymotrypsin, added 2 minutes after thrombin, did not deaggregate normal control platelets: it did deaggregate the delta storage pool deficient platelets and the platelets from patient VR. In investigating the possibility that released ADP might be responsible for the stability of the aggregates of control platelets under these conditions, it was found that the addition of ADP prevented deaggregation of the delta storage pool deficient platelets, but did not affect the instability of VR's platelets that are insensitive to ADP. Furthermore, the addition of apyrase (to degrade ADP released from the control platelets) made it possible to deaggregate the control platelets with the combination of inhibitors. The conclusion from these experiments is that released ADP helps to stabilize platelet aggregates when the release reaction has occurred. This stabilizing effect is specific for ADP since serotonin and epinephrine did not stabilize the aggregates[118]. Other investigators have shown that prostaglandin E_1, an inhibitor of the effect of ADP on platelets, accelerates thrombolysis by tissue plasminogen activator[119,120].

Thus, without fibrin, a stable platelet aggregate requires: GPIIb/IIIa availability, the release of the dense granule contents (principally ADP), and probably the release of proteins from the alpha granules. It is not known whether there is also a role for GMP-140, the granule membrane glycoprotein that becomes expressed on the surface of platelets that have undergone a release reaction.

PLATELETS AND FIBRINOLYSIS

An arterial thrombus that forms *in vivo* has both aggregated platelets that have undergone the release reaction, and fibrin that has formed when thrombin acts on fibrinogen. If an arterial thrombus is to be disrupted, two things must be achieved: the fibrin must be lysed and the platelets must be deaggregated. There are a number of relationships between these processes that affect the stability of thrombi and the efficacy of fibrinolytic agents, both those produced normally and those introduced in thrombolytic therapy[121]. Platelets impair clot lysis in several ways: by clot retraction which consolidates the thrombus and makes the fibrin in it less available to fibrinolytic agents such as tissue plasminogen activator and streptokinase; by providing factor XIII which cross links fibrin and reduces its sensitivity to lysis by plasmin; and by releasing plasminogen activator inhibitor, PAI-1, and alpha$_2$-antiplasmin. Platelets can also be activated by plasmin[122,123]. On the other hand, platelets can also localize plasminogen in several ways and thus promote clot lysis. *In vitro* experiments have shown that platelets can exert both profibrinolytic and antifibrinolytic effects, so prediction of the *in vivo* situation is difficult and as yet unresolved[121]. The thrombolytic agents may paradoxically contribute to platelet-mediated reocclusion by their platelet activating effects. Thus, it is not surprising that evidence from experiments with animals and from clinical trials indicates that the combination of various inhibitors of platelet reactions with fibrinolytic agents increases the success of reperfusion of thrombosed atherosclerotic or injured blood vessels[121,124-128].

Which inhibitors of platelets will be most effective in arterial thrombosis? This depends on the reactions that play the major part in the formation of a thrombus. Some thrombi form largely under the influence of TXA_2 and can be inhibited with non-steroidal anti-inflammatory drugs such as aspirin. The beneficial effect of aspirin in carefully selected male patients with unstable angina probably provides an example of a situation in which TXA_2 formation is important in thrombus formation[129,130]. ADP may initiate and help to stabilize some thrombi. Other thrombi may be most dependent on thrombin for their formation and, if so, heparin or other inhibitors of thrombin would be expected to be inhibitory[131,132].

Further understanding of platelet reactions in thrombosis may make it possible to deal more effectively with the thrombi that are responsible for clinical complications of atherosclerosis.

REFERENCES

1. Ross, R. 1986. The pathogenesis of atherosclerosis - an update. New Engl. J. Med. 314: 488-500.

2. Moore, S. 1981. Injury mechanisms in atherogenesis. In: Moore, S., ed., Vascular injury and atherosclerosis. New York: Marcel Dekker, 131-148.

3. Duguid, J.B. 1946. Thrombosis as a factor in the pathogenesis of coronary atherosclerosis. J. Pathol. Bacteriol. 58: 207-212.

4. Jorgensen, L., Rowsell, H.C., Hovig, T. and Mustard, J.F. 1967. Resolution and organization of platelet-rich mural thrombi in carotid arteries of swine. Am. J. Pathol. 51: 681-693.

5. Woolf, N. 1987. Thrombosis and atherosclerosis. In: Bloom, A.L. and Thomas, D.P., eds. Haemostasis and thrombosis. Second Edition. New York: Churchill Livingstone, 651-678.

6. Faxon, D.P., Sanborn, T.A., Weber, V.J., *et al.* 1984. Restenosis following transluminal angioplasty in experimental atherosclerosis. <u>Arteriosclerosis</u> 4: 189-195.

7. Davies, M.J. 1987. Thrombosis in acute myocardial infarction and sudden death. <u>Cardiovasc. Clin.</u> 18: 151-159.

8. Packham, M.A. and Mustard, J.F. 1986. The role of platelets in the development and complications of atherosclerosis. <u>Semin. Hematol.</u> 23: 8-26.

9. Willerson, J.T., Golino, P., Eidt, J., Campbell, W.B. and Buja, L.M. 1989. Specific platelet mediators and unstable coronary artery lesions. Experimental evidence and potential clinical implications. <u>Circulation</u> 80: 198-205.

10. Kroll, M.H. and Schafer, A.I. 1989. Biochemical mechanisms of platelet activation. <u>Blood</u> 74: 1181-1195.

11. Siess, W. 1989. Molecular mechanisms of platelet activation. <u>Physiol. Rev.</u> 69: 58-178.

12. Mustard, J.F., Packham, M.A. and Kinlough-Rathbone, R.L. 1990. Platelets, blood flow, and the vessel wall. <u>Circulation</u> 81 (Suppl I): I-24-I-27.

13. Holmsen, H. 1989. Physiological functions of platelets. <u>Ann. Med.</u> 21: 23-30.

14. Baumgartner, H.R. and Muggli, R. 1976. Adhesion and aggregation: Morphological demonstration and quantitation *in vivo* and *in vitro*. <u>In</u>: Gordon, J.L., ed., Platelets in biology and pathology. Oxford: Elsevier/North Holland, 23-60.

15. Packham, M.A. and Mustard, J.F. 1984. Platelet adhesion. <u>In</u>: Spaet, T.H., ed. Progress in hemostasis and thrombosis. Vol. 7. New York: Grune & Stratton, 211-288.

16. Packham, M.A. 1988. The behaviour of platelets at foreign surfaces. <u>Proc. Soc. Exp. Biol. Med.</u> 189: 261-274.

17. Tracy, P.B. and Mann, K.G. 1986. Platelet involvement in coagulation. <u>In</u>: Holmsen, H., ed., Platelet responses and metabolism. Vol. I. Boca Raton: CRC Press, 297-324.

18. Nemerson, Y. 1988. Tissue factor and hemostasis. <u>Blood</u> 71: 1-8.

19. Nurden, A.T., George, J.N. and Phillips, D.R. 1986. Platelet membrane glycoproteins: Their structure, function, and modification in disease. <u>In</u>: Phillips, D.R. and Shuman, M., eds., Biochemistry of platelets. 159-224.

20. Plow, E.F., Srouji, A.H., Meyer, D., Marguerie, G. and Ginsberg, M.H. 1984. Evidence that three adhesive proteins interact with a common recognition site on activated platelets. <u>J. Biol. Chem.</u> 259: 5388-5391.

21. Staatz, W.D., Rajpara, S.M., Wayner, E.A., Carter, W.G. and Santoro, S.A. 1989. The membrane glycoprotein Ia-IIa (VLA-2) complex mediates the Mg^{++}-dependent adhesion of platelets to collagen. <u>J. Cell Biol.</u> 108: 1917-1924.

22. Coller, B.S., Beer, J.H., Scudder, L.E. and Steinberg, M.H. 1989. Collagen-platelet interactions: Evidence for a direct interaction of collagen with platelet GPIa/IIa and an indirect interaction with platelet GPIIb/IIIa mediated by adhesive proteins. <u>Blood</u> 74: 182-192.

23. Ruggeri, Z.M. and Zimmerman, T.S. 1987. von Willebrand factor and von Willebrand disease. <u>Blood</u> 70: 895-904.

24. Phillips, D.R. 1985. Receptors for platelet agonists. In: George, J.N., Nurden, A.T. and Phillips, D.R., eds., Platelet membrane glycoproteins. New York: Plenum Press, 145-169.

25. Asch, A.S., Barnwell, J., Silverstein, R.L. and Nachman, R.L. 1987. Isolation of the thrombospondin membrane receptor. J. Clin. Invest. 79: 1054-1061.

26. McGregor, J.L., Catimel, B., Parmentier, S., Clezardin, P., Dechavanne, M., Leung, L.L.K. 1989. Rapid purification and partial characterization of human platelet glycoprotein IIIb. Interaction with thrombospondin and its role in platelet aggregation. J. Biol. Chem. 264: 501-506.

27. Silverstein, R.L., Leung, L.L.K. and Nachman, R.L. 1987. Thrombospondin: A versatile multifunctional glycoprotein. Arteriosclerosis 6: 245-253.

28. Cazenave, J.-P., Blondowska, D., Richardson, M., Kinlough-Rathbone, R.L., Packham, M.A. and Mustard, J.F. 1979. Quantitative radioisotopic measurement and scanning electron microscopic study of platelet adherence to a collagen-coated surface and to subendothelium with a rotating probe device. J. Lab. Clin. Med. 93: 60-70.

29. Hanson, S.R., Harker, L.A. 1985. Studies of suloctidil in experimental thrombosis in baboons. Thromb. Haemost. 53: 423-427.

30. Badimon, L., Badimon, J.J., Turitto, V.T., Rand, J. and Fuster, V. 1989. Functional behaviour of vessels from pigs with von Willebrand disease. Values of platelet deposition are identical to those obtained on normal vessels. Arteriosclerosis 9: 184-188.

31. Sakariassen, K.S., Nievelstein, P.F.E.M., Coller, B.S. and Sixma, J.J. 1986. The role of platelet membrane glycoproteins Ib and IIb-IIIa in platelet adherence to human artery subendothelium. Br. J. Haematol. 63: 681-691.

32. de Groot, P.G. and Sixma, J.J. 1987. Role of von Willebrand factor in the vessel wall. Semin. Thromb. Hemost. 13: 416-424.

33. Badimon, L., Badimon, J.J., Turitto, V.T. and Fuster, V. 1989. Role of von Willebrand factor in mediating platelet-vessel wall interaction at low shear rate; the importance of perfusion conditions. Blood 73: 961-967.

34. Levine, J.D., Harlan, J.M., Harker, L.A., Joseph, M.L. and Counts, R.B. 1982. Thrombin-mediated release of factor VIII antigen from human umbilical vein endothelial cells in culture. Blood 60: 531-534.

35. Ribes, J.A., Francis, C.W. and Wagner, D.D. 1987. Fibrin induces release of von Willebrand factor from endothelial cells. J. Clin. Invest. 79: 117-123.

36. Stel, H.V., Sakariassen, K.S., de Groot, P.G., van Mourik, J.A. and Sixma, J.J. 1985. von Willebrand factor in the vessel wall mediates platelet adherence. Blood 65: 85-90.

37. Bastida, E., Escolar, G., Ordinas, A. and Sixma, J.J. 1987. Fibronectin is required for platelet adhesion and for thrombus formation on subendothelium and collagen surfaces. Blood 70: 1437-1442.

38. Wayner, E.A., Carter, W.G., Piotrowicz, R.S. and Kunicki, T.J. 1988. The function of multiple extracellular matrix receptors in mediating cell adhesion to extracellular matrix: Preparation of monoclonal antibodies to the fibronectin receptor that specifically inhibit cell adhesion to fibronectin and react with platelet glycoproteins Ic-IIa. J. Cell Biol. 107: 1881-1891.

39. Houdijk, W.P.M. and Sixma, J.J. 1985. Fibronectin in artery subendothelium is important for platelet adhesion. Blood 65: 598-604.

40. Ginsberg, M., Pierschbacher, M.D., Ruoslahti, E., Marguerie, G., Plow, E. 1985. Inhibition of fibronectin binding to platelet by proteolytic fragments and synthetic peptides which support fibroblast adhesion. J. Biol. Chem. 260: 3931-3936.

41. Piotrowicz, R.S., Orchekowski, R.P., Nugent, D.J., Yamada, K.Y. and Kunicki, T.J. 1988. Glycoprotein Ic-IIa functions as an activation-independent fibronectin receptor on human platelets. J. Cell Biol. 106: 1359-1364.

42. Nievelstein, P.F.E.M. and Sixma, J.J. 1988. Glycoprotein IIB-IIIA and RGD(S) are not important for fibronectin-dependent platelet adhesion under flow conditions. Blood 72: 82-88.

43. Turitto V.T., Weiss, H.J., Baumgartner, H.R. 1984. Platelet interaction with rabbit subendothelium in von Willebrand's disease: Altered thrombus formation distinct from defective platelet adhesion. J. Clin. Invest. 74: 1730- 1741.

44. Weiss, H.J., Hawiger, J., Ruggeri, Z.M., Turitto, V.T., Thiagarajan, P. and Hoffmann, T. 1989. Fibrinogen-independent platelet adhesion and thrombus formation on subendothelium mediated by glycoprotein IIb-IIIa complex at high shear rate. J. Clin. Invest. 83: 288-297.

45. Nieuwenhuis, H.K., Akkerman, J.W.N., Houdijk, W.P.M. and Sixma, J.J. 1985. Human blood platelets showing no response to collagen fail to express surface glycoprotein Ia. Nature 318: 470-472.

46. Kehrel, B., Balleisen, L., Kokott, R., et al. 1988. Deficiency of intact thrombospondin and membrane glycoprotein Ia in platelets with defective collagen-induced aggregation and spontaneous loss of disorder. Blood 71: 1074-1078.

47. Knapp, W., Dörken B., Rieber, P., Schmidt, R.E., Stein, H. and von dem Borne, A.E.G.K. 1989. CD antigens. Blood 74: 1448-1450.

48. Tandon, N.N., Kralisz, U. and Jamieson, G.A. 1989. Identification of glycoprotein IV (CD36) as a primary receptor for platelet-collagen adhesion. J. Biol. Chem. 264: 7576-7583.

49. Nakano, T., Hanasaki, K. and Arita, H. 1989. Possible involvement of cytoskeleton in collagen-stimulated activation of phospholipases in human platelets. J. Biol. Chem. 264: 5400-5406.

50. Groves, H.M., Kinlough-Rathbone, R.L., Richardson, M., Moore, S. and Mustard, J.F. 1979. Platelet interaction with damaged rabbit aorta. Lab. Invest. 40: 194-200.

51. O'Neill, W., Timmis, G.C., Bourdillon, P.D., et al. 1986. A prospective randomized clinical trial of intracoronary streptokinase versus coronary angioplasty for acute myocardial infarction. New Engl. J. Med. 314: 812-818.

52. Ellis, S.G., Roubin, G.S., King, S.B. III, Douglas, J.S. Jr. and Cox, W.R. 1989. Importance of stenosis morphology in the estimation of restenosis risk after elective percutaneous transluminal coronary angioplasty. Am. J. Cardiol. 63: 30-34.

53. Phillips, D.R., Charo, I.F., Parise, L.V. and Fitzgerald, L.A. 1988. The platelet membrane glycoprotein IIb-IIIa complex. Blood 71: 831-843.

54. Kunicki, T.J. 1989. Platelet membrane glycoproteins and their function: An overview. Blut 59: 30-34.

55. Hawiger, J. 1987. Formation and regulation of platelet and fibrin hemostatic plug. Hum. Pathol. 18: 111-122.

56. Mustard, J.F., Packham, M.A., Kinlough-Rathbone, R.L., Perry, D.W. and Regoeczi, E. 1978. Fibrinogen and ADP-induced platelet aggregation. Blood 52: 453-466.

57. Peerschke, E.I., Zucker, M.B., Grant, R.A., Egan, J.J., Johnson, M.M. 1990. Correlation between fibrinogen binding to human platelets and platelet aggregability. Blood 55: 841-847.

58. Bennett, J.S. and Vilaire, G. 1979. Exposure of platelet fibrinogen receptors by ADP and epinephrine. J. Clin. Invest. 64: 1393-1401.

59. Marguerie, G., Ginsberg, M. and Plow, E.F. 1987. Glycoproteins: The fibrinogen receptor. In: Holmsen, H., ed., Platelet responses and metabolism. Vol III. Response-metabolism relationships. Boca Raton: CRC Press, 285-296.

60. Mustard, J.F., Kinlough-Rathbone, R.L., Packham, M.A., Perry, D.W., Harfenist, E.J. and Pai, K.R.M. 1979. Comparison of fibrinogen association with normal and thrombasthenic platelets on exposure to ADP or chymotrypsin. Blood 54: 987-993.

61. Plow, E.F., Pierschbacher, M.D., Ruoslahti, E., Marguerie, G. and Ginsberg, M.H. 1987. Arginyl-glycyl-aspartic acid sequences and fibrinogen binding to platelets. Blood 70: 110-115.

62. Andrieux, A., Hudry-Clergeon, G., Ryckewaert, J.-J., et al. 1989. Amino acid sequences in fibrinogen mediating its interaction with its platelet receptor, GPIIbIIIa. J. Biol. Chem. 264: 9258-9265.

63. D'Souza, S.E., Ginsberg, M.H., Burke, T.A., Lam, S.C.-T. and Plow, E.F. 1988. Local-ization of an Arg-Gly-Asp recognition site within an integrin adhesion receptor. Science 242: 91-93.

64. D'Souza, S.E., Ginsberg, M.H. and Plow, E.F. 1990. A discrete region of the platelet integrin, GPIIb-IIIa involved in ligand recognition. FASEB J. 4: A892.

65. Harfenist, E.J., Packham, M.A., Kinlough-Rathbone, R.L., Cattaneo, M. and Mustard, J.F. 1987. Effect of calcium ion concentration on the ability of fibrinogen and von Willebrand factor to support the ADP-induced aggregation of human platelets. Blood 70: 827-831.

66. Suzuki, H., Kinlough-Rathbone, R.L., Packham, M.A., Tanoue, K., Yamazaki, H. and Mustard, J.F. 1988. Immunocytochemical localization of fibrinogen on washed human platelets. Lack of requirement for fibrinogen during adenosine diphosphate-induced responses and enhanced fibrinogen binding in a medium with low calcium levels. Blood 71: 850-860.

67. Legrand, C., Dubernard, V. and Nurden, A.T. 1989. Studies on the mechanism of expression of secreted fibrinogen on the surface of activated human platelets. Blood 73: 1226-1234.

68. Courtois, G., Ryckewaert, J.-J., Woods, V.L., Ginsberg, M.H., Plow, E.F. and Marguerie, G.A. 1986. Expression of intracellular fibrinogen on the surface of stimulated platelets. Eur. J. Biochem. 159: 61-67.

69. Suzuki, H., Kinlough-Rathbone, R.L., Packham, M.A., Tanoue, K., Yamazaki, H. and Mustard, J.F. 1988. Immunocytochemical localization of fibrinogen during thrombin-induced aggregation of washed human platelets. Blood 71: 1310-1320.

70. Fujimoto, T. and Hawiger, J. 1982. Adenosine diphosphate induces binding of von Willebrand factor to human platelets. Nature 297: 154-156.

71. De Marco, L., Girolami, A., Zimmerman, T.S. and Ruggeri, Z.M. 1986. von Wille-brand factor interaction with the glycoprotein IIb/IIIa complex. Its role in platelet

function as demonstrated in patients with congenital afibrinogenemia. J. Clin. Invest. 77: 1272-1277.

72. Pietu, G., Cherel, G., Marguerie, G. and Meyer, D. 1984. Inhibition of von Willebrand factor-platelet interaction by fibrinogen. Nature 308: 648-649.

73. Gralnick, H.R., Williams, S.B. and Coller, B.S. 1984. Fibrinogen competes with von Willebrand factor for binding to the glycoprotein IIb/IIIa complex when platelets are stimulated with thrombin. Blood 64: 797-800.

74. Schullek, J., Jordan, J. and Montgomery, R.R. 1984. Interaction of von Willebrand factor with human platelets in the plasma milieu. J. Clin. Invest. 73: 421-428.

75. Gartner, T.K. and Bennett, J.S. 1985. The tetrapeptide analogue of the cell attachment site of fibronectin inhibits platelet aggregation and fibrinogen binding to activated platelets. J. Biol. Chem. 260: 11891-11894.

76. Haverstick, D.M., Cowan, J.F., Yamada, K.M. and Santoro, S.A. 1985. Inhibition of platelet adhesion to fibronectin, fibrinogen, and von Willebrand factor substrates by a synthetic tetrapeptide derived from the cell-binding domain of fibronectin. Blood 66: 946-952.

77. Plow, E.F., Pierschbacher, M.D., Ruoslahti, E., Marguerie, G.A. and Ginsberg, M.H. 1985. The effect of Arg-Gly-Asp-containing peptides on fibrinogen and von Willebrand factor binding to platelets. Proc. Natl. Acad. Sci. USA 82: 8057-8061.

78. Cook, J.J., Huang, T.-F., Rucinski, B., et al. 1989. Inhibition of platelet hemostatic plug formation bytrigramin, a novel RGD-peptide. Am. J. Physiol. 256: H1038-H1043.

79. Kloczewiak, M., Timmons, S., Bednarek, M.A., Sakon, M. and Hawiger, J. 1989. Platelet receptor recognition domain on the c chain of human fibrinogen and its synthetic peptide analogues. Biochemistry 28: 2915-2919.

80. Coller, B.S., Folts, J.D., Smith, S.R., Scudder, L.E. and Jordan, R. 1989. Abolition of in vivo platelet thrombus formation in primates with monoclonal antibodies to the platelet GPIIb/IIIa receptor. Correlation with bleeding time, platelet aggregation, and blockade of GPIIb/IIIa receptors. Circulation 80: 1766-1774.

81. Mickelson, J.K., Simpson, P.J., Cronin, M., et al. 1990. Antiplatelet antibody [7E3 F(ab')$_2$] prevents rethrombosis after recombinant tissue-type plasminogen activator-induced coronary artery thrombolysis in a canine model. Circulation 81: 617-627.

82. Coller, B.S., Scudder, L.E., Berger, H.J. and Iuliucci, J.D. 1988. Inhibition of human platelet function in vivo with a monoclonal antibody with observations on the newly dead as experimental subjects. Ann. Intern. Med. 109: 635-638.

83. Cattaneo, M., Kinlough-Rathbone, R.L., Lecchi, A., Bevilacqua, C., Packham, M.A. and Mustard, J.F. 1987. Fibrinogen-independent aggregation and deaggregation of human platelets: Studies in two afibrinogenemic patients. Blood 70: 221-226.

84. De Marco, L., Girolami, A., Zimmerman, T.S. and Ruggeri, Z.M. 1986. von Willebrand factor interaction with the glycoprotein IIb/IIIa complex. Its role in platelet function as demonstrated in patients with congenital afibrinogenemia. J. Clin. Invest. 77: 1272-1277.

85. Soria, J., Soria, C., Borg, J.Y., et al. 1985. Platelet aggregation occurs in congenital afibrinogenaemia despite the absence of fibrinogen or its fragments in plasma and platelets, as demonstrated by immunoenzymology. Br. J. Haematol. 60: 503-514.

86. Bauman, J.E., Reimers, H.J. and Joist, J.H. 1987. Deaggregation of *in vitro*-degranulated human platelets: Irreversibility of aggregation may be agonist-specific rather than related to secretion *per se*. Thromb. Haemost. 58: 899-904.

87. Kinlough-Rathbone, R.L., Packham, M.A., Cattaneo, M. and Mustard, J.F. 1990. Studies with thrombin-degranulated human platelets: Conditions influencing their responsiveness and the stability of aggregates. (in preparation)

88. Stenberg, P.E., McEver, R.P., Shuman, M.A., Jacques, Y.V. and Bainton, D.F. 1985. A platelet alpha-granule membrane protein (GMP-140) is expressed on the plasma membrane after activation. J. Cell Biol. 101: 880-886.

89. Berman, C.L., Yeo, E.L., Wencel-Drake, J.D., Furie, B.C., Ginsberg, M.H. and Furie, B. 1986. A platelet alpha granule membrane protein that is associated with the plasma membrane after activation. Characterization and subcellular localization of platelet activation-dependent granule-external membrane protein. J. Clin. Invest. 78: 130-137.

90. Hattori, R., Hamilton, K.K., Fugate, R.D., McEver, R.P. and Sims, P.J. 1989. Stimulated secretion of endothelial von Willebrand factor is accompanied by rapid redistribution to the cell surface of the intracellular granule membrane protein GMP-140. J. Biol. Chem. 264: 7768-7771.

91. Bonfanti, R., Furie, B.C., Furie, B. and Wagner, D.D. 1989. PADGEM (GMP140) is a component of Weibel-Palade bodies of human endothelial cells. Blood 73: 1109-1112.

92. George, J.N., Pickett, E.B., Saucerman, S., *et al.* 1986. Platelet surface glycoproteins. Studies on resting and activated platelets and platelet membrane microparticles in normal subjects, and observations in patients during adult respiratory distress syndrome and cardiac surgery. J. Clin. Invest. 78: 340-348.

93. Palabrica, T.M., Furie, B.C., Konstam, M.A., *et al.* 1989. Thrombus imaging in a primate model with antibodies specific for an external membrane protein of activated platelets. Proc. Natl. Acad. Sci. USA 86: 1036-1040.

94. Johnston, G.I., Pickett, E.B., McEver, R.P. and George, J.N. 1987. Heterogeneity of platelet secretion in response to thrombin demonstrated by fluorescence flow cytometry. Blood 69: 1401-1403.

95. Metzelaar, M.J., Sixma, J.J. and Nieuwenhuis, H.K. 1989. Detection of activated platelets with monoclonal antibodies (abstr). Thromb. Haemost. 62: 164.

96. Zimmerman, G.A., McIntyre, T.M., Prescott, S.M. and McEver, R.P. 1989. Granule membrane protein 140 (GMP-140) mediates neutrophil adhesion to activated human endothelium. Blood 74 (suppl 1): 46a.

97. Hamburger, S.A. and McEver, R.P. 1990. GMP-140 mediates adhesion of stimulated platelets to neutrophils. Blood 75: 550-554.

98. Larsen, E., Celi, A., Gilbert, G.E., *et al.* 1989. PADGEM protein: A receptor that mediates the interaction of activated platelets with neutrophils and monocytes. Cell 59: 305-312.

99. Bevilacqua, M.P., Stengelin, S., Gimbrone, M.A. Jr. and Seed, B. 1989. Endothelial leukocyte adhesion molecule 1: An inducible receptor for neutrophils related to complement regulatory proteins and lectins. Science 243: 1160-1165.

100. Jamieson, G.A. and Okumura, T. 1978. Reduced thrombin binding and aggregation in Bernard-Soulier platelets. J. Clin. Invest. 61: 861-864.

101. Phillips, D.R. and Agin, P.P. 1977. Platelet plasma membrane glycoproteins. Identification of a proteolytic substrate for thrombin. Biochem. Biophys. Res. Commun. 75: 940-947.

102. Workman, E.F. Jr., White, G.C. II and Lundblad, R.L. 1977. Structure-function relationships in the interaction of α-thrombin with blood platelets. J. Biol. Chem. 252: 7118-7123.

103. Hallam, T.J. and Rink, T.J. 1989. Receptor-mediated Ca^{2+} entry: Diversity of function and mechanism. TIPS 10: 8-10.

104. Jones, G.D. and Gear, A.R.L. 1988. Subsecond calcium dynamics in ADP- and thrombin-stimulated platelets: A continuous-flow approach using indo-1. Blood 71: 1539-1543.

105. Gerrard, J.M., McNicol, A., Klassen, D., Israels, S.J. 1989. Protein phosphorylation and its relation to platelet function. In: Meyer, P. and Marche, P., eds. Blood cells and arteries in hypertension and atherosclerosis. New York: Raven Press, 93-114.

106. Packham, M.A., Kinlough-Rathbone, R.L., Reimers, H.-J., Scott, S., Mustard, J.F. 1977. Mechanisms of platelet aggregation independent of adenosine diphosphate. In: Silver, M.J., Smith, J.B. and Koksis, J.J., eds. Prostaglandins in hematology. New York: Spectrum Publications, 247-276.

107. Packham, M.A., Bryant, N.L., Guccione, M.A., Kinlough-Rathbone, R.L. and Mustard, J.F. 1989. Effect of the concentration of Ca^{2+} in the suspending medium on the responses of human and rabbit platelets to aggregating agents. Thromb. Haemost. 62: 968-976.

108. Fisher, G.J., Bakshian, S. and Baldassare, J.J. 1985. Activation of human platelets by ADP causes a rapid rise in cytosolic free calcium without hydrolysis of phosphatidylinositol-4, 5-bisphosphate. Biochem. Biophys. Res. Commun. 129: 958-964.

109. Vickers, J.D., Kinlough-Rathbone, R.L., Packham, M.A. and Mustard, J.F. 1990. Inositol phospholipid metabolism in human platelets stimulated by ADP. Eur. J. Biochem. (in press).

110. Sage, S.O. and Rink, T.J. 1987. The kinetics of changes in intracellular calcium concentration in fura-2-loaded human platelets. J. Biol. Chem. 262: 16364-16369.

111. Kinlough-Rathbone, R.L., Cazenave, J.-P., Packham, M.A. and Mustard, J.F. 1980. Effect of inhibitors of the arachidonate pathway on the release of granule contents from rabbit platelets adherent to collagen. Lab. Invest. 42: 28-34.

112. Murray, R. and FitzGerald, G.A. 1989. Regulation of thromboxane receptor activation in human platelets. Proc. Natl. Acad. Sci. USA 86: 124-128.

113. Lalau Keraly, C., Kinlough-Rathbone, R.L., Packham. M.A., Suzuki, H. and Mustard, J.F. 1988. Conditions affecting the responses of human platelets to epinephrine. Thromb. Haemost. 60: 209-216.

114. Lanza, F., Beretz, A., Stierle, A., Hanau, D., Kubina, M. and Cazenave, J.-P. 1988. Epinephrine potentiates human platelet activation but is not an aggregating agent. Am. J. Physiol. 255: H1276-H1288.

115. Kinlough-Rathbone, R.L. and Mustard, J.F. 1986. Synergism of agonists. In: Holmsen, H., ed. Platelet responses and metabolism. Vol. I. Responses. Boca Raton: CRC Press, 193-207.

116. Kinlough-Rathbone, R.L., Mustard, J.F., Perry, D.W., et al. 1983. Factors influencing the deaggregation of human and rabbit platelets. Thromb. Haemost. 49: 162-167.

117. Kinlough-Rathbone, R.L., Perry, D.W., Packham, M.A. and Mustard, J.F. 1985. Deaggregation of human platelets aggregated by thrombin. Thromb. Haemost. 53: 42-44.

118. Cattaneo, M., Canciani, M.T., Lecchi, A., et al. 1990. Released adenosine diphosphate stabilizes thrombin-induced human platelet aggregates. Blood 75: 1081-1086.

119. Vaughan, D.E., Plavin, S.R., Schafer, A.I. and Loscalzo, J. 1989. PGE$_1$ accelerates thrombolysis by tissue plasminogen activator. Blood 73: 1213-1217.

120. Lam, S.C.-T., Dieter, J.P., Strebel, L.C., Muscolino, G., Feinberg, H. and Le Breton, G.C. 1989. Rapid dissociation of platelet-fibrin clots in vitro by a combination of fibrinolytic and antiplatelet agents. Blood 74 (suppl 1): 295a.

121. Coller, B.S. 1990. Platelets and thrombolytic therapy. New Engl. J. Med. 322: 33-42.

122. Schafer, A.I., Maas, A.K., Ware, J.A., Johnson, P.C., Rittenhouse, S.E. and Salzman, E.W. 1986. Platelet protein phosphorylation, elevation of cytosolic calcium, and inositol phospholipid breakdown in platelet activation induced by plasmin. J. Clin. Invest. 78: 73-79.

123. Shebuski, R.J., Bloom, J.C., Sellers, T.S., et al. 1989. Attenuation of the inhibitory effect of prostacyclin on platelet function after tissue plasminogen activator or streptokinase infusion in the rabbit. Fibrinolysis 3: 115-123.

124. Fitzgerald, D.J. 1989. Platelet inhibition with an antibody to glycoprotein IIb/IIIa. Circulation 80: 1918-1919.

125. Verstraete, M. 1989. Thrombolysis: An approach still on the move. Drugs 37: 116-122.

126. Schumacher, W.A. and Heran, C.L. 1989. Effect of thromboxane antagonism on recanalization during streptokinase-induced thrombolysis in anesthetized monkeys. J. Cardiovasc. Pharmacol. 13: 853-861.

127. Golino, P., Ashton, J.H., McNatt, J., et al. 1989. Simultaneous administration of thromboxane A$_2$- and serotonin S$_2$-receptor antagonists markedly enhances thrombolysis and prevents or delays reocclusion after tissue-type plasminogen activator in a canine model of coronary thrombosis. Circulation 79: 911-919.

128. Terres, W., Beythien, C., Kupper, W. and Bleifeld, W. 1989. Effects of aspirin and prostaglandin E$_1$ on in vitro thrombolysis with urokinase. Evidence for a possible role of inhibiting platelet activity in thrombolysis. Circulation 79: 1309-1314.

129. Lewis, H.D. Jr., Davis, J.W., Archibald, D.G., et al. 1983. Protective effects of aspirin against acute myocardial infarction and death in men with unstable angina. Results of a Veterans Administration Cooperative Study. New Engl. J. Med. 309: 396-403.

130. Cairns, J.A., Gent, M., Singer, J., et al. 1985. Aspirin, sulfinpyrazone, or both in unstable angina. Results of a Canadian multicenter trial. New Engl. J. Med. 313: 1369-1375.

131. Eidt, J.F., Allison, P., Noble, S., et al. 1989. Thrombin is an important mediator of platelet aggregation in stenosed canine coronary arteries with endothelial injury. J. Clin. Invest. 84: 18-27.

132. Heras, M., Chesebro, J.H., Penny, W.J., Bailey, K.R., Badimon, L. and Fuster, V. 1989. Effects of thrombin inhibition on the development of acute platelet-thrombus deposition during angioplasty in pigs. Heparin versus recombinant hirudin, a specific thrombin inhibitor. Circulation 79: 657-665.

EICOSANOID METABOLISM AND ENDOTHELIAL CELL ADHESION MOLECULE

EXPRESSION: EFFECTS ON PLATELET/VESSEL WALL INTERACTIONS

Michael R. Buchanan, Maria C. Bertomeu,
Stephanie J. Brister and Thomas A. Haas

Departments of Pathology and Surgery
McMaster University
Hamilton, Ontario, Canada

Platelet interactions with the blood vessel wall following injury influence both the acute thrombotic event (platelet/vessel wall adhesion) and the chronic pathogenesis of arteriosclerosis (chronic vessel wall thickening and occlusion). Platelets are thought not to interact with the healthy intact endothelium, but only to adhere to vessel wall components exposed following endothelial cell damage and vessel wall injury (1-3). When platelets adhere to the injured vessel wall, they release a number of constituents which facilitate further platelet activation (3), activation of the coagulation cascade (1), and/or are mitogenic, facilitating smooth muscle cell proliferation and hyperplasia (4). Endothelial cells also synthesize and release a number of constituents which counterbalance the platelet responses, thereby enhancing fibrinolysis, inhibiting coagulation and preventing platelet aggregation and adhesion (5-7). Recently, with the development of endothelial cell culture techniques and a number of molecular biologic tools, we have learned that platelets and other blood cells actively interact not only with the injured vessel wall, but also with the intact endothelium itself. In this chapter, we will focus on the latter interaction and specifically on endothelial cell fatty acid metabolism and its possible consequence on the reactivity of adhesion molecule receptors on the surface of the intact endothelium which in turn may influence platelet/vessel wall interactions.

Fatty Acid Metabolism by Vascular Endothelial Cells

The study of fatty acid metabolism by the vessel wall has focussed predominantly on the metabolism of arachidonic acid via the cyclo-oxygenase pathway following endothelial cell stimulation. A number of biological stimuli, such as thrombin, endotoxin and cytokines, induce the synthesis and/or release of prostacyclin (prostaglandin I_2, PGI_2) from endothelial cells (8-11). PGI_2 is synthesized from arachidonic acid by the cyclo-oxygenase pathway following the liberation of esterified arachidonic acid from the endothelial cell membrane phospholipids. Until recently, PGI_2 was thought to be the major cyclo-oxygenase derived fatty acid metabolite produced by endothelial cells, although some reports have suggested that endothelial cells also synthesize small amounts of thromboxane A_2, TxA_2 (12,13).

PGI_2 is a potent inhibitor of platelet aggregation and a potent vasodilator of the vessel wall (14). PGI_2 is thought to inhibit platelet aggregation by elevating cyclic adenosine monophosphate (cAMP), thereby rendering the platelets hypo-responsive to stimuli (15,16). The activation of adenylate cyclase by PGI_2 (to elevate cAMP) is thought to be mediated through a PGI_2 receptor and to be representative of the general mechanism of action of PGI_2 on a variety of cells (15,16). This effect is opposite to that

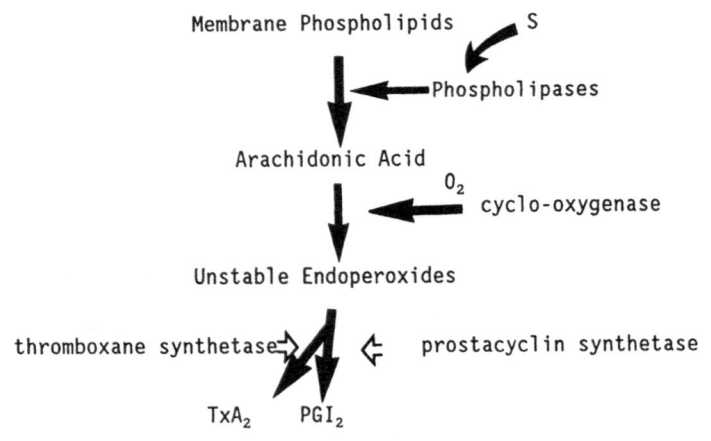

Figure 1. Metabolism of arachidonic acid via the cyclo-oxygenase pathway in stimulated (S) endothelial cells.

initiated by the platelet-derived cyclo-oxygenase arachidonic acid metabolite, TxA_2. As a consequence, it has been postulated that the relative amounts of TxA_2 and PGI_2 play important opposing roles in the regulation of platelet/endothelial cell interactions following blood cell and/or vessel wall injury (14). Thus, PGI_2 appears to influence platelet/vessel wall interactions by rendering the platelets hypo-responsive to stimuli rather than inhibiting platelet/vessel wall interactions directly, such as influencing a specific adhesion receptor on the platelet. In support of this, physiological concentrations of PGI_2 do not inhibit platelet adhesion either to intact endothelial cells or to the injured vessel wall (9,17).

When blood and vascular wall cells are stimulated, arachidonic acid liberated from the membrane phospholipids is also metabolized by another enzyme, lipoxygenase which is located within the cytosol of the cells. Lipoxygenase, in the presence of molecular oxygen, converts arachidonic acid into the unstable hydroperoxy-eicosatetraenoic acid, which is hydrolysed by a cytosolic-associated peroxidase, to hydroxy-eicosatetraenoic acid or HETE. Early studies by Greenwald et al (18) suggested such a pathway also existed in the endothelium since vessel wall segments synthesized a "12-HETE" like substance. More detailed studies, however, demonstrated that both endothelial cells and vascular smooth muscle cells can only synthesize HETE when arachidonic acid is added exogenously (19,21). However, under these conditions, the HETE produced has the first double bond in the 15 position (15-HETE), not the twelfth position like the 12-HETE platelet product (19,20). The position of the hydroxyl group in the 15-HETE indicates that the lipoxygenase enzyme within endothelial cells is an omega-6 lipoxygenase and therefore different from the lipoxygenase enzymes found in leukocytes (which synthesize 5-HETE) and platelets (which synthesize 12-HETE). In summary, while endothelial cells are capable of metabolizing free arachidonic acid (added exogenously) into two major metabolites, PGI_2 via the cyclo-oxygenase pathway at the plasma membrane level, and 15-HETE via the lipoxygenase pathway within the cytosol, their endogenous stores of arachidonic acid are metabolized only to PGI_2 and only after stimulation (21).

A number of recent studies have demonstrated that endothelial cells are also metabolically active under basal conditions (11,22,23). In particular, triglycerides are rapidly turned over in unstimulated endothelial cells (11,22). As a consequence, linoleic acid is continuously released from this triglyceride pool and metabolized by the endothelial cell lipoxygenase into the monohydroxide, 13-hydroxyoctadecadienoic acid or 13-HODE (Figure 2) (11,23). Interestingly, 13-HODE synthesis is rapidly halted when endothelial cells are stimulated by thrombin, endotoxin, cytokines or biologics (9-11,23).

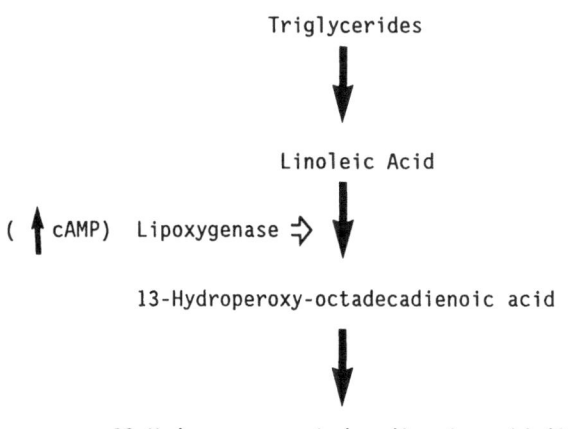

Triglycerides

Linoleic Acid

(↑cAMP) Lipoxygenase ⇒

13-Hydroperoxy-octadecadienoic acid

13-Hydroperoxy-octadecadienoic acid (13-HODE)

Figure 2. Metabolism of linoleic acid (18:2 n - 6) via the lipoxygenase pathway in unstimulated endothelial cells.

The levels of intracellular 13-HODE in cultured endothelial cells can be elevated by increasing the intracellular levels of cAMP both in vitro and in vivo (11,24). This effect appears to be due to an increase in linoleic acid turnover in the endothelial cell triglyceride pool (11). Other investigators have reported that, following stimulation, endothelial cells can also metabolize linoleic acid into 13-HODE and 9-HODE via the cyclo-oxygenase pathway (25). These studies, however, must be interpreted with caution, since these investigations measured the production of these metabolites using exogenous linoleic acid as both the stimulus and substrate, they used short incubation times, and concentrations of linoleic acid were far in excess of those found endogenously. These studies provide no evidence of 9- or 13-HODE synthesis via the cyclo-oxygenase pathway, when measuring these products generated from steady state endogenous stores in intact cells. This point is rather important since most two double bond 18 carbon fatty acids such as linoleic acid are poor substrates for the cyclo-oxygenase enzyme. The observations suggest instead that the production of 9- and 13-HODE via the cyclo-oxygenase pathway is most likely due to the swamping of the enzyme with very high concentrations of the substrate.

When we analyzed the lipoxygenase products derived from endogenous substrate stores of the vessel wall in both human endothelial cells and intact or injured animal tissue, and in relation to the endogenous substrate concentrations over chronic steady state conditions, we only detected 13-HODE (9,21,24,26). Recently, we examined the fatty acid extracts of segments of saphenous veins, internal mammary arteries and gastroepiploic arteries obtained fresh from patients undergoing elective coronary artery bypass surgery. While we noted that there were marked differences between the amounts of 13-HODE produced in veins and arteries, all human vessel wall segments produced only 13-HODE. There were no detectable amounts of either 15-HETE or 12-HETE (27). HPLC tracings of internal mammary artery extracts divided in two segments and analyzed by reverse phase HPLC with and without known amounts of internal standards for 13-HODE and 15-HETE are shown in Figure 3.

We suggest, therefore, that while a number of studies provided evidence that vessel wall cells, presumably both endothelial cells and smooth muscle cells, are capable of metabolizing 9-HODE and 13-HODE from linoleic acid via both the cyclo-oxygenase and lipoxygenase pathways and can synthesize 15-HETE from free arachidonic acid via the lipoxygenase pathway, our studies indicate that only 13-HODE, synthesized via the lipoxygenase pathway, is produced from their endogenous stores. Thus, vessel wall cells synthesize only two major fatty acid metabolites, PGI_2 derived from arachidonic acid via the cyclo-oxygenase pathway and only following cell stimulation, and 13-HODE derived from linoleic acid via the lipoxygenase pathway and only under basal conditions.

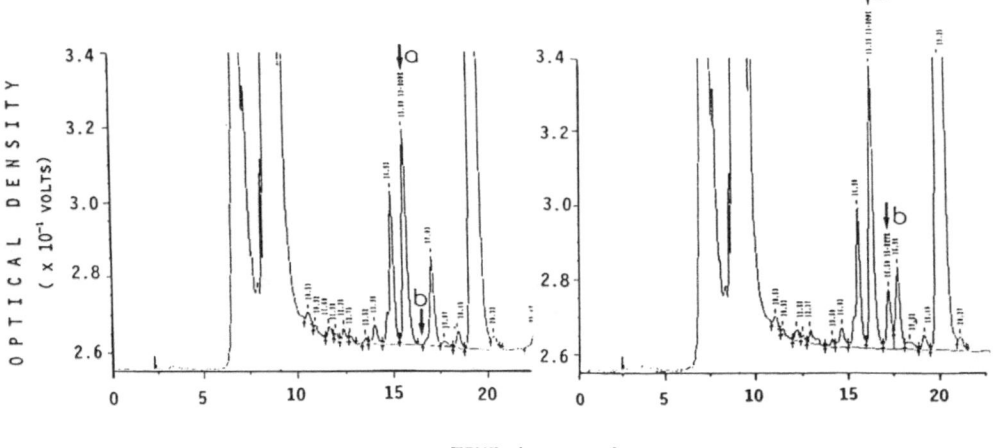

TIME (MINUTES)

Figure 3. HPLC tracing of an internal mammary artery extract without (left) and with (right) internal standards of (a) 13-HODE (25 ng) and (b) 15-HETE (8 ng).

Role of 13-HODE in Regulating Adhesive Molecule Expression

Little is known about the endogenous regulators in cells which influence the expression of adhesion molecules on the external surfaces of endothelial cells and circulating blood cells. Recent studies by a number of investigators have demonstrated that there is a consistent and significant relationship between the synthesis of the lipoxygenase-derived linoleic and arachidonic acid metabolites [i.e., 13-HODE and (5-, 12-, 15-HETE)] and blood cell/endothelial cell adhesion. Thus, our laboratory has reported that the intracellular levels of 13-HODE in tumor cells and endothelial cells inversely correlates with tumor cell and endothelial cell adhesivity, whereas intracellular levels of 12-HETE directly correlate with platelet and tumor cell adhesivity (10,11,28-30). Similar observations have been obtained by adding exogenous 13-HODE or 12-HETE to stimulated tumor cells (31). Exogenous 13-HODE blocks enhanced adhesion of tumor cells associated with increased glycoprotein IIb/IIIa receptor expression whereas 12-HETE increases tumor cell adhesion. These data suggested to us that 13-HODE and the HETEs down and up regulate adhesion molecule expression on endothelial cells and other cells.

The results of recent studies in our laboratory provide evidence that 13-HODE specifically down-regulates vitronectin receptor expression on the endothelial cell surface. We found that 13-HODE and the vitronectin receptor are located in the same vesicles within unstimulated endothelial cells (Figure 4), i.e., in endothelial cells which were not adhesive (32). In contrast, when the endothelial cells were stimulated, 13-HODE was no longer detectable in the vesicles, the vitronectin receptor was expressed on the surface of the endothelial cell and platelet/endothelial cell adhesion was increased, i.e., when 13-HODE and the vitronectin receptor dissociated (32). These observations are similar with the observations of Singer et al. (33) who found that the laminin and fibronectin receptors were also located in specific granules in other cells under unstimulated conditions.

We propose, therefore, that in unstimulated, non-adhesive endothelial cells, the vitronectin receptor is physically associated with 13-HODE in vesicles inside the endothelial cells. The intracellular positioning of the vitronectin receptor would prevent it from recognizing its extracellular adhesive ligands such as vitronectin, fibronectin and von Willebrand factor (34). When 13-HODE synthesis is halted, such as following endothelial cell stimulation, 13-HODE and the vitronectin receptor dissociate, and the receptor relocates in the phospholipid membrane, undergoing conformational changes which unmask its adhesive sites on the β chain. The possibility

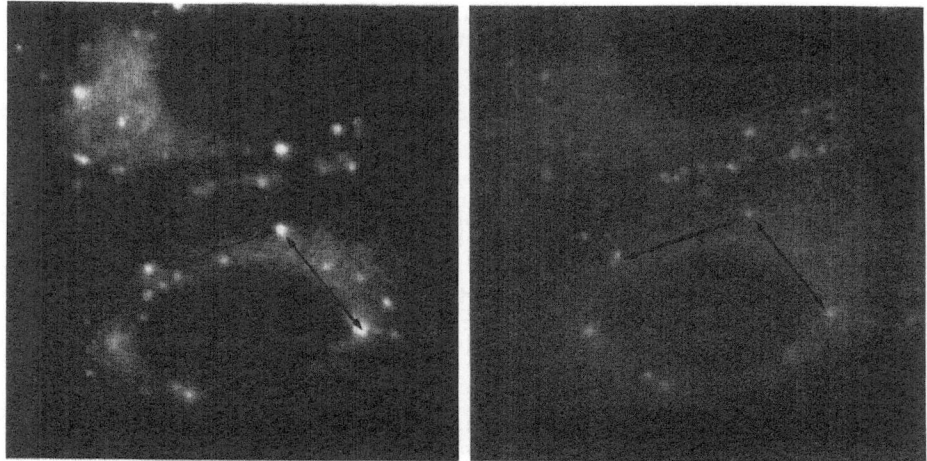

Figure 4. Histoimmunofluorescent micrographs of 13-HODE (left) and the vitronectin receptor (right) in unstimulated endothelial cells. The identical staining pattern of the antigens in the same cells under fixed and permeabilized conditions indicate the identical location of 13-HODE and the vitronectin receptor in the same vesicles (as indicated by the arrows) in non-adhesive cells.

that both the α and β chains of the vitronectin receptor can undergo such conformational changes which alter their abilities to recognize specific ligands and contribute to cell-cell adhesion when associated with specific fatty acids is supported by the recent observations of Conforti et al. (34). These investigators demonstrated that the vitronectin receptor's ability to recognize different ligands is altered when the vitronectin receptor is inserted into liposomes containing different combinations of phospholipids. Further studies are presently ongoing to determine whether 13-HODE alters the ability of the vitronectin receptor as well as other endothelial cell receptors to recognize the adhesive ligand counterparts by inducing specific conformational changes or simply by blocking the RGD recognizing site on these molecules.

REFERENCES

1.   Ross R. Atherosclerosis: A problem of the biology of arterial wall cells and their interaction with blood components. Arteriosclerosis 1: 293-311, 1981.

2.   Ross R. The pathogenesis of atherosclerosis - an update. N. Engl. J. Med. 314: 488-500, 1986.

3.   Packham, M.A. and Mustard, J.F. The role of platelets in the development and complications of atherosclerosis. Semin. Hematol. 23: 8-26, 1986.

4.   Nilsson, J.   Growth factors and the pathogenesis of atherosclerosis. Atherosclerosis 62: 185-99, 1986.

5.   Buchanan, M.R. Mechanisms of pathogenesis of arterial thrombosis: potential sites of inhibition by therapeutic compounds. Semin. Thromb. Haemost. 14: 33-40, 1988.

6.   Collen, D. Potential approaches for therapeutic intervention of thrombosis by fibrinolytic agents. Semin. Thromb. Haemost. 14: 95-99, 1988.

7.   Turpie, A.G.G. Clinical studies: evidence for intervention with specific antiplatelet drugs in arterial thromboembolism. Semin. Thromb. Haemost. 14: 41-49, 1988.

8.   Rossi, V., Breviario, F., Ghezzi, P., Dejana, E. and Mantovani, A. Prostacyclin synthesis induced in vascular cells by interleukin-1. Science 229: 174-176, 1985.

9.   Buchanan, M.R., Haas, T.A., Lagarde, M. and Guichardant, M.   13-Hydroxy-octadecadienoic acid is the vessel wall chemorepellent factor, LOX. J. Biol. Chem. 260: 16056-16059, 1985.

10.   Bastida, E., Almirall, L., Ordinas, A., Buchanan, M.R., Bertomeu, M.C. and Haas, T.A. Effects of antiplatelet drugs and cytokines on endothelial cell 13-HODE synthesis and tumor cell/endothelial cell adhesion. Thromb. Haemost. 62: 138, 1989.

11.   Haas, T.A., Bertomeu, M.C., Bastida, E. and Buchanan, M.R.   Cyclic AMP regulation of endothelial cell triglyceride turnover, 13-hydroxyoctadecadienoic acid (13-HODE) synthesis and endothelial cell thrombogenicity. Biochem. Biophys. Acta. 1031: 174-178, 1990.

12.   Ingerman-Wolenski, C., Silver, M.I., Smith, J.B. and Macarack, E.   Bovine endothelial cells in culture produce thromboxane as well as prostacyclin. J. Clin. Invest. 67: 1291-1296, 1981.

13.   Ali, A.E., Banet, J.C. and Eling, T.E.   Prostaglandin and thromboxane production by fibroblasts and vascular endothelial cells. Prostaglandins 20: 667-668, 1988.

14.   Moncada, S. and Vane, J.R. Arachidonic acid metabolites and the interaction between platelet and blood vessel wall. New Engl. J. Med. 300: 1142-1147, 1979.

15.   Gorman, R.R., Bunting, S. and Miller, D.V. Modulation of human platelet adenylate cyclase by prostacyclin (PGX). Prostaglandins 13: 377-388, 1977.

16.   Hajjar, D.P., Weksler, B.B., Falcone, D.J., Hefton, D.M., Tack-Goldman, K. and Minick, C.R. Prostacyclin modulates cholesteryl ester hydrolyte activity by its effects on cyclic adenosine monophosphate in rabbit aorta smooth muscle cells. J. Clin. Invest. 70: 479-488, 1982.

17.   Weiss, H.J. and Turitto, V.T. Prostacyclin (prostaglandin I$_2$) inhibits platelet adhesion and thrombus formation in subendothelium. Blood 53: 200-244, 1979.

18.   Greenwald, J.E., Bianchine, J.R. and Wong, L.K.   The production of the arachidonate metabolite HETE in vascular tissue. Nature 281: 588-589, 1979.

19.   Hopkins, N.K., Oglesby, T.D., Bundy, G.L. and Gorman, R.R. Biosynthesis and metabolism of 13 hydroperoxy 5,8,11,15 eicosatetraenoic acid by human umbilical vein endothelial cells. J. Biol. Chem. 259: 14048-14058, 1984.

20.   Kuhn, H., Ponicke, K., Hall, W., Weisner, R., Scheme, T. and Forster, W. Metabolism of (1-^{14}C)-arachidonic acid by cultured calf aortic endothelial cells and evidence for the presence of lipoxygenase pathway. Prostaglandins Leukotrienes Med. 17: 291-303, 1988.

21.   Buchanan, M.R., Crozier, G. and Haas, T.A. Fatty acid metabolism and the vascular endothelial cell. New thoughts about old data. Haemost. 18: 360-375, 1988.

22.   Denning, G.M., Figard, P.H., Kaduce, T.L. and Spector A. Role of triglycerides in endothelial cell arachidonic acid metabolism. J. Lipid. Res. 24: 993-1011, 1983.

23.   Buchanan, M.R., Butt, R.W., Magas, Z., Van Ryn, J., Hirsh, J. and Nazir, D.J. Endothelial cells produce a lipoxygenase derived chemorepellent which influences platelet/endothelial cell interactions. Effect of aspirin and salicylate. Thromb. Haemost. 53: 306-311, 1985.

24. Weber, E., Haas, T.A., Hirsh, J. and Buchanan, M.R. Relationship between vessel wall 13-HODE production and subendothelial basement membrane thromboresistance: Influence of salicylate. Thromb. Haemost. 58: 316, 1987.

25. Kaduce, T.L., Figard, P.H., Laifur, R. and Spector, H.H. Formation of 9-hydroxyoctadecadienoic from linoleic acid in endothelial cells. J. Biol. Chem. 264: 6823-6830, 1988.

26. Bertomeu, M.C., Crozier, G.L., Haas, T.A., Fleith, M. and Buchanan, M.R. Dietary fats, vessel wall 13-HODE synthesis and vessel wall thrombogenecity. Thromb. Haemost. 62: 24, 1989.

27. Brister, S.J., Bertomeu, M.C., Haas, T.A., Austin, J. and Buchanan, M.R. Coronary artery bypass grafts: Variations in vessel wall thrombogenecity with 13-HODE synthesis, patient age and vessel wall type. Adv. Prost. Leuk. Res. 20: 1990, in press.

28. Buchanan, M.R. and Bastida, E. Endothelium and underlying membrane reactivity with platelets, leukocytes and tumor cells: regulation by the lipoxygenase-derived fatty acid metabolites 13-HODE and HETEs. Med. Hypoth. 27: 317-325, 1988.

29. Bastida, E., Bertomeu, M.C., Haas, T.A., Almirall, L., Lauri, D., Orr, F.W. and Buchanan, M.R. Regulation of tumor cell adhesion by intracellular 13-HODE:15-HETE ratio. J. Lipid. Med.: 1990, in press.

30. Van Ryn McKenna, J. and Buchanan, M.R. Relative effects of flurbiprofen on platelet 12-hydroxyeicosatetraenoic acid and thromboxane A_2 production: influence on collagen-induce platelet aggregation and adhesion. Prostaglandins Leuko. Essent. Fatty Acids 36: 171-174, 1989.

31. Grossi, I.M., Fitzgerald, L.A., Umbarger, L.A., Nelson, K.K., Diglio, C.A., Taylor, J.D. and Koon, K.V. Bidirectional control of membrane expression and/or activation of the tumor cell IRGp IIb/IIIa receptor and tumor cell adhesion by lipoxygenase products of arachidonic acid and linoleic acid. Cancer Res. 49: 1029-1037, 1989.

32. Bertomeu, M.C., Haas, T.A., Eltringham-Smith, L. and Buchanan, M.R. Relationship between endothelial cell vitronectin receptor expression and 13-HODE: Histoimmunofluorescent localization. Proc. VIth Intl. Sym. Vascular Wall Cells, Paris, August, 1990.

33. Singer, I.I., Scott, S. Kawka, W. and Dazizis. Adhesomes: specific granules containing receptors for laminin, C3bi/fibrinogen, fibronectin, and vitronectin in human polymorphonuclear leukocytes and monocytes. J. Cell. Biol. 109: 3169-3182, 1989.

34. Conforti, G., Zanetti, I., Pasquali-Ronchetti, D., Qualino, Jr. P., Negroz, P. and Dejana, R. Modulation of vitronectin receptor binding by membrane lipid composition. J. Biol. Chem. 265: 4011-4019, 1990.

SECTION 6

LIPIDS IN ATHEROGENESIS

RECEPTOR-MEDIATED LOW DENSITY LIPOPROTEIN METABOLISM

Wolfgang J. Schneider

Lipid and Lipoprotein Research Group
 and Department of Biochemistry
327 Heritage Medical Research Centre
University of Alberta
Edmonton, Alberta T6G 2S2

INTRODUCTION

The concept of receptor-mediated metabolism of lipoproteins emerged 17 years ago from studies on human skin fibroblasts grown in culture[1]. These experiments were designed to elucidate the normal function of low density lipoprotein (LDL), about which little was known at that time. Biochemical studies showed that a specific cell surface receptor, the LDL receptor, mediates the binding, uptake and degradation of LDL, thus supplying almost all cells in the body with cholesterol. Detailed insight into the molecular mechanisms underlying this complex process was obtained from studies with fibroblasts derived from patients with the phenotype of homozygous familial hypercholesterolemia (FH). As in many other biological systems, the expression of a disease state in a defined cellular system was essential to the discovery of the causal factor: FH is now one of the best characterized genetic diseases at the molecular level. As will be outlined below, several groups of mutations occur naturally in the structural gene for the LDL receptor which disrupt its normal function and lead to severe hypercholesterolemia, myocardial infarctions and premature atherosclerosis. Thus, the important role of lipoprotein receptors in normal physiology is underscored by the dramatic consequences of their functional absence.

THE LDL RECEPTOR - PHYSIOLOGICAL ASPECTS

The LDL Receptor Pathway

This pathway describes how cells, through receptor-mediated endocytosis of LDL, control their internal cholesterol concentration. The cholesterol derived from the lysosomal hydrolysis of LDL cholesterol esters mediates a complex array of feedback control mechanisms that protects the cell from overaccumulation of cholesterol. First, the LDL-derived cholesterol suppresses the activities of the enzymes 3-hydroxy-3-methylglutaryl CoA synthase (HMG-CoA synthase) and HMG-CoA reductase, rate controlling enzymes in cholesterol biosynthesis, thereby turning off cholesterol synthesis in the cell[2]. Second, the cholesterol activates a cholesterol-esterifying enzyme called acyl-CoA: cholesterol acyltransferase (ACAT) so that excess cholesterol can be stored as droplets of cholesteryl esters[3]. Third, the cholesterol suppresses the synthesis of new LDL receptors, preventing further cellular entry of LDL and thus overloading with cholesterol[4].

Atherosclerosis, Edited by A. I. Gotlieb *et al.*
Plenum Press, New York, 1991

The overall effect of this regulatory system is to coordinate the intracellular and extracellular sources of cholesterol in order to maintain a constant level of cholesterol within the cell, while the external supply in the form of lipoproteins fluctuates. Human fibroblasts and other mammalian cells are able to subsist in the absence of lipoproteins because they can synthesize cholesterol from acetyl CoA. In contrast, when LDL is accessible, the cells primarily use the LDL receptor to take up LDL and keep their own cholesterol synthesis suppressed[5,6,7].

Familial Hypercholesterolemia: Clinical Consequences of the Genetic Disruption of the LDL Receptor Pathway

The delineation of the LDL receptor pathway in fibroblasts was greatly facilitated by studies on fibroblasts from a large number of patients with the phenotype of homozygous FH. In 1974, it was recognized that mutations affecting the function of the LDL receptor are responsible for FH[1,8]. Now, 16 years later, at least 18 different alleles at the LDL receptor locus have been identified in over 110 fibroblast lines from patients with the clinical features of FH homozygotes.

FH is characterized by three cardinal features: (i) a selective elevation in the plasma level of LDL; (ii) deposition of cholesterol in abnormal sites, in particular in tendons (formation of xanthomas) and in arteries (forming atheromas); and (iii) inheritance as an autosomal dominant trait with a gene dosage effect. This means that individuals who inherit two mutant alleles (FH homozygotes) are more severely affected than those with one mutant allele (FH heterozygotes)[9,10]. Heterozygotes occur at a frequency of about 1 in 500 persons, while homozygotes occur at a frequency of about 1 in one million persons among European and American populations. Through the delineation of the molecular defects in certain patients (see below), it is now clear that the severely affected offspring of a marriage between two heterozygotes can be either a true homozygote or a heteroallelic genetic compound[9,11,12,13,14]. Thus, the term *homozygote* in most cases is not strictly correct in a genetic sense, yet it is a convenient one that is generally applied to patients with two mutant alleles at the LDL receptor locus.

Familial hypercholesterolemia is the outstanding example of a single-gene mutation that regularly produces atherosclerosis. The deposition of LDL-cholesterol within the intima of major arteries in homozygotes becomes rapidly progressive, and leads to remarkably uniform clinical findings. Myocardial infarction, angina pectoris, and sudden death usually occur in homozygotes before the age of 15[15].

The clinical features of heterozygous FH are more variable and less severe than those of the homozygous form. Seventy-five percent of heterozygous men experience a myocardial infarction before 60 years of age; in normal men, this risk is 15%[16,17]. Despite the expression of the same genetic abnormality and similarly elevated plasma LDL levels, heterozygous women suffer from coronary heart disease less often and at a later age than do heterozygous men. The risk that female heterozygotes will develop coronary artery disease by age 60 is 45% as compared with 10% in unaffected females[16,17].

MOLECULAR DISSECTION OF THE LDL RECEPTOR

The complete amino acid sequence of the human LDL receptor was derived from the nucleotide sequence of the full-length receptor cDNA. The strategy for the isolation of this cDNA is outlined in Ref. 18. The amino acid sequence, in combination with biochemical experiments[19], revealed that the mature receptor can be divided into five domains (see Fig. 1).

First Domain: Ligand Binding

The mature receptor (after cleavage of a 21-residue signal sequence) consists of 839 residues. The ligand binding domain (292 residues) is located at the NH_2-terminus of the receptor. These 292 amino acids, 42 of which are cysteines, exist in a tightly folded conformation that is held together by multiple disulfide bonds. Analysis of the amino acid sequence and exon/intron mapping data[20] suggest that the first domain is made up of seven repeats, each consisting of 40 amino acids. Within these 40 residues, 6 cysteines are spaced 4-7 amino acids apart and are located exactly in register for all repeats. At the COOH-terminus of each repeat unit, there is a cluster of negatively charged amino acids. These sequences are complementary to positively charged sequences in the receptor's ligands, apoE[21] and apo B[22]. It has been speculated[22,23] that the negatively charged clusters of amino acids within the cysteine-rich repeat sequences constitute binding sites for apo E and apoB. Studies by Mahley and associates[25,26] and others[27,28], strongly support this notion. In addition, the disulfide bridges may be important in stabilizing the binding site structure, in particular when the receptor recycles via the endosome with its acidic environment[19].

Second Domain: EGF Precursor Homology

This region, consisting of about 400 residues, is homologous to a portion of the extracellular domain of the precursor to epidermal growth factor (EGF)[18,19]. The growth factor is a peptide of 53 amino acids, which is derived by proteolytic cleavage from a large, membrane-bound precursor of 1217 residues[29,30]. The portion of the precursor facing the cytoplasm contains multiple homologous but not identical repeats of the EGF sequence, as well as unrelated spacer sequences. The region of the LDL receptor that exhibits EGF precursor homology is flanked by 3 cysteine-rich repeats, which are highly homologous to four repeat sequences in the EGF precursor[20,29,31]. Repeats A, B, and C (Fig. 1) of the LDL receptor are also homologous to certain proteins of the blood clotting system, such as Factor IX, Factor X, and protein C[20,31].

All observations regarding the significance of these homologies seem to suggest that the homologous regions arose by a duplication of an ancestral gene. These genes might have differentiated further by, e.g., acquisition of exons from other genes that provided specific functions such as nutrient delivery (LDL) or that signaled of cell proliferation through the secretion of a peptide (EGF).

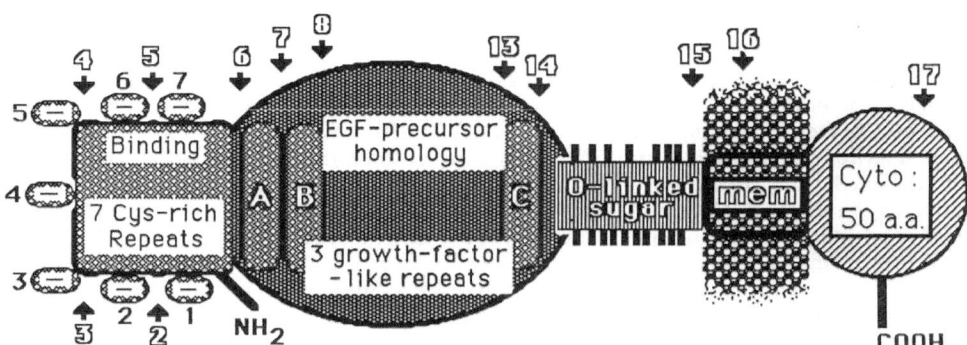

Figure 1. Schematic representation of the human LDL receptor. The five domains of the mature protein (839 amino acids) are indicated, and their salient features are highlighted. Regions in the EGF-precursor homology domain (A,B,C) that are homologous to cysteine-rich regions in the ligand binding domain are shaded similarly. The numbered arrows denote the approximate positions of introns discussed in the text. The numbered (1-7), negatively charged clusters on the binding domain indicate 7 cysteine-rich repeats and their carboxyterminal negative charges. Eighteen O-linked carbohydrate chains on the third domain are represented by black bars. From Ref. 38, with permission.

Third Domain: 0-linked Sugars

There is a sequence of 58 amino acid residues that contains 18 serines or threonines immediately external to the membrane[18]. This region contains a large number of carbohydrate chains attached to the hydroxyl groups of Ser and Thr[19,32]. The elongation of these sugar chains during posttranslational processing of the receptor brings about the dramatic increase in apparent molecular weight that is observed in biosynthetic studies of radiolabeled receptor (see below). Carbohydrate chains in 0-linkage are also present on other parts of the receptor; these chains do not appear to be clustered as they are in the 0-linked sugar domain[33]. The role of 0-linked carbohydrate in the function of the LDL receptor, if any, is unknown.

Fourth Domain: Membrane-spanning Region

Twenty-two uncharged and hydrophobic amino acids constitute the only membrane-spanning portion of the receptor[19]. There is a single intramembraneous cysteine residue in the human receptor, which is replaced by an alanine in the bovine receptor[19]. Since the two receptors function similarly, it appears likely that this cysteine residue is not involved in a disulfide bridge, but rather exists in a reduced state.

Fifth Domain: Cytoplasmic Region

At the COOH-terminus of the LDL receptor, 50 amino acid residues project into the cytoplasm. The intracellular location of this tail region was demonstrated in proteolysis experiments by the use of an anti-peptide antibody directed against the COOH-terminal sequence[19]. As outlined below, the cytoplasmic region is crucial for clustering of LDL receptors in coated pits, a prerequisite for subsequent internalization of receptor/LDL complexes. Localization to coated pits may occur through interaction of the cytoplasmic tail with clathrin itself or with some clathrin-associated protein on the intracellular side of the plasma membrane. The intracellular regions of other coated pit receptors have been investigated for common features that might identify a mediator site on the LDL receptor, but no obvious structural element has been identified[23]. Thus, while the function of the cytoplasmic tail of the LDL receptor is established, the mechanism of its action has yet to be elucidated.

THE BIOSYNTHETIC PATHWAY OF LDL RECEPTOR IN NORMAL HUMAN FIBROBLASTS

The biosynthesis of LDL receptors can be conveniently studied by metabolic labeling of cultured cells with 35S-methionine, followed by immunoprecipitation of receptor protein with a monoclonal anti-receptor antibody[34]. Experimental procedures are detailed in Refs. 13 and 35. The normal human LDL receptor is initially synthesized as a precursor with an apparent M_r of 120,000, as estimated by SDS polyacrylamide gel electrophoresis (SDS-PAGE). It contains asparagine-linked (N-linked) high mannose oligosaccharide chains and N-acetylgalactosamine (GalNAc) residues attached to serine and threonines by O-linkage[32]. Between 30 and 60 min after synthesis, the LDL receptor precursor undergoes a sudden shift in apparent M_r from 120,000 to 160,000 in SDS-PAGE[13]. This change in apparent M_r is due to the elongation of the O-linked chains[32] which temporally coincides with the maturation of the N-linked carbohydrate moiety[32]. The calculated molecular weight of the protein portion of the receptor is 93,102, and including the mature carbohydrate its M_r is 115,000, yet the O-linked carbohydrate which forms a cluster in a Ser/Thr rich region of the receptor[19] causes it to migrate to the position of a 160 kDa-protein in SDS-PAGE[32].

In other studies, it was shown that the 120 kDa precursor is localized intracellularly[14], and that the addition of N-linked carbohydrate chains is required neither for transport to the surface nor for the receptor's ability to bind LDL and to recycle. The role of O-linked carbohydrate in the receptor's cellular itinerary is still under investigation. A deletion of the clustered O-linked carbohydrates does not appear to impair the function of the LDL receptor[33]. However, there is a small number of isolated chains of O-linked carbohydrate in addition to the clustered chains, and their significance for receptor function needs to be delineated[33].

MOLECULAR DEFECTS IN LDL RECEPTORS OF PATIENTS WITH FAMILIAL HYPERCHOLESTEROLEMIA

At least 18 mutant alleles at the LDL receptor locus have been described[14,15,36,37] (reviewed in Ref. 38). Fibroblasts from over 110 FH homozygotes were analyzed in terms of LDL receptor structure and function; these studies revealed that the mutations could be divided into four classes (for review, see Ref. 23). The mutant alleles are described in the following.

Class 1 Mutations: No Detectable Precursor

These alleles, designated R-O, fail to express receptor protein as measured by functional assays (binding of ^{125}I-LDL) or as identified immunologically with a variety of monoclonal and polyclonal antibodies directed against the LDL receptor. R-O alleles are the most frequent of the mutant alleles and probably account for about 40% of all mutations at the LDL receptor locus. This class most likely includes nonsense mutations that introduce termination codons early in the protein coding region, but point mutations in the promoter that block transcription, mutations in intron/exon junctions that lead to abnormal splicing of the mRNA, and large deletions may also be included in this class of mutant alleles. Examples are described in Refs. 13 and 14.

Class 2 Mutations: Precursor Not Processed

These alleles specify receptor precursors that are produced in normal or reduced amounts, have normal (120 kDa) or abnormal (100 kDa, 135 kDa) molecular weight, but do not undergo any apparent increase in molecular weight after synthesis. Receptors encoded by these alleles never reach the cell surface[14].

Most of the mutant receptor precursors in this class have an apparent M_r of 120,000 (allele designation, R-120) but precursors with apparent molecular weights of 100,000 (R-100 allele) and 135,000 (R-135 allele) have also been identified as members of this class[14]. These M_r abnormalities are likely due to alterations in the length of the protein chain and do not result from changes in the carbohydrate moieties[14].

Variants of the class 2 mutations were observed in a black American family, in Afrikaners, and in WHHL rabbits[36]. In these variants, the receptor is produced as a precursor of apparently normal M_r that is processed to the mature form, but much more slowly than the normal precursor. Eventually, about 5-10% of the receptors reach the cell surface as 160 kDa mature proteins. Once on the cell surface, these receptors display a reduced ability to bind LDL; that means that the mutation has disrupted both transport and binding capacity of the receptor[36]. Recently, a mutation in the WHHL rabbit receptor gene has been identified that causes the deletion of four amino acids in the LDL binding domain. While this might explain the inability of the mutant receptors to bind LDL normally, the reason for their delayed processing is less obvious.

Class 3 Mutations: Precursor Processed, Abnormal Binding

Receptors in this class of mutants reach the cell surface at normal rates, but are able to bind less than 15% of the normal amount of LDL[14,15,34]. Most of the class 3 receptors have normal M_r in SDS-PAGE (allele designation, R-160 b-). Receptors with apparent M_r of 140,000 (R-140 b-)[14] and 210,000 (R-210 b-)[13] have been described as well. All mature receptor proteins in this class originate as precursors with apparent molecular weights that are 40,000 less, i.e., 120,000, 100,000, and 170,000, respectively. It appears that the abnormal molecular weights are due to alterations in the amino acid sequence and not to changes in carbohydrate content[32].

The structure of the LDL binding domain (7 repeats of a 40 amino acid sequence) suggests an explanation for the abnormally sized, binding deficient receptors in class 3. The DNA encoding the repeat structure might be susceptible to deletion or duplication following mispairing and recombination of homologous regions during meiosis. Such deletions or duplications would change size and binding capacity of the receptors.

Class 4 Mutations: Precursor Processed, Receptor Binds LDL but Does Not Cluster in Coated Pits

These mutations have been termed internalization-defective[38]. Clustering of receptors in coated pits is a prerequisite for subsequent internalization of ligand, and this process is disrupted in class 4 mutations. To date, the molecular defects in five patients with this phenotype have been elucidated. The original example was patient J.D.[38]. While he inherited a R-O (class 1) allele from his mother, he inherited a gene from his father that produces a cell surface receptor of normal size and normal LDL binding capacity, but which is unable to carry the bound LDL into the cell (R-160 i⁻ allele)[39]. Thus J.D. is a compound heterozygote with regards to his genotype at the LDL receptor locus.

It was suspected early that the R-160 i⁻ allele produces a receptor that is altered in its intracellular portion of 50 amino acid residues, because this region of the receptor is believed to mediate the clustering of receptors into coated pits through interaction with cytoplasmic coated pit specific proteins[39]. The molecular defect, indeed, was found to reside in the intracellular domain[40]. A point mutation has converted the codon for tyrosine 807 into one for cysteine. Subsequently, four other patients with internalization defects have been identified and their genes analyzed. In one of them, the internalization-defective allele produces a receptor that lacks the intracellular domain and the membrane-spanning region, and the truncated protein is secreted from these cells[41]. The secreted receptor has an apparent M_r of 150,000 (allele designation, R-150 i⁻, sec). Another is a patient from Japan[42], the offspring of consanguineous parents, apparently homozygous for a R-150 i⁻, sec allele. Two more internalization-defective alleles were identified and the mutant genes analyzed[37]. In one (FH patients 682 and 683), the cytoplasmic domain consists of only two residues due to a point mutation that changes a tryptophan codon into a termination codon[37]. The other mutant gene (found in patient FH 763) contains a four-base duplication, producing a frameshift that alters the reading frame. The cytoplasmic tail of this mutant receptor is made up of six of the normal plus eight additional amino acids instead of the normal 50[37].

OUTLOOK

Delineation of mutations at the LDL receptor locus in FH has confirmed results of cell biological investigations, and afforded an understanding of structure-function relationships of the key player in systemic cholesterol homeostasis. In addition, this fascinating experimental system has established concepts of a more general nature.

It is clear that, given a normal LDL receptor gene, clustering of LDL receptors in coated pits is essential for endocytosis and regulatory functions of the ligand. This fact sets receptor-mediated endocytosis apart from other cellular uptake processes. Whether or not similar mechanisms operate in the metabolism of lipoproteins such as high density lipoproteins and chylomicron remnants is still an unsettled issue, but increasing biochemical, physiological, and genetic evidence indicates a different mode of cellular interaction for lipoproteins other than LDL.

ACKNOWLEDGEMENTS

W.J. Schneider is a Heritage Medical Scientist of the Alberta Heritage Foundation for Medical Research and an MRC Scientist. Special thanks go to Yolanda Gillam for expert help in preparation of this manuscript.

REFERENCES

1. Brown, M.S. and Goldstein, J.L. 1974. Familial hypercholesterolemia-defective binding of lipoproteins to cultured fibroblasts associated with impaired regulation of 3-hydroxy-3-methylglutaryl coenzyme A reductase activity. Proc. Natl. Acad. Sci. 71: 788.

2. Goldstein, J.L. and Brown, M.S. 1990. Regulation of the mevalonate pathway. Nature 343: 425.

3. Goldstein, J.L., Dana, S.E. and Brown, M.S. 1974. Esterification of low density lipoprotein cholesterol in human fibroblasts and its absence in homozygous familial hypercholesterolemia. Proc. Natl. Acad. Sci. USA 71: 4288.

4. Brown, M.S. and Goldstein, J.L. 1975. Regulation of activity of low density lipoprotein receptor in human fibroblasts. Cell 6: 307.

5. Brown, M.S. and Goldstein, J.L. 1976. Receptor-mediated control of cholesterol metabolism. Science 191: 150.

6. Goldstein, J.L. and Brown, M.S. 1976. LDL pathway in human fibroblasts. Receptor-mediated mechanisms for regulation of cholesterol metabolism. Curr. Topics Cell Reg. 11: 147.

7. Goldstein, J.L. and Brown, M.S. 1977. Low density lipoprotein pathway and its relation to atherosclerosis. Ann. Rev. Biochem. 46: 897.

8. Goldstein, J.L. and Brown, M.S. 1974. Binding and degradation of low-density lipoproteins by cultured human fibroblasts. Comparison of cells from a normal subject and from a patient with homozygous familial hypercholesterolemia. J. Biol. Chem. 249: 5153.

9. Fredrickson, D.S., Goldstein, J.L. and Brown, M.S. 1978. The familial hypercholesterolemias. In: The Metabolic Basis of Inherited Disease. Wyngaarden, J.B. and Fredrickson, D.S. (eds), McGraw-Hill, New York.

10. Goldstein, J.L. and Brown, M.S. 1978. Hypercholesterolemia. Pathogenesis of a receptor disease. John Hopkins Med. 143: 8.

11. Khachadurian, A.K. 1964. Inheritance of essential familial hypercholesterolemia. Am. J. Med. 37: 402.

12. Goldstein, J.L., Dana, S.E., Brunschede, G.Y. and Brown, M.S. 1975. Genetic heterogeneity in familial hypercholesterolemia. Evidence for 2 different mutations affecting function of low density lipoprotein receptor. Proc. Natl. Acad. Sci. USA 72: 1092.

13. Tolleshaug, H., Goldstein, J.L., Schneider, W.J. and Brown, M.S. 1982. Posttranslational processing of the LDL receptor and its genetic disruption in familial hypercholesterolemia. Cell 30: 715.

14. Tolleshaug, H., Hobgood, K.K., Brown, M.S. and Goldstein, J.L. 1983. The LDL receptor locus in familial hypercholesterolemia. Multiple mutations disrupt transport and processing of membrane receptor. Cell 32: 941.

15. Goldstein, J.L. and Brown, M.S. 1983. Familial hypercholesterolemia. In: The Metabolic Basis of Inherited Disease. Stanbury, J.B., Wyngaardon, J.B., Fredrickson, D.S., et al. (eds), McGraw-Hill, New York.

16. Slack, J. and Nevin, N.C. 1968. Hyperlipidemic xanthomatosis. I. Increased risk of death from ischemic heart disease in first degree relatives of 53 patients with essential hyperlipidemia and xanthomatosis. J. Med. Genet. 5: 4.

17. Stone, N.J., Levy, R.I., Fredrickson, D.S. and Verber, J. 1974. Coronary artery disease in 116 kindred with familial type II hyperlipoproteinemia. Circulation 49: 476.

18. Yamamoto, T., Davis, C.G., Brown, M.S., Schneider, W.J., Casey, M.L., Goldstein, J.L. and Russell, D.W. 1984. The human LDL receptor. A cysteine-rich protein with multiple ALU sequences in its messenger RNA. Cell 39: 27.

19. Russell, D.W., Schneider, W.J., Yamamoto, T., Luskey, K.L., Brown, M.S. and Goldstein, J.L. 1984. Domain map of the LDL receptor. Sequence homology with the epidermal growth factor precursor. Cell 37: 577.

20. Sudhof, T.C., Goldstein, J.L., Brown, M.S. and Russell, D.W. 1985. Cassette of 8 exons shared by genes for LDL receptor and EGF precursor. Science 228: 815.

21. Innerarity, T.L., Weisgraber, K.H., Arnold, K.S., Rall, S.C. Jr. and Mahley, R.W. 1984. Normalization of receptor binding of apolipoprotein E2. Evidence for modulation of the binding site conformation. J. Biol. Chem. 259: 7261.

22. Knott, T.J., Rall, S.C. Jr., Innerarity, T.L., Jacobson, S.F., Urdena, M.S., Levy-Wilson, B., Powell, L.M., Pease, R.J., Eddy, R., Nakai, H., Byers, M., Priestly, L.M., Robertson, E., Rall, L.B., Betsholtz, C., Shows, T.B., Mahley, R.W. and Scott, J. 1985. Human apolipoprotein B. Structure of carboxyl terminal domains, sites of gene expression, and chromosomal location. Science 230: 37.

23. Goldstein, J.L., Brown, M.S., Anderson, R.G.W., Russell, D.W. and Schneider, W.J. 1985. Receptor-mediated endocytosis. Concepts emerging from the LDL receptor system. Ann. Rev. Cell Biol. 1: 1.

24. Mahley, R.W. and Innerarity, T.L. 1983. Lipoprotein receptors and cholesterol homeostasis. Biochim. Biophys. Acta 737: 197.

25. Innerarity, T.L. and Mahley, R.W. 1978. Enhanced binding by cultured human fibroblasts of Apo-E containing lipoproteins as compared with low density lipoproteins. Biochemistry 7: 1440.

26. Pitas, R.E., Innerarity, T.L. and Mahley, R.W. 1980. Cell surface receptor binding of phospholipid-protein complexes containing different ratios of receptor-active and receptor-inactive E apolipoprotein. J. Biol. Chem. 255: 5454.

27. Basu, S.K., Goldstein, J.L., Anderson, R.G.W. and Brown, M.S. 1976. Degradation of cationized low density lipoprotein and regulation of cholesterol metabolism in homozygous familial hypercholesterolemia fibroblasts. Proc. Natl. Acad. Sci. USA 73: 3178.

28. Schneider, W.J., Beisiegel, U., Goldstein, J.L. and Brown, M.S. 1982. Purification of the low density lipoprotein receptor. An acidic glycoprotein in 164,000 molecular weight. J. Biol. Chem. 257: 2664.

29. Scott, J., Urdea, M., Quiroga, M., Sanchez-Pescador, R., Fong, N., Selby, M., Rutter, W.J. and Bell G.I. 1983. Structure of a mouse submaxillary messenger RNA encoding epidermal growth factor and seven related proteins. Science 221: 236.

30. Gray, A., Dull, T.J. and Ullrich, A. 1983. Nucleotide sequence of epidermal growth factor cDNA predicts a 128,000 molecular weight protein precursor. Nature 303: 722.

31. Doolittle, R.F., Feng, D.-F. and Johnston, M.S. 1984. Computer-based characterization of epidermal growth factor precursor. Nature 307: 558.

32. Cummings, R.D., Kornfeld, S., Schneider, W.J., Hobgood, K.K., Tolleshaug, H., Brown, M.S. and Goldstein, J.L. 1983. Biosynthesis of N-linked and O-linked oligosaccharides of the low density lipoprotein receptor. J. Biol. Chem. 258: 15261.

33. Davis, C.G., Elhammer, A., Russell, D.W., Schneider, W.J., Kornfeld, S., Brown, M.S. and Goldstein, J.L. 1986. Deletion of clustered O-linked carbohydrates does not impair function of low density lipoprotein receptor in transfected fibroblasts. J. Biol. Chem. 261: 2828.

34. Beisiegel, U., Schneider, W.J., Goldstein J.L., Anderson, R.G.W. and Brown, M.S. 1981. Monoclonal antibodies to the low density lipoprotein receptor as probes for the study of receptor mediated endocytosis and the genetics of familial hypercholesterolemia. J. Biol. Chem. 256: 11923.

35. Schneider, W.J., Goldstein, J.W. and Brown, M.S. 1985. Purification of the LDL receptor. Methods Enzymol. 109: 405.

36. Schneider, W.J., Brown, M.S. and Goldstein, J.L. 1983. Kinetic defects in the processing of low density lipoprotein receptor in fibroblasts from WHHL rabbits and a family with familial hypercholesterolemia. Mol. Biol. Med. 1L353.

37. Lehrman, M.A., Goldstein, J.L., Brown, M.S., Russell, D.W. and Schneider, W.J. 1985. Internalization-defective LDL receptors produced by genes with nonsense and frameshift mutations that truncate the cytoplasmic domain. Cell 41: 735.

38. Schneider, W.J. 1989. The low density lipoprotein receptor, Biochim. Biophys. Acta 988: 303.

39. Goldstein J.L., Brown, M.S. and Stone, N.J. 1977. Genetics of LDL receptor. Evidence that mutations affecting binding and internalization are allelic. Cell 12: 629.

40. Davis, C.G., Lehrman, M.A., Russell, D.W., Anderson, R.G.W., Brown, M.S. and Goldstein J.L. 1986. The JD mutation in familial hypercholesterolemia. Amino acid substitution in cytoplasmic domain impedes internalization of LDL receptors. Cell 45: 15.

41. Lehrman, M.A., Schneider, W.J., Sudhoff, T., Brown, M.S., Goldstein, J.L. and Russell, D.W. 1985. Mutations in LDL receptor. ALU-ALU recombination deletes exon encoding transmembrane and cytoplasmic domains. Science 227: 140.

42. Miyake, Y., Tajima, S., Yamamura, T. and Yamamoto, A. 1981. Homozygous familial hypercholesterolemia mutant with a defect in internalization of low density lipoprotein. Proc. Natl. Acad. Sci. USA 78: 5151.

THE MOLECULAR BASIS FOR LIPOPROTEIN INTERACTION

WITH VASCULAR TISSUE

Göran Bondjers*, Eva Hurt Camejo, Olle Wiklund,
Urban Olsson, Sven-Olof Olofsson and German Camejo

Wallenberg Laboratory for Cardiovascular Research
University of Göteborg
Göteborg, Sweden

ABSTRACT

Local and focal retention of lipoproteins rather than increased influx appears to be the basis for lipoprotein deposition in arterial tissue during atherogenesis. Binding of low density lipoproteins to arterial chondroitin sulphate rich proteoglycans is mediated by specific hydrophilic peptide sequences in apoB, characterized by a high frequency of basic amino acids. One of these sequences is considered to be involved in the interaction between the lipoproteins and the LDL receptor. Consequently, it might be postulated that the receptor-mediated cellular uptake of LDL could be inhibited in the presence of an excess of arterial proteoglycans. However, LDL which has been precipitated by proteoglycan and subsequently resolubilized is taken up more avidly than native LDL both in macrophages and in smooth muscle cells. This appears to be due to a selection by the proteoglycans of a more reactive fraction of LDL. This fraction has a smaller size and less surface phospholipids. As smaller LDL particles also have a higher transfer rate into the arterial tissue they may be particularly atherogenic. Low density lipoproteins appear to be taken up with higher affinity by arterial macrophages than by smooth muscle cells. The selective transfer of the LDL to macrophages can be inhibited by alpha-tocopherol suggesting that oxidative modification may be involved in this process. In fact, binding of LDL to arterial proteoglycans also appears to increase the susceptibility of the lipoproteins to oxidative modification, as well as the susceptibility to proteolytic degradation. Thus, the interaction of LDL with proteoglycans might be involved in several of the key elements of the atherogenic process.

INTRODUCTION - THE CELL POPULATION IN ATHEROSCLEROSIS

The existence of a relationship between high levels of plasma lipoproteins and atherosclerosis is a central dogma in prevailing hypotheses about atherogenesis. As an increased arterial lipid content is one of the most obvious differences between the atherosclerotic and the normal arterial tissue, it has been suggested that plasma lipoproteins may be deposited in the arterial tissue in proportion to the increased levels

---

*Correspondence: Göran Bondjers, Wallenberg Laboratory for Cardiovascular Research, Sahlgrenska sjukhuset, S-413 45 Göteborg, Sweden

in plasma. However, it appears that this view is over-simplified and can neither explain the focality of atherosclerosis nor the cellular accretion of lipids.

The cellular population of the atherosclerotic lesion is more heterogenous than we thought a few years ago. Using monoclonal antibodies specific for different cell types, it has been possible to establish the presence of three subendothelial cell types (Jonasson *et al.* 1986; Tsukada *et al.*, 1986):

1. Smooth muscle cells, which have been the focus of interest for numerous investigators during the last decade, as summarized by several reviews in this volume. The smooth muscle cell is the major cellular inhabitant of the non-atherosclerotic intima.
2. Monocyte-derived macrophages, frequent around the necrotic core of the human plaque, as well as in diet-induced experimental atherosclerosis.
3. Lymphocytes, primarily T lymphocytes, particularly common in the periphery of the human plaque, but also present in diet-induced experimental atherosclerosis (Hansson and Bondjers, studies in progress).

The growth and differentiation of one category of cells may be controlled by the others through different mechanisms discussed by other contributors to this volume.

The characteristic cell in the atherosclerotic lesion is the foam cell, a lipid-containing cell rarely found in normal arterial tissue. Both smooth muscle cells and macrophages may transform into foam cells (Jonasson *et al.*, 1986; Tsukada *et al.*, 1986; Rosenfeld *et al*,. 1987). Lymphocytes, on the other hand, do not appear to be involved in such processes. In view of cellular mechanisms for the control of lipoprotein uptake, it is not immediately obvious why foam cells would develop. In this presentation we will discuss some of the factors which may be involved in lipid accretion in smooth muscle cells and macrophages in vascular tissue. To set the scene for this discussion, we have to elaborate the role of the extracellular tissue in the atherosclerotic lesion in some detail.

LIPOPROTEIN DEPOSITION IN ARTERIAL TISSUE

The rate of transfer for lipoproteins through the endothelial cell layer appears to be inversely proportional to size. Therefore, high density lipoproteins enter the subendothelial space more readily than low density lipoproteins, and these more readily than very low density lipoproteins (Stender and Zilversmit, 1981). Even within the lipoprotein density classes, the inverse relationship between transfer rates and size is apparent. Thus, smaller LDL particles enter the arterial wall more readily than larger LDL particles (Nordestgard and Zilversmit, 1989). The transfer of different lipoprotein particles do not appear to have any closer correlation with the actual concentration of lipoprotein, neither in the normal intima nor in the atherosclerotic lesion. Thus, the arterial LDL concentrations appear to be higher than those of HDL (Smith and Staples, 1982). This suggests that factors besides influx of lipoproteins are significant. A very clear demonstration of this point was obtained from a study of lipoprotein deposition after arterial injury. In arterial tissue devoid of endothelial lining the influx of LDL was severalfold higher than in tissue covered with endothelium. Nevertheless, the actual lipoprotein concentrations in intimal thickenings underneath an intact endothelium were similar to those in intimal thickenings devoid of endothelium. Actually, studies *in vivo* as well as in an *in vitro* perfusion system in our laboratory indicate that the lipoprotein pool in arterial tissue is saturated by the influx of lipoproteins during a 24 hour period (Olsson *et al.*, 1990; Wiklund *et al.*, 1987). These data clearly indicate that retention of lipoproteins might be of greater significance than influx in determining the actual lipoprotein concentrations that vascular cells are exposed to. Such retention may be both specific and selective and lead to considerable differences between lipoprotein concentrations in plasma and in arterial tissue. In addition, exposure to various components in the arterial tissue may induce modifications in lipoprotein characteristics affecting their interaction with both macrophages and smooth muscle cells.

Once the lipoprotein has been transferred from plasma to the subendothelial space, it will encounter the extracellular matrix of the arterial intima. Elastin, collagen and proteoglycans all appear to have the capacity to bind lipoproteins (review with references: Camejo, 1982). In particular, the arterial proteoglycans have been implicated in the reaction with lipoproteins. We have described a system for evaluating the reaction between LDL and arterial proteoglycans in quantitative terms (Camejo et al., 1988; 1990). Results from our studies with this model indicate that HDL does not bind to arterial proteoglycans, VLDL binds to a minor extent, whereas LDL binds with high affinity. The binding of LDL appears to be mediated by specific peptide sequences in apoB (Camejo et al., 1988; Olsson et al., 1991). These sequences were characterized by a number of basic amino acids, surrounded by hydroxylated amino acids. The minimal recognition site in the area with the highest affinity had the sequence: ARG-LEU-THR-ARG-LYS-ARG-GLY-LEU-LYS. This sequence is quite similar to those in a number of other glycosaminoglycan binding proteins (Cardin and Weintraub, 1989), including FGF and PDGF (Fager et al., 1990). It might be speculated that these proteins compete for binding sites on the proteoglycans. If so, the stimulation of smooth muscle proliferation in hyperlipidemia might be related to such competition between LDL and PDGF.

As apoB is found both in VLDL and in LDL, the discrepancy in binding capacity between these lipoprotein fractions may appear surprising. During the lipolytic cascade, one LDL particle is formed from one VLDL particle (review with references: Eisenberg, 1988). Even if initially surprising, the difference in binding to proteoglycans may concur with the fact that LDL but not VLDL binds to the LDL receptor even though apoB is the ligand for this receptor (review with references: Brown and Goldstein, 1983). It is unclear, however, what modifications in the lipoprotein structure during the lipolytic degradation of VLDL to LDL may be involved in the acquisition of reactivity with the proteoglycans or the LDL receptor. It should be noted though that the putative ligands for the LDL receptor and for arterial proteoglycans appear to coincide (Camejo et al., 1988, Yang et al., 1986), suggesting that changes in conformation in this area affect both reactions.

LDL-PROTEOGLYCAN INTERACTION AND FOAM CELL FORMATION

As the receptor-binding and proteoglycan-binding regions of apoB appeared to overlap, we hypothesized that proteoglycan binding might affect LDL interaction with its cellular receptor. In previous experiments we had observed that LDL after exposure to the arterial tissue appeared to be taken up almost exclusively by macrophages, with very low uptake in smooth muscle cells (Wiklund et al., 1987). Could these results be due to inhibition of LDL binding to the receptor for native LDL, with increased time of exposure to macrophage receptors for modified LDL?

To test this hypothesis, LDL was exposed to proteoglycans, leading to the formation of an LDL-proteoglycan precipitate. The lipoproteins were then re-eluted from the precipitate by increasing the salt concentration and used in cell experiments to test whether the eluted lipoproteins might be changed in their binding to the LDL receptor. We reasoned that the re-eluted lipoproteins might be similar to those lipoproteins found in the arterial tissue after primary association followed by dissociation from the proteoglycans there.

In fact, we observed significant differences in LDL accumulation in smooth muscle cells, when the lipoproteins had been exposed to proteoglycans. However, contrary to our hypothesis the uptake of lipoproteins had increased (Bondjers et al., 1988; 1989). The uptake of native and proteoglycan-exposed LDL appeared to be mediated by the same receptor, the LDL receptor. It also appeared that both native LDL and proteoglycan-treated LDL regulated the expression of this receptor. To evaluate the relationships between the processes involved in uptake of normal and proteoglycan-treated LDL further, LDL receptor-negative fibroblasts were incubated with proteoglycan-treated LDL. If a specific receptor for proteoglycan-treated LDL existed, it appeared reasonable to assume that this receptor would be expressed also in these cells. However, no evidence of a high-affinity uptake of LDL (native or proteoglycan-treated) was observed in these cells (Bondjers et al., 1988; 1989). On the other hand, the non-saturable, low-affinity uptake was significantly higher for the proteoglycan-treated LDL. From these experiments it was not easy to conclude, however, how large a proportion of the increase in lipoprotein uptake induced by exposure to proteoglycans might be explained by low-affinity uptake of the lipoproteins.

In other experiments, the effects of the presence of proteoglycans during the incubations with low density lipoproteins was evaluated. We had previously demonstrated that low density lipoproteins and arterial proteoglycans formed associations under physiological salt concentrations. Thus, the apparent molecular weight of LDL was higher in the presence of arterial proteoglycans at physiological salt concentrations (unpublished observations). When smooth muscle cells were incubated with low density lipoproteins in the presence of proteoglycans, the binding and internalization of the lipoproteins was actually inhibited, contrary to the observations with eluted lipoproteins. This supports the hypothesis that the localization of the ligand for the LDL receptor on the lipoprotein particle co-localizes with the ligand for the arterial proteoglycans.

We concluded from the results of these experiments that binding of LDL to arterial proteoglycans, followed by elution of the lipoproteins from this complex leads to an increased uptake of LDL into smooth muscle cells, and that most of that increased uptake could be mediated by non-specific pathways. As such pathways are not subject to feed-back regulation. This effect could explain the formation of smooth muscle foam cells where down-regulation of lipoprotein uptake in response to lipid accumulation otherwise might be anticipated.

The possibility that macrophages might behave differently from smooth muscle cells was tested in parallel experiments. In short, these experiments showed an increased uptake of proteoglycan-treated LDL compared with native LDL, and the uptake appeared to be mediated by similar mechanisms (Bondjers et al., 1988; 1989; Hurt Camejo et al., 1990). Therefore, exposure to proteoglycan-treated LDL might induce foam cell transformation also in macrophages, via pathways not subjected to feed-back regulation of the uptake.

CHANGES IN LDL CONFORMATION OR SELECTION OF LDL PARTICLES?

Two alternative explanations to our observations of an enhanced uptake might be discussed. The first explanation would be that exposure of LDL to the proteoglycans in itself might induce changes in LDL structure leading to increased cellular lipoprotein uptake. This possibility was supported by a number of observations from Dr. Camejo's laboratory indicating changes in physico-chemical characteristics of the lipoproteins following exposure to the proteoglycans (review with references: Camejo, 1982) and changes in sensitivity to proteolytic degradation (Camejo et al., 1991). An alternative explanation, however, would be that the proteoglycans select a sub-population of lipoprotein particles with specific characteristics including an increased affinity for proteoglycans and an increased cellular binding and uptake. To evaluate this alternative we decided to precipitate LDL sequentially, i.e., first precipitating one fraction of LDL and then removing the LDL-proteoglycan complex. The supernate would then be exposed again to proteoglycans leading to the formation of another LDL-proteoglycan precipitate, which would be removed and so on. With this method four LDL fractions, comprising about 90% of the total LDL, were isolated. The different fractions were exposed to macrophages, and large differences in cellular uptake between the fractions were observed (Hurt Camejo et al., 1990). The LDL fraction that was precipitated first was taken up much more readily than the second, which was taken up more readily than the third, which was taken up more readily than the fourth. These observations suggest that the arterial proteoglycans select a lipoprotein fraction with a tendency to be internalized more efficiently than LDL as a whole. This might be explained if the ligand for the LDL receptor and that for the arterial proteoglycan are localized in the same segment of apoB, and this segment is exposed to a variable degree, as discussed above. The variation in exposure might be related to the position of the LDL particle in the lipolytic cascade.

When we compared other characteristics of the different fractions of LDL precipitated by repeated exposure to proteoglycans, some differences were obvious. Thus, the order of precipitation was parallelled by an increase in size as well as an increase in surface lipids, primarily phospholipids. Therefore, it appeared that the proteoglycans have higher affinity for a lipoprotein fraction which is smaller and internalized in cells more rapidly. These observations focus our attention to the size of low density lipoproteins as a factor of major significance for lipoprotein deposition and foam cell formation during atherogenesis. Small low density lipoproteins would be transferred more rapidly into the arterial tissue, would be specifically retained by arterial proteoglycans, and finally internalized to transform macrophages and smooth muscle cells into foam cells.

LDL BINDING TO PROTEOGLYCANS IN THE ARTERIAL WALL IN SITU

The role of LDL binding to proteoglycans for cellular lipid deposition had primarily been tested in cell culture systems so far. Therefore, an uncertainty remained whether such processes might be significant in the arterial tissue. To clarify this, we decided to use the possibility of inhibiting LDL-proteoglycan interaction with basic polypeptides, as suggested by our studies with synthetic polypeptides (Camejo *et al.* 1988). In order to achieve an increased inhibitory capacity, polylysine was used in our *in vitro* perfusion system, during exposure of arterial tissue to LDL. It was obvious that the polylysine had reduced the tissue uptake of LDL by 20-40% (Wiklund *et al.* 1990). These observations reinforced our impression that LDL binding to proteoglycans is a significant step in cellular lipid accretion during atherogenesis.

OXIDATIVE MODIFICATION OF LDL

The significance of LDL modification for foam cell formation has been subject to considerable discussion (reviews with references: Brown and Goldstein, 1983; Steinberg, 1988). LDL which has been modified by various treatments are taken up via high-affinity receptors on macrophages, the scavenger receptors. Two such receptors were recently cloned and characterized (Kodama *et al.*, 1990). Recently, much of the interest concerning modifications of lipoproteins has been focused on oxidative modifications. Such modifications were first induced by exposing LDL to endothelial cells (Henriksen *et al.*, 1981), smooth muscle cells (Heinecke *et al.*, 1984) or macrophages (Parthasarathy *et al.*, 1986). Such modification might be inhibited by the presence of an antioxidant such as vitamin E (Steinbrecher *et al.*, 1984). However, the significance of modified lipoproteins for foam cell formation in atherogenesis is still unclear as most studies on such processes have been performed in cell culture. To shed further light on the role of oxidative modification for lipoprotein uptake in the intact arterial tissue, we decided to use our *in vitro* perfusion system. We wanted to investigate whether oxidative modification of LDL might be significant for the selective transfer of LDL to macrophages and whether the presence of alpha tocopherol as an anti-oxidant might inhibit this process. A significant reduction of lipoprotein uptake in macrophages, isolated from arterial tissue after perfusion was observed in the presence of the antioxidant (Wiklund *et al.* 1990). This effect was not exerted in the perfusion medium as suggested by control experiments where the uptake of lipoproteins in mouse peritoneal macrophages was studied. Thus, these studies suggest that an oxidative modification of LDL within the arterial tissue might lead to an increased uptake of lipoproteins into tissue macrophages, and by inference to foam cell formation.

AN INTERRELATIONSHIP BETWEEN LDL-PROTEOGLYCAN INTERACTION AND OXIDATION?

Our studies with the *in vitro* perfusion system had suggested that both LDL binding to proteoglycans and oxidative modification of LDL might be significant for cellular uptake of lipoproteins during atherogenesis. We had also demonstrated that LDL binding to proteoglycans leads to an increased susceptibility to enzymatic degradation. In view of these observations, it appeared logical to test whether the effects we had observed were interrelated. Thus, it could be hypothesized that binding of LDL to proteoglycans might lead to increased susceptibility to oxidative modification, followed by increased uptake in macrophages. To test this hypothesis, LDL was complexed with proteoglycans, and exposed to Cu(II) in cell culture medium. The formation of thiobarbituric acid-reacting substances (TBARS), diene conjugation and lipid composition during incubation at 37°C were followed (Camejo et al., 1991). The formation of TBARS and diene conjugation increased dramatically after exposure of the lipoproteins to proteoglycans, indicating an increased sensitivity to oxidative modification. As a consequence of these changes, cellular uptake of LDL increased after oxidative modification, if the lipoproteins were exposed to proteoglycans compared with native lipoproteins. Thus, binding of LDL to proteoglycans might increase oxidative modification through two different routes. Retention of lipoproteins by proteoglycans might increase the residence time in the intima, with an increased accumulated exposure to free radicals generated by the arterial cells. In addition, the binding of the lipoproteins to the proteoglycans in itself might increase the susceptibility to oxidative modification.

CONCLUSIONS

The atherosclerotic lesion is primarily made up by extracellular components. A major constituent of the extracellular space are proteoglycans, macromolecules composed of glycosaminoglycans and proteins. Through charge interactions, these macromolecules bind various proteins including lipoproteins and growth factors. The interactions between lipoproteins and proteoglycans lead to a selection of a more reactive species of lipoprotein particles, and in addition to an increased sensitivity to oxidative modification. Working together, these processes lead to an increased cellular uptake of lipoproteins and ultimately foam cell formation.

REFERENCES

Bondjers, G., Wiklund, O., Olofsson S.-O., Fager, G., Hurt, E. and Camejo, G. 1988. Low density lipoprotein interaction with arterial proteoglycans: Effects on lipoprotein interaction with arterial cells. In: "Hyperlipidaemia and Atherosclerosis". Suckling, K.E. and Groot, P.E. (eds), Academic Press, London p. 135.

Bondjers, G., Camejo, G., Fager, G., Olofsson, S.-O. and Wiklund, O. 1989. Functional interrelationships between the smooth muscle and macrophage cell populations of the atherosclerotic plaque. In: "Hypertension and Atherosclerosis". C. Dal Palú and R. Ross (eds), Excerpta Medica, Amsterdam p. 75.

Brown, M.S. and Goldstein, J.L. 1983. Lipoprotein metabolism in the macrophage: Implications for the cholesterol deposition in atherosclerosis. Ann. Rev. Biochem. 52: 223.

Camejo, G. 1982. The interaction of lipids and lipoproteins with the intercellular matrix of arterial tissue: Its possible role in atherogenesis. Adv. Lipid Res. 19: 1.

Camejo, G., Olofsson, S.-O., Lopez, F., Carlsson, P. and Bondjers, G. 1988. Identification of apo B-100 segments mediating the interaction of low density lipoproteins with arterial proteoglycans. Arteriosclerosis 8: 368.

Camejo, G., Rosengren, B., Olsson, U., Lopez, F., Olofsson, S.-O., Westerlund, C. and Bondjers, G. 1991. Molecular basis of the association of arterial proteoglycans with low density lipoproteins: Its effect on the structure of the lipoprotein particle. Eur. Heart J. (in press).

Cardin, A.D. and Weintraub, H.J.R. 1989. Molecular modelling of protein-glycosaminoglycan interaction. <u>Arteriosclerosis</u> 9: 21.

Eisenberg, S. 1988. Regulation of the apoB100 cascade. <u>In</u>: "Hyperlipidemia and Atherosclerosis. K.E. Suckling and P.H.E. Groot (eds), Academic Press, London, p. 65.

Fager, G., Camejo, G., Olsson, U., Östergren-Lundén, G. and Bondjers, G. 1990. The long platelet-derived growth factor a-chain transcript encodes a peptide with an amino acid sequence that specifically binds to heparin-like glycosaminoglycans. <u>In Vitro</u> (in press).

Heinecke, J.W., Rosen, H. and Chait, A. 1984. Iron and copper promote modification of low density lipoprotein by human arterial smooth muscle cells in culture. <u>J. Clin. Invest.</u> 74: 1890.

Henriksen, T., Mahoney, E.M. and Steinberg, D. 1981. Enhanced macrophage degradation of low density lipoprotein previously incubated with cultured endothelial cells: Recognition of receptors for acetylated low density lipoproteins. <u>Proc. Natl. Acad. Sci. USA</u> 78: 6499.

Hurt Camejo, E., Camejo, G., Rosengren, B., Lopez, F., Wiklund, O. and Bondjers, G. 1990. Differential uptake of proteoglycan-selected subfractions of low density lipoprotein by human macrophages. <u>J. Lipid Res.</u> (in press).

Jonasson, L., Holm, J., Bondjers, G., Skalli, O. and Hansson, G.K. 1986. Regional accumulations of T cells, macrophages and smooth muscle cells in the human atherosclerotic plaque. <u>Arteriosclerosis</u> 6: 131.

Kodama, T., Freeman, M., Rohrer, L., Zabrecky, J., Matsudaira, P. and Krieger, M. 1990. Type I macrophage scavenger receptor contains alpha-helical and collagen-like coiled coils. <u>Nature</u> 343: 6258.

Nordestgard, B.G. and Zilversmit, D.B. 1989. Comparison of arterial intimal clearances of LDL from diabetic and nondiabetic cholesterol-fed rabbits. Differences in intimal clearance explained by size differences. <u>Arteriosclerosis</u> 9: 176.

Olsson, G., Wiklund, O. and Bondjers, G. 1990. Arterial low density lipoprotein uptake and turnover. Significance of endothelial integrity. <u>Atherosclerosis</u> (submitted).

Olsson, U., Camejo, G. and Bondjers, G. 1990. Molecular parameters that control the association of low density lipoprotein apoB-100 to chondroitin sulfate. <u>Biochim. Biophys. Acta</u> (submitted).

Parthasarathy, S., Printz, D.J., Boyd, D., Joy, L. and Steinberg, D. 1986. Macrophage oxidation of low density lipoprotein generates a modified form recognized by the scavenger receptor. <u>Arteriosclerosis</u> 6: 505.

Rosenfeld, M.E., Tsukada, T., Gown, A.M. and Ross, R. 1987. Fatty streak initiation in Watanabe heritable hyperlipidemic and comparably hypercholesterolemic fat-fed rabbits. <u>Arteriosclerosis</u> 7: 9.

Smith E.B. and Staples, E.M. 1982. Intimal and plasma protein concentrations and endothelial function. <u>Atherosclerosis</u> 41: 295.

Steinberg, D. 1988. Metabolism of lipoproteins and their role in the pathogenesis of atherosclerosis. <u>Atherosclerosis Reviews</u> 18: 1.

Steinbrecher, U.P., Parthasarathy, S., Leake, D.S., Witztum, J.L. and Steinberg, D. 1984. Modification of low density lipoprotein by endothelial cells involves lipid peroxidation and degradation of low density lipoprotein phospholipids. <u>Proc. Natl. Acad. Sci. USA</u> 81: 3883.

Stender, S. and Zilversmit, D.B. 1981. Transfer of plasma lipoprotein components and of plasma proteins into aortas of cholesterol-fed rabbits. Molecular size as a determinant of plasma lipoprotein influx. <u>Arteriosclerosis</u> 1: 38.

Tsukada, T., Rosenfeld, M., Ross, R. and Gown, A.M. 1986. Immunocyto-chemical analysis of cellular components in atherosclerotic lesions. Use of monoclonal antibodies with the Watanabe and fat-fed rabbit. <u>Arteriosclerosis</u> 6: 601.

Wiklund, O., Björnheden, T., Olofsson, S.-O. and Bondjers, G. 1987. Influx and cellular degradation of low density lipoproteins in rabbit aorta determined in an *in vitro* perfusion system. <u>Arteriosclerosis</u> 7: 565.

Wiklund, O., Mattsson, L., Björnheden, T., Camejo, G. and Bondjers, G. 1990. Uptake and degradation of low density lipoproteins in atherosclerotic rabbit aorta; dependence of LDL modification. <u>J. Lipid Res.</u> (in press).

Wiklund, O., Camejo, G., Mattsson, L., Lopez, F. and Bondjers, G. 1990. Cation in polypeptides as modulators of *in vitro* association of LDL with isolated arterial proteoglycans, cellular B, E receptor and arterial tissue. <u>Arteriosclerosis</u> (in press).

Yang C.-Y., Chen, S.-H., Gianturco, S.H., Bradley, W.A., Sparrow, J.T., Tanimura, M., Li, W.-H., Sparrow, D.A., DeLoof, H., Rosseneu, M., Le, F.-S., Gu, Z.-W., Gotto, Jr. A.M. and Chan, L. 1986. Sequence, structure, receptor-binding domains and internal repeats of human apolipoprotein B-100. <u>Nature</u> 323: 738.

CHOLESTEROL, IS THERE A CONSENSUS?

Louis Horlick

Department of Medicine
University Hospital
University of Saskatchewan
Saskatoon, Saskatchewan

The Canadian Cholesterol Consensus Conference was held in March 1988[1]. It was the fourth conference of its kind to be held within a five-year span. The American[2], British[3], European[4] and Canadian Consensus Conferences examined the evidence relating cholesterol to coronary heart disease and found it convincing, and made recommendations for reducing cholesterol levels for the population as a whole, and more specifically for individuals at high risk. Now, less than two years later what appears to have been an international consensus is in danger of falling apart, with potentially disastrous effects for an effective prevention campaign.

The critics include a small group of physicians[5] and some writers for the popular magazines and lay press[6]. They question the importance of cholesterol as a risk factor for coronary heart disease; they impugn the evidence that lowering cholesterol will reduce risk of coronary heart disase; they question the effectiveness of dietary control in lowering cholesterol levels; and they claim that existing evidence is inadequate for making recommendations for women and the elderly. In addition, a small group of medical economists are suggesting that National programs for lowering cholesterol levels would be inordinately expensive[7,8]. Additional controversies have arisen about the wisdom of instituting population strategies versus high risk strategies for cholesterol control, and about the intensity of intervention programs and the cut points for dietary and pharmacological intervention[9].

What has been the effect of all of this on physicians' attitudes and behaviour? In a survey of Saskatchewan physicians carried out in 1989, we found that they considered cholesterol an important risk factor, along with smoking and hypertension, but were not very confident about their ability to deal effectively with it. More than 90% regularly measured the blood cholesterol levels of their adult patients. Their decision points for dietary and pharmacological intervention for blood cholesterol control were quite similar to those described in the 1983 NIH surveys of American physicians[10]. A portion of our Saskatchewan cohort was then enrolled in a continuing education program based on the National Cholesterol Education Program guidelines[11], and the entire population was resurveyed 6-9 months later. There were no basic changes in attitudes about cholesterol, but there were significant changes in their definition and management of hypercholesterolemia. They were now prepared to intervene at lower levels of cholesterol with diet, to persevere longer with dietary treatment, and to intervene with drugs at lower cut points. There was no difference in this regard between the controls and those who had received the special educational package. This suggests an important

Atherosclerosis, Edited by A. I. Gotlieb *et al.*
Plenum Press, New York, 1991

secular trend among physicians and reflects the strong influence of the N.C.E.P. program, and of the medical and lay press and media. The majority of physicians appear to be convinced of the validity of the Consensus recommendations and are acting on them.

In Canada, the greatest impediment to the implementation of the national preventive strategy recommended by the Canadian Consensus Conference[1] appears to be the excessive caution and timidity of our provincial and federal health planners. In this, they are aided and abetted by a small but influential group of health economists who are not convinced that attempts to treat asymptomatic hypercholesterolemia are likely to yield significant medical benefits, and who believe that the costs of any such program could not be supported by our health care system. They[9] are advocating cut off points for dietary and drug treatments that would leave the bulk of our population unprotected from the ravages of hypercholesterolemia[9]. As a result, a clear cut national policy has not yet emerged, and there is confusion in the minds of some physicians and lay people as to what, if anything, should be done about cholesterol.

It seems essential therefore to review once again the strength of the evidence supporting the Consensus recommendations. The weight of factual evidence supporting the causal relationship of blood cholesterol levels and coronary heart disease is immense, and comprises animal experimentation in many species including primates, extensive epidemiological surveys over several decades, controlled clinical intervention studies, and some fundamental discoveries in cellular biology. I will cite only a few of the most important of these studies.

The Framingham Heart Study[12] provided some of the first evidence that elevated cholesterol was an important risk factor for coronary heart disease. The 14 year follow up on 2,282 men and 2,845 women reported in 1971, showed a clear curvilinear relationship between blood cholesterol levels and incidence of CHD over the range of cholesterol levels between 150 and 300 mg/dl. This fact was greatly reinforced by the findings of the Multiple Risk Factor Intervention Study (MRFIT)[13]. In the initial cohort of 361,662 men followed for an average period of 6 years, risk increased steadily, particularly above levels of 200 mg/dl. It doubled between 200 and 250 mg/dl, and doubled again between 250 and 300 mg/dl. Recent data from rural China[14], where cholesterol levels are very low, suggests that the curve bottoms out at levels below 150 mg/dl. Of great interest was the finding that the majority of cardiovascular events occurred at levels of cholesterol between 200 and 250 mg/dl, levels formerly considered to be "normal" for our population. While there is general acceptance of the relationship of cholesterol levels and CHD incidence, there is a tendency among the critics to overlook the very large number of individuals with cholesterol levels between 200 and 250 mg/dl who are at risk.

What is the strength of the evidence that reducing blood cholesterol levels will reduce the risk of CHD? The evidence comes mainly from the Lipid Research Clinics Primary Prevention Trial (LRCPPT)[15], and the Helsinki Heart Study[16]. In the LRCPPT more than 3,800 hypercholesterolemic, middle aged men, who were refractory to lipid lowering diet, were randomized to either of a placebo or cholestyramine (a bile acid sequestrant). Over a seven year period, the treatment group experienced a fall in total cholesterol which was 9% greater than that of the controls and a reduction in coronary events of 19% (CHD death and non-fatal MI). Critics are insistent in pointing out that the actual reduction in events between the treated and placebo groups was from 9.8% to 8.1% or only 1.7% and question its clinical significance. Because of side effects, many subjects took less than the prescribed dose of 24 gm of cholestyramine per day. Subjects who took the full dose of medication reduced their cholesterol levels by 25% and experienced 50% fewer coronary events than the placebo group. The LRCPPT results indicated that a 1% reduction in blood cholesterol was associated with a 2% reduction in CHD events. The Helsinki Heart Study comprised 2,051 middle aged, hyperlipidemic men (cholesterol > 260 mg/dl) who were treated with the fibric acid derivative, gemfibrozil, and compared with an equal number of men randomized to placebo. The gemfibrozil-treated group showed a fall in total and LDL cholesterol of 8%, and a rise in HDL cholesterol of 10% and a reduction in CHD events of 34%. At the end of the period of observation, the event rate curves were still diverging with a 50% difference in CHD event rates. In this study, a fall in cholesterol of 1% was associated with a 3-4% reduction in CHD events. Further analyses of the Helsinki Heart data indicate that individuals with combined hyperlipidemia (increase in both cholesterol and triglycerides) and those with initially low HDL levels were the most likely to achieve benefit from gemfibrozil treatment. Critics tend to concentrate on the high initial cholesterol levels in this group of male, middle aged subjects and question the generalizability of the results. They also harp on the fact that there was no difference in all causes mortality and that the major difference was in the number of non-fatal MI's. Analysis of the data from these two trials lends strong support to the thesis that lowering cholesterol results in a reduced incidence of CHD events. These trials were not designed to test the effects of intervention on total mortality. In order to do that, they would have had to comprise many more subjects and lasted many more years. There is, however, some evidence from three recently reported trials that intervention can influence total mortality. These studies are (1) The Coronary Drug Project[17], (2) The Oslo Study Diet and Antismoking Trial[18] and (3) The Stockholm Ischemic Heart Disease Study[19]. In all of these studies, a long term follow up was required to determine the effect on all causes mortality. This was 9 years for the CDP study and the Oslo study, and 5 years for the Stockholm study. The differences in all causes mortality were 11%, 40% and 26%, respectively.

All of the studies described above have depended on clinical end points. During the last decade, we have also had three studies with angiographic end points which have demonstrated that cholesterol lowering is associated with a slowing of progression, or actual regression of atherosclerotic disease in the coronary arteries. These studies were (1) The NHLBI Type II Coronary Intervention Trial[20], (2) The Leiden Intervention Trial[21], and (3) The Cholesterol Lowering Atherosclerosis Study (CLAS)[22]. The 116 patients in the NHLBI Type II trial were randomized to placebo or cholestyramine and angiograms were performed at the start of the study and five years later. Post-treatment angiograms indicated progression of disease in 49% of the placebo group and 32% of the cholestyramine group. The 39 patients in the Leiden study were placed on a vegetarian type diet high in polyunsaturated fat and low in cholesterol for two years. Forty-six percent of the subjects showed no progression of their lesions and this correlated with low initial TC/HDL ratios or a fall in the ratio during treatment. In the CLAS, 162 men who had recently undergone coronary artery bypass surgery were randomized into two groups, one of which received dietary therapy only, and the other received diet and a combination of colestipol and nicotinic acid in large doses. The latter group showed very significant falls in total and LDL cholesterol and a rise in HDL cholesterol. Atherosclerotic regression was evident in 16% of the group who received drug therapy as compared to 3.6% on diet alone. Sixty-one percent of the drug treated group showed either regresssion or no change compared with 39% of the placebo group. The data from these three trials strongly supports the thesis that lowering cholesterol arrests or reverses the disease process in affected individuals.

The largest proportion of individuals at risk lie in the range of "moderate" risk (5.2-6.2 mmols/L or 200-240 mg/dl) and it seems unreasonable to expose this group of individuals to expensive and potentially hazardous drug therapy. The critics contend that dietary therapy is ineffective in lowering cholesterol to the extent that is required to substantially affect risk of CHD.

Yusuf et al.[23] have examined all the randomized trials of cholesterol lowering, both primary and secondary, diet and drug treated. Twenty-two trials involving 40,000 individuals over a time span of one to seven years were included. Overall 8.4% of 19,813 subjects in the treated group suffered a coronary event (non-fatal myocardial infarction or death) compared with 10.7% of 20,506 subjects in the control group. This represents a 23% risk reduction with 95% confidence limits of -18% to -28% with a p < .0001. The reduction in coronary artery disease was directly related to the degree and duration of cholesterol lowering. For a given reduction in cholesterol level over a standard period of time, roughly similar results were observed in the diet and drug interventions in both primary and secondary trials with respect to fatal and non-fatal events.

Although dietary trials have shown an average fall of about 10% in plasma cholesterol levels, the actual degree of fall in any one individual is unpredictable. For example, in the National Diet Heart Study[24] although the average fall in cholesterol was 10% sustained over the one year period of the study, 35% of the population showed a fall of more than 15% which was sustained. Thus a large part of the population at risk might be expected to respond to a diet low in total fat, saturated fatty acids and cholesterol. Even if we accept the lower estimate of an average 10% reduction in cholesterol, we would anticipate a 20% reduction in CHD events.

The critics contend that current Consensus guidelines which are based mainly on studies in men should not be extended to women or the elderly until the results of specific clinical trials become available. There is a considerable body of evidence which supports the application of the Consensus recommendations on cholesterol control to women. For example, coronary risk was increased in women with total cholesterol levels above 265 mg/dl in the Framingham[25] and Donolo-Tel Aviv[26] studies, and in women with total cholesterol above 235 mg/dl in the Lipid Research Clinics Program Follow-Up Study[27]. There was a strong inverse correlation between HDL cholesterol and CHD; a decrease of 10 mg/dl was associated with a 50% increase in CHD risk among the Framingham women and a 42% increase in risk among the women in the LRC Follow-Up Study. There is no reason to believe that the pathologic process which underlies CHD is any different in women than in men, and there is no good reason to deprive women who are at risk of the benefits of cholesterol reduction.

Recent studies indicate that both total cholesterol and LDL cholesterol are predictive of disease in the elderly. Analysis of the Framingham data indicates that both LDL and total cholesterol are coronary risk factors for elderly individuals. LDL is associated with increased risk at all ages through age 82, while HDL is negatively correlated with CHD through ages 49-82. In the most recent analysis of the Framingham data[28], the investigators found that although the relative risk of CHD declines with increased age, the attributable risk of cholesterol increases with age. In effect, an elevated level of cholesterol contributes to more cases of heart disease in older individuals than in younger ones. It is obvious that clinical judgment must play an important role in determining how the Consensus guidelines are applied to the elderly, but they should not be deprived of the benefits of cholesterol reduction simply on account of chronological age.

A major objection raised by the critics is that cholesterol reduction is not cost effective and will damage our health promotion efforts by diverting large amounts of money from more useful endeavours. Currently about 1.5 million Americans suffer a heart attack each year and approximately 300,000 bypass operations are done annually. A conservative estimate of the cost to the health care system would be 34.5 billion dollars. If we add lost output due to disability, and the cost to society of premature death of these individuals, the cost could be as high as 100 billion dollars annually[29]. Extrapolating to the Canadian population, we would be looking at between 2 and 3 billion dollars in direct health care costs alone. If we were able to lower serum cholesterol levels by 10% through dietary modification alone, this should translate into a reduction in CHD events of 20% and an annual saving of 200-300 million dollars per year. This would largely offset the costs of a national cholesterol control program.

The Canadian Cholesterol Consensus Conference (CCCC)[1] made important recommendations for a National effort to reduce cholesterol levels. The keystone was the recommendation that Canadian food guidelines be changed to reduce the total fat content of the diet to 30% of calories, and that there be a reduction in saturated fats to less than 10% of calories. This was to be coupled to a vigorous health promotion program which would not only promote a heart healthy diet but would also be aimed at promoting increased levels of physical activity, reducing weight and eliminating smoking and other risk factors. In addition to the population strategy, the CCCC also made recommendations for detection and treatment of individuals at moderate and high risk. A task force commissioned by the Ontario Government[9] has issued a report which, while agreeing with the population strategy approach of the CCCC, takes serious issue with the recommendations for moderate and high risk individuals. CCCC recommended that individuals at moderate risk (cholesterol 5.2-6.2 mmol/L) should receive dietary advice from their physicians (AHA level 1) and, if they did not experience a fall in cholesterol, should be referred for dietary therapy. The Ontario recommendations do not specifically address this largest group of individuals at risk, and recommend diet therapy at levels in excess of 6.2 only for males with at least one risk factor and females with two risk factors. Their reasons for this recommendation are that they believe that individuals with cholesterol levels below 6.2 mmol would be covered by the anticipated change in Canada's Food Rules, and that diet therapy is basically not very effective in individuals with asymptomatic moderate hypercholesterolemia. I believe that the evidence which I have presented above deals adequately with that objection.

We must not allow wrangling over definitions and cut points to deflect us from the primary mission of the Cholesterol Consensus. The reexamination of the data supporting the Consensus recommendations shows them to be as valid today as they were when issued in 1988. Although the Federal Government has not yet decided on a national strategy for Canada, a number of encouraging events have occurred. The Federal Government, in collaboration with several Provincial Governments and the Heart and Stroke Foundations, has begun a comprehensive nation-wide survey of risk factors modelled on the Nova Scotia Heart Health Study. Surveys have been completed in British Columbia, Saskatchewan, and Manitoba, and are scheduled to begin shortly in Ontario and Quebec. When this program is completed, Canada will have the most comprehensive national data bank on risk factors. This may permit us to explain the marked variation in coronary heart disease rates within Canada, and will form the basis for intervention campaigns. Nova Scotia, which led the country in its risk factor survey, has already begun a number of targetted programs for children and adolescents, and for economically and educationally underprivileged groups, which are known to have a high incidence of CHD.

A recent survey of cholesterol lowering programs in several countries was presented at the Tenth International Symposium on *Drugs Affecting Lipid Metabolism* which was held in Houston, Texas in November of 1989[30]. Probably the most ambitious and far reaching program in existence is the National Cholesterol Education Program of the U.S.A. This is a cooperative program involving the Federal Government (through the National Heart Lung and Blood Institute), the American Heart Association and many other private agencies. It is modelled on the highly successful National High Blood Pressure Education Program. Its initial effort has been focussed on high risk strategy with the production of detailed guidelines for the identification and treatment of individuals at moderate and high risk. These have been widely disseminated among physicians and health care workers. In addition, the public have been urged to "know your cholesterol". A population based strategy will be proclaimed in March 1990, and additional panels of experts are at work on identification of children and adolescents at risk, and on laboratory methodology for improving the accuracy of lipid measurement. The American people have become very diet conscious, and surveys suggest that at least 50% of them have modified their diets to reduce fat and cholesterol intake.

Australia, through its National Heart Foundation, has implemented screening for risk factors, national risk factor prevalence surveys and developed educational programs to combat hypercholesterolemia. Australia has experienced perhaps the most dramatic fall in CHD mortality of any country (50% reduction in the past 20 years). A recently begun Food Approval Program sponsored by the National Heart Foundation aims to educate the public to choose foods wisely, and to influence the food industry to formulate reduced fat containing products.

Reports from the United Kingdom, Germany, France and Spain all indicate that active programs to reduce cholesterol and other risk factors are under way. A Handbook for CHD Prevention, based on the European Consensus Recommendations and providing algorithms for the treatment of hyperlipidemias, has been distributed to over 45,000 physicians. Of all the European countries, the U.S.S.R. is the least advanced in programming for cholesterol and other risk factor reduction. Diagnostic facilities are scarce and given the state of the Soviet economy, there is unlikely to be any dietary strategy for quite some time. There is a clear need for a unified European population strategy on cholesterol and other risk factors.

REFERENCES

1. Canadian Consensus Conference on Cholesterol; final report. 1988. Can. Med. Assoc. J. 139(suppl.): 1-8.

2. Consensus Conference: Lowering blood cholesterol to prevent heart disease. 1985. J.A.M.A. 253: 2080-2086.

3. Report of the British Cardiac Society Working Group on Coronary Disease Prevention. 1987. British Cardiac Society, London.

4. Study Group. European Atherosclerosis Society, Strategies for the prevention of coronary heart disease. 1987. Eur. Heart J. 8: 77-88.

5. Corday, E. and Ryden, L. 1989. Why some physicians have concerns about the cholesterol awareness program. J. Amer. Coll. Cardiol. 13: 497-502.

6. Moore, T.J. 1989. The Cholesterol Myth. Atlantic Monthly (September): 37-70.

7. Oster, G. and Epstein, A.M. 1987. Cost effectiveness of antihyperlipidemic therapy in the prevention of coronary heart disease. The case for cholestyramine. J.A.M.A. 258: 2381-2387.

8. Taylor, W.C., Pass, T.M., Shepard, D.S. and Komaroff, A.L. 1987. Cholesterol reduction and life expectancy. A model incorporating multiple risk factors. Ann. Int. Med. 106: 605-614.

9. Detection and Management of Asymptomatic Hypercholesterolemia. 1989. A Policy Document by the Toronto Working Group on Cholesterol Policy. Prepared for the Task Force On the Use and Provision of Medical Services. Ontario Ministry of Health.

10. Schucker, B., Wittes, J., Cutler, J. et al. 1987. Changes in physicians perspectives on cholesterol and heart disease; results from two national surveys. J.A.M.A. 258: 3521-3526.

11. National Cholesterol Education Program Expert Panel, National Heart, Lung and Blood Institute. NCEP report of the Expert Panel on Detection, Evaluation, and Treatment of High Blood Cholesterol in Adults. Bethesda, MD. October 1987.

12. Kannel, W.B., Castelli, W.P., Gordon, T. and McNamara, P.M. 1971. Serum cholesterol, lipoproteins, and the risk of coronary heart disease. The Framingham Study. Ann. Int. Med. 74: 1-12.

13. Multiple Risk Factor Intervention Trial Research Group. 1982. Multiple risk factor intervention trial: Risk factor changes and mortality results. J.A.M.A. 248: 1465-1477.

14. Lewis, B. 1989. The contribution of diet change to CHD prevention. Proceedings of the Xth International Symposium on Drugs Affecting Lipid Metabolism, Houston, TX, Nov. 1989 (in press).

15. The Lipid Research Clinics Program: The Lipid Research Clinics Coronary Primary Prevention Trial results, I. 1984. J.A.M.A. 251: 351-364.

16. Frick, M.H., Elo, O., Haapa, K., Heinonen, O.P., Heinsalmi, P., Helo, P., Huttunen, J.K., Kaitaniemi, P., Koskinen, P. and Manninen, V. 1987. Helsinki Heart Study. Primary prevention trial with Gemfibrozil in middle-aged men with dyslipidemia:

Safety in treatment, changes in risk factors, and incidence of coronary heart disease. N. Engl. J. Med. 317: 1237-1245.

17. Canner, P.L., Berge, K.G., Wenger, N.K., Stamler, J., Prineas, R.J. and Friedwald, W. 1986. Fifteen-year mortality in Coronary Drug Project patients. Long term benefit with Niacin. J. Am. Coll. Cardiol. 8: 1245-1255.

18. Hjermann, I., Holme, I. and Leren, P. 1986. Oslo Study Diet and Antismoking Trial. Results after 102 months. Am. J. Med. 80: 7-11.

19. Carlson, L.A. and Rosenhamer, G. 1988. Reduction of mortality in the Stockholm Ischemic Heart Disease Secondary Prevention Study by combined treatment with clofibrate and nicotinic acid. Acta Med. Scand. 223: 405-418.

20. Brensike, J.F., Levy, R.I., Kelsey, S.F., Passamani, E.R., Richardson, J., Loh, I.K., Stone, N.J., Aldrich, R.F., Battaglini, J.W., Moriarty, D.J., et al. 1984. Effects of therapy with cholestyramine on progression of coronary arteriosclerosis. Results of the NHLBI Type II Coronary Intervention Study. Circulation 69: 313-324.

21. Arntzenius, A.C., Kromhout, O., Barth, J.D., Reiber, J.H., Bruschke, A.V., Buis, B., van Gent, C.M., Kempen-Voogd, N., Strikwerda, S. and van der Veide, D.A. 1985. Diet, lipoproteins and the progression of coronary atherosclerosis. The Leiden Intervention Trial. N. Engl. J. Med. 312: 805-811.

22. Blankenhorn, D.H., Nessim, S.A., Johnson, R.L., San Marco, M.E., Azen, S.P., Cashen-Hemphill, L. 1987. Beneficial effects of combined colestipol-niacin therapy on coronary atherosclerosis and coronary venous bypass grafts. J.A.M.A. 257: 3233-3240.

23. Yusuf, S., Wittes, J. and Friedman, L. 1988. Overview of results of randomized clinical trials in heart disease. 1. Treatment following myocardial infarction. J.A.M.A. 260: 2088-2093.

24. National Diet-Heart Research Group. 1968. The National Diet-Heart Study Final Report. Circulation 37(suppl 1): 111-428.

25. Eaker, E.D. and Castelli, W.P. 1987. Differential risk for coronary heart disease in women in the Framingham Study. In: Coronary Heart Disease in Women. Eaker, E., Packard, B., Wenger, N. et al. (eds). Haymarket-Doyma Inc., New York, pp. 33-41.

26. Livshits, G., Weisbort, J., Mesulam, N. and Brunner, D. 1989. Multivariate analysis of the twenty year follow-up of the Donolo-Tel Aviv prospective coronary artery disease study and the usefulness of high density lipoprotein percentage. Am. J. Cardiol. 63: 676-681.

27. Bush, T.L., Criqui, M.H., Cowan, L.D. et al. 1987. Cardiovascular disease mortality in women. Results from the Lipid Research Clinics follow up study, In: Coronary Heart Disease in Women. Eaker, E., Packard, B., Wenger, N. et al. (eds), Haymarket-Doyma Inc., New York, pp. 106-111.

28. Castelli, W.P., Wilson, P.W., Levy, D. and Anderson, K. 1989. Cardiovascular risk factors in the elderly. Am. J. Cardiol. 63: 12H-19H.

29. American Heart Association. 1989. 1989 Heart Facts. Dallas.

30. National and International Cholesterol Campaigns. 1989. Proc. Xth Int. Symposium on Drugs Affecting Lipid Metabolism. Nov. 1989 (in press).

CONTRIBUTORS

Dr. Goran Bondjers
Professor of Cardiovascular
    Biology
Director, Wallenberg Laboratory
    for Cardiovascular Research
Göteborgs University
Sahlgren's Hospital
S-143 45 Gothenburg
Sweden

Dr. Michael R. Buchanan
Associate Professor of
    Pathology
Department of Pathology
1280 Main Street W.
McMaster University
Hamilton, ON
L8N 3Z5 Canada

Dr. Tucker Collins
Assistant Professor of
    Pathology
Department of Pathology
Brigham and Women's Hospital
Harvard Medical School
75 Francis Street
Boston, MA 02115
USA

Dr. Paul DiCorleto
Staff Member
Department of Atherosclerosis
    Research
Cleveland Clinic Foundation
9500 Euclid Avenue
Cleveland, OH 44106
USA

Dr. F. Fedoroff
Professor
Department of Anatomy
College of Medicine
University of Saskatchewan
Saskatoon, SK
S7N 0W0 Canada

Dr. Giulio Gabbiani
Professor and Acting Chairman
Department of Pathology
University of Geneva, CMU
1 rue Michel Servet
1211 Geneva 4, Switzerland

Dr. Avrum I. Gotlieb
Professor of Pathology
Director, Vascular Research Laboratory
Department of Pathology and
Banting and Best Diabetes Centre
Toronto General Hospital Research
    Centre
University of Toronto
200 Elizabeth Street, CCRW 1-857
Toronto, ON
M5G 2C4 Canada

Dr. David Harrison
Professor of Medicine
Division of Cardiology
Department of Medicine
Emory University
School of Medicine
Drawer LL
Atlanta, GA 30322
USA

Dr. Louis Horlick
Professor of Medicine
Department of Medicine
University Hospital
University of Saskatchewan
Saskatoon, SK
S7N 0X0 Canada

Dr. Fred Keeley
Associate Professor
Research Institute
Hospital for Sick Children
University of Toronto
555 University Avenue
Toronto, ON
M5G 1X8 Canada

Dr. B. Lowell Langille
Associate Professor of Pathology
Vascular Research Laboratory
Toronto General Hospital Research
      Centre
University of Toronto
200 Elizabeth Street, CCRW 1-858
Toronto, ON
M5G 2C4 Canada

Dr. Peter Libby
Associate Professor, Medicine
Harvard Medical School
Director, Vascular Medicine and
      Atherosclerosis Unit
Department of Medicine
Brigham and Women's Hospital
75 Francis Street
Boston, MA 02115
USA

Dr. Joseph Madri
Associate Professor of Pathology
Department of Pathology
School of Medicine
Yale University
310 Cedar Street
New Haven, CT 06510-8023
USA

Dr. Marian Packham
Professor of Biochemistry
Department of Biochemistry
University of Toronto
1 King's College Circle
Toronto, ON
M5S 1A8 Canada

Dr. Michael Reidy
Associate Professor
Department of Pathology
University of Washington, SJ-60
Seattle, WA 98195
USA

Dr. Michael Sawdey
The Research Institute of Scripps
      Clinic
10666 North Torrey Pines Road
La Jolla, CA 92037
USA

Dr. Wolfgang Schneider
Professor of Biochemistry
Lipid and Lipoprotein Research Group
Faculty of Medicine
University of Alberta HMRC-327
Edmonton, AB
T6G 2F2 Canada

Dr. Stephen M. Schwartz
Professor of Pathology
Department of Pathology
University of Washington, SJ-60
Seattle, WA 98195
USA

Dr. David M. Stern
Associate Professor
Department of Physiology
Columbia University
College of Physicians and Surgeons
630 W. 168th Street
New York, NY 10032
USA

Dr. Thomas N. Wight
Professor of Pathology
Department of Pathology
School of Medicine
University of Washington
Seattle, WA 98195
USA

acetylcholine, 153-155,157
acetyl *CoA*, 238
acetylsalicylic acid, 92
A-chain, 14
  alternative splicing in
    vascular cells, 7,143-146
  genes, 140,141,143-145,148
actin, 53,69,71
  α-*SM* actin mRNA expression,
    53-55
  ß-, 53,54
  bundles, 7,72,77,176
  ɣ-, 53,54
  isoforms, 53,54
  microfilaments, 71-73,78
  polymerization, 215
  reorganization, 69
  total, 53,54,71,79
α-actinin, 71,72,129
acute hypertension, 43,153,156
acute myocardial infarction, 17,
  187
acute vasomotion, 35,44
acyl-*CoA*: cholesterol acyltransferase (*ACAT*), 237
adenosine, 155
ADP, 188,209,210,212,213,
  215-217
ATP, 209,215
  -stimulated prostacyclin
    production, 94
adenovirus *E4* gene, 141
E4TFI, 141
adenyl cyclase, 92,94,227
adherens junctions, 73
adhesion molecule,
  receptors, 227
  suspension of, 230
ACTH, 193
adventitia, 4
afibrinogenemia, 211,213
aging process, 12
agonist-induced vasodilatation,
  35
alanine, 240
albumin, 163,175,177

alternative splicing of *PDGF A*-
  chain, 145
Altschul, 1-4
amiloride, 93,94
amino acid analysis, 103
angina pectoris, 238
  unstable, 217
angiogenic response, 12,192
angioplasty, 59,127,128,212
angiotensin II, 153
animal model systems, 13,15,17
anisomycin, 163
anticoagulation factors, 67
antifibronectin antisera, 119
α₂-antiplasmin, 188,209,217
anti-platelet serum, 8
antithrombin III, 118,174
aorta, 11-15,36,40,41,44,154
  abdominal, 40,72
  atheromatous, 162,164
  de-endothelialized and re-
    endothelialized, 120
  endothelium-dependent
    relaxations in rabbit,
    154,155
  hypoxic endothelial cells
    from, 182
  optimum wall stress level,
    102
  perfusion fixed, 72
  thoracic, 40,68,69,72,102,
    104-106,156
  wall thickness of, 11
aortic arch, 63,64
aortic diameter, 40
aortic organ culture model, 69
aortic thickening,
  experimental, 53,54
ApC formation, 178
apoB, 239,247,249,251
apoE, 239
AP-1,
  consensus sequence, 141-143
  binding sites, 142,195,196
AP-2, 142
apyrase, 216

arachidonic acid, 92,209,213-215
  metabolism, 227-230
arterial diameter, 37,43
arterial endothelium repair, 59-
  65
arterial injury, 13,59,92,249
arterial lipid content, 247
arterial proteoglycans, 247,249-
  251
  LDL binding to, 249
arterial smooth muscle cells, 53
  cytoskeleton of, 53,54
  proliferation of, 44
arterial thrombi and thromboemboli, 209,210
arterial tissue, 247
  lipoprotein deposition in, 248
arterial wall, 248,251
  injury, 209
  proteoglycan accumulation
    in, 115
  response to tangential
    stress, 101,102,107
  thickening of, 43,44,227
arteries, 33-46
  gastroepiploic, 229
  iliac, 154
  internal mammary, 229,230
  umbilical, 40
arteriolar wall thickening, 44
arteriosclerosis, 3,4,128,131,
  132,227
asparagine-linked high mannose
  oligosaccharide chains, 240
aspirin, 215,217
atherectomy, 128
atherogenesis, 7,37,43,45,46,
  131,139,161-166,197,247,
  251,252
atherogenic agents, 2
atherogenic lesions, 3
atheroma, 16,163,165,238
  cells, 147
atheromatosis, 53,54
atheromatous plaque, 53,54,139
atherosclerosis, 2,35,37,43,46,
  101,115,130,139,141,209,
  217,247,248
  and cholesterol lowering,
    257
  diet-induced, 248
  endothelial regulation of
    vasomotor tone in, 153-
    157
  and monocyte adhesion, 89-
    94
  natural history of, 7-17
  and *PAI-1* activity, 197,198
  premature, 237,238
  proteoglycan accumulation
    in, 115,118,122

atherosclerosis *(continued)*
  spontaneous, 68
atherosclerotic lesion, 17,11-
  14,16,17,67,68,72,131,140,
  142,162,197,209,248,252
atherosclerotic plaque, 7,8,11-
  17,89,90,94,120,122,140,
  163,197,209,211,212
atherosclerotic regression, 257
atherosclerotic vessels, 2
autocrine and paracrine regulatory loops, 161
autocrine/paracrine factors, 90,
  130
autoradiography, 13,60,119
avidin-gold localization, 132

balloon catheter injury or denudation, 8,13-15,17,60-65,
  77,115,123,130,131,133,143,
  157
baroreceptor-mediated changes in
  sympathetic drive, 43
barrier function of endothelial
  monolayers, 173,176,177,182
basement membrane, 7,128,129
  HS-PG in, 115,117
basic amino acids, 247,249
basic polypeptides, 251
Baumgartner flow chamber, 211
B cell proliferation, 163
B-chain,
  genes, 140,141,143,144
  localization, 15,16
  promoters, 141,142
benign leiomyomas, 13
Bernard-Soulier syndrome, 211,
  214
biglycan, 118
blood cell/endothelial cell
  adhesion, 230
blood cell injury, 228
blood flow, 35,38,40,43,46,127,
  153,209,210,212
  chronic changes in rates
    of, 37
  lung, 40
  placental, 40
blood pressure, 101-105,109
bone marrow, 164
bovine aortic endothelial cells
  (*BAEs* or *BAEC*), 90,92,128,
  129,156,188,191,192,195
  and *B*-chain promoter, 141
  plasminogen activator and
    PAI-1, 130
  sheet migration on laminin
    substrates, 129
bovine aortic smooth muscle
  cells (*BASMC*), 128-132
  migration, 130,131

bradykinin-induced phosphoinositide turnover, 94
branch ostia, 44,46
bronchial circulation, 40

Cajal's gold chloride, 1
calcium, 41
  ionophore *A23187*, 92,154, 155,213
Ca²⁺/calmodulin-dependent protein kinase, 215
capping of receptors, 41
18 carbon fatty acids, 229
carbon monoxide poisoning, 1,2
cardiovascular tissues, 33
  responses to mechanical loads, 33
carotid artery, 38-40,53
  denuded, 59-64
  human lesions, 140
  proximal internal, 46
  rabbit, 59-61
  rat injury model, 54
CAT, 141
catecholamines, 9,14
caveolae, 7
cDNA probes, 102,141-145,174, 177,178
cell adhesion or attachment, 71, 108,119
cell adhesion molecule, 10
cell-cell adhesion, 230,231
cell contact mechanism, 15
cell contractility, 17,109
cell death, 8,13
cell elongation, 78
cell injury, 7,9
cell migration, 69,71,73-77
cell motility, 71,74,130
cell-substrate adhesion, 35
cell surface coagulant properties, 173
cell surface matrix binding proteins, 128,129,131,133
cell surface receptors, 187
cell translocation, 69,78
cellular lipid accretion, 248, 251
centrioles, 73
centrosomes, 73-78
  function in endothelial regeneration, 74,75
  redistribution of, 74-78
cephalin, 175
cerebral artery, 17
c-fms, 140
chloramine *T*, 189
cholesterol, 2,3,237,238,255-260
  blood levels, 255,256
  clefts, 17
  diet, 162-164

cholesterol *(continued)*
  feeding, 90,154,157
  lowering, 255,257-259
  plasma levels, 258
  serum levels, 3,258
  synthesis, 237,238
  total, 256-258
cholesteryl esters, 237
cholestyramine, 256,257
chondroitin sulphate proteoglycan (*CS-PG*), 115,116,118-120,247
chromaffin cells, 1
chromium, 211
chromosome 7, 142
chromosome 22, 140
chronic medial responses to shear force, 37
chylomicron remnants, 242
chymotrypsin, 216.
circulatory function, 33
circumferential stress, 41,43
classical lesion, 12,16,17
clathrin, 240
coagulation,
  cascade, 17,92,190,227
  factors, 69
  mechanism, 173
coated pits, 240,242
colcemid, 74
colchicine-treated cells, 73, 109-112
colestipol, 257
collagen, 17,37-39,139,188,210-212,215,216,249
  DS-PG in fibrils of, 115, 117,119,121-123
  interstitial, 129
  receptor, 210
  responses to vessel wall perturbation, 101-112
  type I,III,IV, 69,128-131
  type V, 128
colony stimulating factors (*CSFs*), 161,164,165
compression, 33
connective tissue proteins, 102, 105,106,112
consensus binding site, 141-143, 148
contact-mediated paracrine stimulation, 162
contractile proteins, 10,11
coronary artery, 12,17,154,155
  bypass surgery, 229,257,258
  disease (*CAD*), 197,238,257
coronary heart disease (*CHD*), 238,255,256,259
coronary microcirculation and atherosclerosis, 154,155
cortisol, 193

creatine phosphate/creatine phosphokinase (*CP/CPK*), 215
c-sis oncogene, 8,140,141
Cu(II), 252
cyanogen bromide, 103
cyclic adenosine monophosphate (*cAMP*), 227,229
   in platelets, 210,216
   regulatory domains, 142
cycloheximide, 163,174,181,193, 194
cyclo-oxygenase,
   activity, 92
   pathway, 227-229
cylindrical symmetry, 41
cysteine protease, 181-183
cysteine residue, 144,188,239, 240,242
cytochalasin *B*, 74,77
cytokines, 67,89,94,127,130,133, 173,179,182,190,192-194, 227,228
   production by vascular wall cells, 161-166
cytoplasmic Ca^{2+}, 214,215
cytoplasmic proteins, 145
cytoskeletal proteins, 69,74
cytoskeleton, 41,129
   change in configuration of, 176
   organization of, 129,133
   reorganization of, 215
   role in endothelial repair, 67-79
   submembranous, 109
cytosol, 228
cytosolic-associated peroxidase, 228
cytosolic Ca^{2+}, 89,92,215

Dahl salt-sensitive rat model, 103,109,111
DEAE-Sephacel ion exchange chromatography, 121
decorin, 119,120
deformability of tissue, 33,35
deletion analysis, 141
delta storage pool deficiency, 216
dense peripheral band (*DPB*), 72, 73,75-79
DNA synthesis, 14,38,54,69
   inhibition of, 119
DNA transfection studies, 195
dermatan sulfate proteoglycan (*DS-PG*), 115,116,118,120
desmin, 11,53,54
desquamation of the cell, 35
dexamethasone, 193,195,196
diabetes, 156
diacylglycerol, 89,92,214,215

diene conjugation, 252
dietary therapy, 255-257,259
diet-induced arterial abnormalities, 162
disulfide bridge, 212,239,240
DOCA-salt model, 103
ductus arteriosus, 40

eicosanoids, 8
elastase, 189
elastic lamina, 154
elastin, 11,37-39,249
   in injured vessels, 121-123
   insoluble, 103,107,108
   responses to vessel wall perturbation, 101-112
electron microscopy, 7,60,69, 116,117,121,128
   scanning, 69,107
   transmission, 69,71,72
ELISA, 128
endarterectomy, 127,128,140,143, 147
endocytosis of *LDL*, 237,242
endoglycosidase, 15
endoplasmic reticulum, 10
endothelial cell growth factor (*ECGF*), 192
ECGF-heparin, 192
endothelial cell-extracellular matrix interactions, 127-133
endothelial cells, 2-4,35-37,41, 43,45,139-143,147,162,164, 209,211,214,248,251
   alterations in morphology, 153,154
   bovine adrenal microvascular, 173
   bovine capillary, 191,192
   brain, 2
   cultured human umbilical vein, 146,147,191-195
   culture of and exposure to hypoxia, 173,174
   cytoskeleton of, 67,69,73
   denudation of, 59-61,64,65, 154
   fatty acid metabolism, 227-229
   function, 67-69,74,89,90, 93,94,173-183
   growth of, 59-65
   high turnover of, 35
   human umbilical vein, 89
   human vascular, 162-164
   hypoxic, 175-177,179-182
   lysates of, 9
   microvascular, 127,128
   migration of, 68,69,71,73, 75,77,121,127-133,164

endothelial cells *(continued)*
  non-adhesive, 230,231
  PAI-1 from, 188
  porcine aortic, 68,70
  proliferation of, 69
  proteoglycan synthesis by,
    119-122
  replication rates of, 59,
    60,62-64
  senescence, 60,65
  shape and cytoskeleton, 176
  transcriptional initiation
    in, 195,196
endothelial denudation, 67,68
endothelial dysfunction, 4,45,
  89,153,157
  macro- and microvascular,
    182,183
endothelial injury or lesion,
  11,35,43,45,53,68,69,79,
  192,197
endothelial integrity, 35,45,67,
  79
endothelial lining of capillaries, 15
endothelial permeability, 79,
  173,174,177,182,183
endothelial regeneration, 68,69,
  74,75
endothelial regrowth, 37
endothelial repair, 67-79
endothelial reponses to shear
  stress, 35
endothelial surface, 17,45
endothelial tubes, 11
endothelin, 153
endothelium, 3,4,9,11,14-17,53,
  54,67-69,73,89,90,94,107,
  127,128,131,139,153-157,
  197,209,211,212,214,227,
  228,248
  bovine aortic, 182
  shear on, 35
  thrombogenicity under
    hypoxia, 182
  vascular, 153
  venous, 188
endothelium-dependent relaxation
  in atherosclerosis, 154
endothelium-derived relaxing
  factor *(EDRF)*, 35,37,153,
  155-157
  chemical identity of, 156
  -mediated vasomotion, 37
  tonic release of, 37
endothelium-mediated dilatation,
  43
endothelium-smooth muscle cell
  interface, 62
endotoxin, 162,163,182,227,
  228

epidermal growth factor *(EGF)*,
  9,10,40,142,191
  precursor homology, 239
epinephrine, 216
epithelial cells,
  plasma membrane of, 115
ethchlorvynol, 77,79
exercise training, 37
exonuclease III deletions, 141
extensive cell kinetic analysis,
  12,13
extracellular matrix *(ECM)*, 10,
  14,15,17,120,121,130
  and integrin expression,
    129,133
  interaction with *PAI-1*,
    189,190
extracellular space, 252
extracellular wall constituents,
  37,44
extrinsic coagulation pathway,
  210

F-actin microfilaments, 35,69,
  71,174
F-actin stress fibers, 35,79,176
FACS analyses, 132
Factor V of intrinsic coagulation pathway, 209
Factor IX, 175,181,239
Factor X, 175,176,179,181,183,239
  activation, 175,179,180,181
  ^{125}I-, 179,180
Factor Xa, 92
  clotting assay, 175,176,180
  Factor Xaα and Xaß, 179,180
Factor XII activation, 17
Factor XIII, 217
familial hypercholesterolemia
  (FH), 237,238,241,242
fatty acid metabolism, 227
fetal calf serum (FCS), 54,55,173
fetal placental circulation, 40
fibric acid derivative, 256
fibrillar elements, 101
fibrin, 9,17,187,209,210,216,217
  clot, 175,188
  degradation, 187
  formation, 190,192
  intravascular deposition
    of, 187,190
  on venous valve cusps, 182
fibrinogen, 129,130,133,209-213,
  215-217
fibrinolysis, 198,217,227
  deficient, 197
  physiologic, 187
fibrinolytic activity, 182,187,
  190,192,197
fibrinolytic agents or substances, 67,216,217

fibroblast growth factor (*FGF*),
  8,9,40,63,65,161,249
  autocrine role of b*FGF*, 69
  basic (*bFGF*), 63-65
    antibody, 63
    and *PA-I* biosynthesis,
      191,192
fibroblasts, 4,9,12,144,145
  adventitial, 11
  biosynthetic pathway of
    LDL receptor in humans,
    240,241
  dermal, 141
  non-migrating, 71
  and *PAI-1*, 190,192
  PDGF-stimulated, 164
fibrofatty atherosclerotic
  plaques, 67,68
fibromuscular proliferation, 7
fibronectin, 9,10,69,128-131,
  133,175,177,178,190,209-
  211,213,216
  receptor, 230
fibronectin-dependent adhesion,
  211
fibrous cap, 13,16,17
firefly luciferase reporter
  gene, 195
flow disturbance, 12
flow-induced arterial growth
  modulation, 40
flow-induced vasodilatation, 35
flow-induced vasomotion, 37
flow resistance, 44,45
flow trajectories, 46
fluorescent methods, 69,73,78
fluorophore, 215
f Met-Leu-Phe (*fMLP*) tripeptide,
  17,155
fms, 16
foam cell, 2,3,90,162,164,165,
  248
  formation of, 249-252
  and hypercholesterolemia,
    154
  macrophages, 17
fodrin, 129
foramen ovale, 40
free radicals, 252
frictional forces, 34

G-actin, 71
gap junctions, 75
gel electrophoresis, 53,55
gemfibrozil, 256,257
gene transcription in *BAE*s, 194
glioma *cDNA* library, 142
glucocorticosteroids, 191,193,
  195
glucose-6-phosphate dehydrogenase (*G-6-PD*), 12,13

glutamine, 173
glycoproteins, 118,144,145,188
  GPIa/IIa, 210,211
  GPIb/IX, 210,211,214
  GPIIb/IIIa, 210-214,216,217
    receptor, 230
  GPIV, 210,211,213
  GPV, 210,214
  membrane, 210
glycosaminoglycan (*GAG*),
  binding proteins, 249
  side chains, 116
GMP-140, 213,214,217
Goldblatt or renal clip model,
  103
Golgi apparatus, 75
G protein (*G*$_p$), 214
  activation, 94
gram-negative sepsis, 188,190,
  192
granulocyte-macrophage colony
  stimulating factor (GM-CSF),
  164,165,195
GRO, 164
growth factors, 67,209
  chemotactic, 139
  mitogenic, 139
growth inhibitor mechanisms, 13
guanidine *HCl*, 189
guanylate cyclase, 153
G0/G1 phase of cell cycle, 54,
  55,120

HA1004 inhibitor, 92
HeLa cells, 195
hematopoietic growth modulation, 162
hemodynamic forces, 67,68,75
hemodynamics, 33,45
hemostasis, 115,187
hemostatic plug formation, 213
heparan sulfates, 15
heparin, 15,53-55,68,90,120-123,
  216,217
heparin-binding growth factors
  (*HBGFs*), 8,9
heparin-like growth inhibitors,
  8,15
heparitinase, 15
hepatocytes, 163,188,189,191,193
hepatoma, 189,190,192,193,195
  FT02B, 195
hexuronic acid, 115
high altitude pulmonary edema, 182
high density lipoproteins (*HDL*),
  242,248,249
HDL cholesterol, 256-258
^3H-inulin, 174,177
hirudin, 216
histamine, 77,128-131
  -stimulated phosphatidyl-
    inositol turnover, 94

H7 inhibitor, 92
³*H*-leucine, 175,177,178
³*H*-sorbitol, 174,177
³*H*-thymidine, 60-62,64
human aortic endothelial cells, 90,91,93,94
human aortic smooth muscle cells (*HAoSMC*), 147
human fibrosarcoma, 191-193
human mammary carcinoma cells, 191,192
human plaque, 13,15,17,248
   atheromatous, 54
   carotid artery, 16
human saphenous vein smooth muscle cells (*HSVSMC*), 147
human skin fibroblasts, 192,237
human teratocarcinoma (*TERA-2*), 147
hydrogen peroxide, 156
hydroperoxy-eicosatetraenoic acid (*HETE*), 228
   12-HETE, 228-230
   15-HETE, 228-230
9-HODE, 229
13-hydroxyoctadecadienoic acid (*13-HODE*), 228-231
hydroxyproline, 103
HMG-CoA reductase, 237
3-hydroxy-3-methylglutaryl CoA synthase (*HMG-CoA synthase*), 237
hypercholesterolemia, 3,4,7,90, 153-157,237,238,241,242,255, 256,259
hyperinsulinemia, 193,197
hyperlipemia, 3,4
hyperlipemic serum, 10
hyperlipidemia, 68,101,197,249, 256,257,259
hyperlipidemic cynomolgous monkeys, 68
hyperlipidemic rabbits, 154
hyperoxia, 79
hyperplasia, 44,106,118,227
hypertension, 11,43,44,68,101- 103,105,106,109,153,157,255
hypertriglyceridemia, 197
hypertrophy, 44,45,109,147
hypoperfusion, 37,43
hypoxemia, 173,182
hypoxia and endothelial cell function, 173-183
hypoxia-induced Factor X activator, 175,176

IGF-1, 9,10
¹²⁵*I*-albumin, 177
¹²⁵*I-BSA*, 174
¹²⁵*I*-Factor X, 175,179,180
¹²⁵*I-LDL* binding, 241,242

¹²⁵*I-TGF-ß*1 binding, 132
immune interferon, 79
immunocytochemical methods, 69, 115
immunofluorescence microscopy, 71-73,77,128,129
immunofluorescent staining, 53
immunoglobulins, 163
immunohistochemistry, 122
indium, 211
inflamed rheumatoid synovia, 163
inflammatory cells, 155,157,187
inflammatory response, 162,163
injury-repair cycle, 78
inositol phosphate production, 89,92
inositol 1,4,5-triphosphate (*IP₃*), 214
inositol 1,3,4,5-tetrakisphosphate (*IP₄*), 215
in situ hybridization, 14-16, 140,163
in situ localization, 71
insulin, 75,191,193
integrin, 129-132,210,212
intercostal arteries of thoracic aorta, 59
interferons, 161,165
   interferon α, 165
   interferon ß, 165
   interferon γ, 165
interlaminar cell, 12
Interleukin 1 (*Il-1*), 161-165, 173,179,182
   IL-1α, 161,162
   IL-1ß, 161,162
   and monocytic cell adhesion, 90
   and *PAI-1*, 192
Interleukin 6 (*IL6*), 163
Interleukin 8 (*IL8*), 164
Interleukin 9 (*IL9*), 164,165
interleukins (*IL*), 161
intermediate filaments, 69,71,73
   pattern of, 53
internal elastic lamina, 13
internalization-defective allele, 242
internal radius of vessel, 101, 107
interstitial stroma, 128
intima, 4,7,8,12,13,15-17,35,43, 53,127,130,132,140,141,155, 197,248,249,252
   proteoglycan content of, 122
intimal injury, 127
intimal lesions, 7,16,115
intimal mesenchymal cell, 15,17
intimal thickenings, 7,8,12,14, 77,127,128,130,155,248
   CS-PG accumulation in, 115

intimal thickenings *(continued)*
  diffuse, 7
  and heparin infusion, 121
  rat, 53
intimal wounds, 69
intracellular pH (pH_i), 93,94
intracellular signalling systems,
  210
intrinsic coagulation pathway,
  210,214
inverted-phase microscope, 69
in vitro organ culture models,
  102,107,110
in vitro perfusion system, 248,
  251
in vitro translation, 102
in vitro wound, 90,92
in vitro wound model system, 68
ion channel activation, 41,94,109
ion exchange chromatography, 118
ischemia, 37,156,182
isoelectric focussing, 175,181

keratan sulfate proteoglycan
  (KS-PG), 115,116

lactoperoxidase method, 174,175
lamellipodia, 68-71,74,75,77,78
laminin, 69,128,129,210
  receptor, 230
laminin binding protein *(LB69)*,
  129
lapse cinemicrophotography, 69,
  76,77
laser ablation, 128
Law of Laplace, 42,44,101,103
leukocyte adhesion,
  inhibitory activity, 164
  molecules, 94
leukocyte-endothelium adhesion,
  37
leukocytes, 8,14,45,67,90,92,
  155,161,164-166,228
  and *GMP-140*, 214
leukotriene *D4*, 17
linoleic acid, 228-230
lipid accumulation, 7,14,16,115,
  120,197
lipid insudation, 45
lipids,
  cellular accretion of, 247
  proteoglycan composition
    and, 118
lipocortin, 215
lipolytic cascade, 249
lipopolysaccharide *(LPS)*, 90,92,
  94,164,190,191,193,194
lipoprotein lipase, 118
lipoproteins, 67,77
  cellular uptake of, 247,
    250-252

lipoproteins *(continued)*
  influx of, 248
  interaction with proteogly-
    cans, 118
  interaction with vascular
    tissue, 247-252
  plasma, 247
  retention of, 247,248,252
liposomes, 231
lipoxygenase, 228-230
low density lipoproteins *(LDL)*,
  89,247-252
  accumulation of, 77
  binding to proteoglycans,
    251
  differential mitogenic
    effect of, 10
  plasma levels of, 238
  significance of size, 251
LDL cholesterol, 237,238,256-258
LDL-proteoglycan interaction,
  251,252
LDL receptor, 237-242,247,249-
  251
  cysteine-rich repeats in,
    239
  cytoplasmic region, 240
  EGF-precursor homology
    domain, 239
  ligand binding domain, 241,
    242
  membrane-spanning region,
    240,242
  molecular defects in *FH*
    patients, 241
  mutations in, 237,238,241,
    242
  O-linked carbohydrate
    chains, 239-241
  biosynthetic pathway, 237,
    238,240
LDL receptor-mediated metabolism,
  237-242
luciferase activity, 195
luminal surface, 41,60,61,127,
  130
lymphocytes, 163-165,248
lymphokine, 165
lysosomal hydrolysis of *LDL*
  cholesterol, 237

macrophages, 7,15-17,45,140,141,
  164,247-252
  and *GMP-140*, 214
  monocyte-derived, 90,248
  subintimal, 155
media, 4,127,130,132
medial muscle cells, 101
medial remodeling, 37
medial thickening, 11
megakaryotic line *K562*, 142

meiosis, 242
melanoma cells, 121
membrane-associated dense bands, 7
membrane phospholipids, 209,214, 227,228
ß-mercaptoethanol, 189
mRNA, 53-55
  cap site of human endothelial cell, 195,197
  elastin, 112
  fibronectin, 131,175,177
  growth factor, 15
  interferon ß, 165
  IL-1, 162,163
  IL-6, 163
  M-CSF, 164
  PAI-1, 190,191,193-195,197,198
  PDGF A-chain, 143,146,147
  PDGF B-chain, 140
  TF, 17
  tissue plasminogen activator, 17
  thrombomodulin, 175,177,183
  translation, 54
  TNF, 163
metal impregnation methods, 1-3
methionine residues, 103,188,189
methionine sulfoxide reductase, 189
methylcholine, 154
micelles of negatively charged phospholipids, 189
microcirculation, 35
microelectrode, 41
microfilaments, 69,71-74,76-78
  central, 72,77-79
microtubules, 69,71-74,76,77,109
microvascular pericyte response, 127
migration assay, 128,129,132
mitogens, 8-10,14-16,41,44,45, 63-65,89,139,140,145,197
mitomycin *C*, 130,131
mitosis, 15
mix-and-match receptor-agonist system, 8
modulation hypothesis, 11
molecular oxygen, 228
molecular sieve chromatography, 118
monoclonal antibodies, 248
  against *CS-PG*, 117,118
  against *LDL* receptor, 240, 241
monoclonality of plaques, 12-14
monocrotaline, 106
monocyte adhesion,
  thrombin-induced, 90-92
monocyte chemoattractant protein-1 (*MCP-1*), 164

monocyte chemotactic and activating factor (*MCAF*), 164
monocyte *CSF* (*M-CSF*), 164,165
monocyte-endothelial adhesion molecules, 165
monocyte/macrophage,
  U937, 195
  uptake, 45
monocytes, 13,14,16,17,89,90,92, 139,163,164,198
  peripheral blood, 90
  U937 cells, 90-92
mononuclear phagocyte, 162,165
mucopolysaccharide elements, 101
multiplication stimulating factor, 75
murine tissues, 142,146
muscularis of vein walls, 188
myeloschisis, 1
myocardial infarction (*MI*), 187, 190,192,237,238
  non-fatal, 257
myogenic relaxation, 44
myometrium, 147
myosin, 71,72,215

Na^+/H^+ antiporter, 93,94
N-acetylgalactosamine (*GalNAc*), 240
neointima, 14,15,130
neoplasm, 14
nephrectomy, 37
nerve growth factor (*NGF*), 12
neuroendocrine regulation, 161
neutrophil activating factor, 164
neutrophils, 73,74
  binding to endothelial cells, 89,90
  lysosomal enzyme release in, 164
niacin therapy, 3
nicotinamide adenine dinucleotide (*NAD*), 3
nicotinic acid, 3,4,257
nitric oxide, 35,153,155,156
nitroglycerine, 154
nitroprusside, 155
nitrosothiol, 156
NK cells, 165
N-linked glycosylation, 144
non-insulin diabetes, 197
non-integrin binding proteins, 129
norepinephrine, 128-131
Northern blot hybridization, 16, 53,55,128,140,143,175,177, 194
nuclear lamina, 109
nuclear membrane, 109
nuclear targeting sequence, 145

nuclear transcription run-on
  experiments, 193
nuclease protection experiments,
  147,195
nucleotide sequence of human
  PAI-1 promoter, 196
nucleus, 74,76,145

occlusive thrombosis, 17,197,209
O-glycosidic linkages, 115
oil red O, 16
1-oleoyl-2-acetyl-*rac*-glycerol,
  92
2'-5'-oligoadenylate synthetase,
  165
oligonucleotides, 142,147
oligosaccharides, 15
  N-linked, 116
  O-linked, 116
omega-3 fatty acids, 157
osteosarcoma cell line *U-20S*,
  143,147
oxidative modification of *LDL*,
  251
oxygen regulated proteins, 182
oxygen tension, 3

parturition, 11,40
pathogenesis of arterial
  diseases, 35,44
penicillin-streptomycin, 173
pericytes, 12,61,127,128
perinatal vascular development,
  40,41
peripheral tissues, 35,38,40
peroxide-treated cholesterol, 2
pertussis toxin-insensitive *G*
  protein, 94
phallacidin, 71
phalloidin, 71
phenylmethylsulfonyl fluoride
  (*PMSF*), 175
  -treated thrombin, 91
phorbol dibutyrate, 92
phorbol esters, 90,92,142
phorbol myristate acetate (*PMA*),
  91,92,195
phosphatidylinositol biphosphate
  (*PIP$_2$*), 214
phosphatidylserine, 215
phospholipase A_2, 215
phospholipase *C*, 93,94,214-216
phytohemagluttinin, 162
pial vessels, 156
placebo, 256,257
plaque rupture, 17
plasma, 210-213,215
  cells, 163
plasmalemma, 7,93
plasmids, 141,195
plasmin, 187-189,216,217

plasminogen, 187
  activator, 37,130,131,216,
    217
plasminogen activator inhibitor
  (*PAI*), 17,187-198,216,217
  activity, 187-190,192,193,
    195,197,198
  tissue- and urokinase-type,
    187
  type one (*PAI-1*), 130,131,
    187-198,209,217
  type two or placental
    (*PAI-2*), 188
PAI-1 biosynthesis regulation,
  190,192,193
PAI-1 gene,
  expression, 193-195
  glucocorticoid-inducibility
    of, 195
PAI-1 promoter activity, 195,197
platelet activating factor
  (*PAF*), 92,212,213
platelet activation-dependent
  granule-external membrane
  (*PADGEM*) protein, 213
platelet adhesion, attachment or
  binding, 17,45,129,211,228
platelet aggregates,
  stabilization of, 216,217
platelet aggregation, 128,212-
  214,227
  inhibition of, 157
platelet antiaggregation
  factors, 67
platelet-collagen adhesion, 211
platelet-derived growth factor,
  (*PDGF*), 8-10,12,14-17,40,
  41,45,67,89-94,128,191,192,
  198,209,249
  B-chain gene structure, 140
  genes, 16,94
  PDGF-A, 8,94,140-148
  PDGF-B, 8,94,140-145,148
  and proteoglycan metabolism,
    120,121
  receptor, 8,14,143,144,145,148
  regulation of expression in
    vascular cells, 140-142
  synergistic effect on cen-
    trosomal redistribution,
    75
  thrombin-induced production,
    89,91-93
platelet/endothelial cell interactions, 228,230
platelet factor *4*, 15,209,216
platelet-fibrin thrombi, 209
platelet releasates, 130,133
platelet release, 8,190
platelets, 8,10,14-17,45,46,60,
  67,130,139

platelets *(continued)*
  and atherosclerosis, 209
  and fibrinolysis, 217
  α-granules of, 188,190,209-
    214,217
  and vasoconstrictors, 157
platelet/vessel wall interaction,
  45,228
point mutation, 241,242
poly(A) sites, 143
polyadenylation signal, 143
polyclonal antibodies, 53
  against *LDL* receptor, 241
polylysine, 251
polymerase chain reaction
  analysis, 146,147
  of *cDNA*, 162
polypeptides, 7,8,14
polypyrimidine tracts, 148
potassium channels, 41
pregnancy, 37,188,190
pressure, 33,34,41-43
  arterial, 35,40
    pulmonary, 40
    systemic, 40
  intraluminal, 40
  perinatal, 40
procoagulant activity, 17
  and hypoxia, 175,179,182
  and *TNF-α*, 192
procollagen, 190
procollagenase gene, 195
proenzymes, 187
  pro-u-PA, 187
profibrinolytic factors, 67
prostacyclin, 35,89,94,210,216,
  227
prostaglandins,
  PGE1, 216
  PGE2, 17,155
  PGI₂, 227-229
prostanoid production, 92
protease inhibitors, 175,178
  leupeptin, 175
  PMSF, 175
protease-protease inhibitor
  system, 131,133
α-1-proteinase inhibitor
  (α-1-PI), 189
protein, 10,12,14,15
  core, 115,116,120
  levels, 130,131
  secretion, 69,193
protein *C*, 174,178,239
  activation, 174,175,178
protein *C*/protein *S* pathway, 177
protein *4.1*, 129
protein kinase *C*, 195,214,215
  activation, 92
protein sequence analysis of
  PDGF, 140,142

protein synthesis experiments,
  54,175,177,178,181,182
  post-transcriptional level,
    54,55
  transcriptional level, 54,
    55
protein synthesis inhibitors,
  163
proteoglycans, 15,108,139
  aortic, 117
  dynamic interaction of,
    115-123
  synthesis of, 119-122
proteolysis,
  experiments on, 240
  extracellular, 187
  plasmin-mediated, 189
  and *PAI-1*, 192
proteolytic degradation, 247,250
prothrombin, 173
prothrombinase, 214
prothrombotic tendency, 173
puerperium, 147
pulmonary artery, 40
  arterial pressure, 40,44
  ligation, 40
  sensitivity to hypertension,
    105
pulmonary gas exchange, 40
pyrogen, 162,163

rabbit reticulocyte lysate, 54
radioligand binding assays, 132
rat carotid balloon de-endothelialization model, 129,131
receptor-*G* protein coupling, 93
recombinant *DNA* technology, 161
recombinant tumor necrosis
  factor, 79
reendothelialization, 59,61,68,
  69,74,130
  microfilaments in, 76-79
reproductive cycles, 37
resistance vasculature, 154
reversible pulmonary edema, 79
RGD-containing peptides, 129
RGD recognizing site, 231
RGD sequences, 211-213
rhabdomyosarcoma, 195
rhodamine phalloidin, 71,76,78,
  174,176
RNA synthesis, 190,192
del Rio Hortega's silver
  carbonate, 1
Russell's viper venom, 175,179,
  180

saphenous veins, 229
saturated fats, 258
scavenger receptors, 251
schizophrenia, 3

SDS PAGE, 118,132,175,176,179-181,183,189,240,241
secondary flow, 46
selectin family of receptors, 214
serine, 240
  proteases, 130,181,187,188
  residues, 115
serotonin, 9,37,128-131,209,216
  receptor, 9
serpin superfamily, 188
serum, 75,190,191
  protein, 90
  triglyceride levels, 190
S/G2 SMC population, 54,55
35S-GTPγS binding, 94
shear, 33,34,46
  at blood-endothelium interface, 37
  distribution in large vessels, 45
  and platelet adhesion, 211, 213
shear strain, 33,35
shear stress, 33-35,37,45,46,101
  biological responses to, 34
  hemodynamic, 72,77
  transduction of, 41
SHR rats, 11
signal peptide, 188
signal transduction,
  mediators, 109
  pathways, 109,195
simian sarcoma virus, 140
  SV40 large *T* antigen, 145
single cell wound model of endothelial injury, 69,70,77
35S-methionine infusion, 54,240
smoking, 255,258
α-smooth muscle (*SM*) actin, 10, 16
  mRNA expression, 53-55
SMC adaptation, 55
SMC differentiation, 53-55
SMC proliferation, 53-55
smooth muscle cells (*SMC*), 4,127, 128,131,133,139,140,143,145, 147,162-166,247,248
  abnormal growth of, 7
  accumulation of, 7,8,11,12, 16
  aortic, 53,54
  arterial, 53,116,120-122
  chemotaxis of, 129
  de-differentiation of, 11, 15
  denudation of, 8,15-17
  differentiation of, 10,12
  growth inhibition of, 9,15, 121
  intimal, 7,14,15,45,141

smooth muscle cells (*continued*)
  as key cells in early lesions, 7
  lipid accretion in, 248
  medial, 35,37,53,54,127,129-132
  migration, 132,139
  neointimal, 130
  origins of, 11,12
  pathologic replication of, 8
  polyploidy in hypertrophy of, 44
  proliferation of, 7,8,10, 12-14,16,17,118,120, 121,139,209,227
  replication of, 7-15
  vascular, 130,131,139,162-165,188,189,191
smooth muscle endothelial interaction, 10
smooth muscle lesion, 7
smooth muscle-specific anti-A actin *cDNA* probe, 11
S-nitroso-*L*-cysteine, 156
S1 nuclease analysis, 141,146, 147
S phase of cell cycle, 53
SP1, 141,143
35S-sulfate incorporation, 121, 122
stenosis,
  arterial, 197
  and platelet binding, 213
  post-therapy, 129
strain, 33,34
  in vessel wall, 107,108
  longitudinal arterial, 42
strain rate, 34
streptokinase, 217
stress, 33,34,41
  mechanical, 41,71,101,110
  response of large arteries to, 101
  tangential wall, 101,102, 107
subendothelial matrix, 75
  proteins, 17
subendothelial muscle cells, 101
subendothelial space or subendothelium, 15,17,67, 68,248,249
  and *PAI-1*, 189
  and platelet deposition, 209-211
  proteins, 17
sulfhydryl groups, 156
superoxide, 164
surface phospholipids, 247
synthetic grafting, 127,129
synthetic polypeptides, 251
systemic arterial wall tension, 43

systemic cholesterol homeostasis, 242

TC/HDL ratios, 257
tenase, 214
tendons, 238
tensile stress, 41-43,46
  wall, 42,44
tension or stretch, 33,35,41
  longitudinal, 43
thalamic syndrome, 1
thiobarbituric acid-reacting
  substances (*TBARS*), 252
3T3 cells, 9,14
threonine, 240
thrombin, 14,37,89-94,142,154,
  157,175,178,188,190-192,
  210-217,227,228
  α-, 90,93
  γ, 91
  receptor, 210,214
thrombin phorbol 12-myristate
  13-acetate (*TPA*), 77
thrombin-stimulated endothelial
  cell functions, 89-94
thromboasthenia, 212
thromboembolism, 43,209
ß-thromboglobulin, 164,209
thrombolysis, 216,217
thrombomodulin, 173-175,177,178,
  182,183
thrombosis, 16,17,43,115,182
  deep vein, 190
  platelet reactions in, 209-217
thrombospondin, 9,10,209,210,213,
  216
thrombotic disease and *PAI-1*
  activity, 189,190
thrombotic occlusion, 7,198,227
thromboxane A_2 (*TxA$_2$*), 17,209,
  215-217,227,228
thrombus formation, 37,173,209-
  212,214,215,217
thymidine cellular labelling, 54
  tritiated, 13
thymocyte costimulation, 162
tight junctions, 73
tissue culture, 53
tissue factor (*TF*), 17,192,197
tissue perfusion, 35,154,155,157
tissue remodeling, 187
tissue-type PA (*t-PA*), 187-190,192
T lymphocytes, 163-165,194,248
toxic lipid peroxidation
  products, 16
transcription factors, 141-143,
  195
transforming growth factor type
  ß (*TGF-ß*), 9,10,14,40,61,69,
  120-122,142,209
  effect on *PAI-1*, 190-195,198

TGF-ß1, 128-133
TGF-ß2, 128,131,132
transmembrane proteins, 41
transmural pressure gradients,
  34
α-tocopherol, 247,251
trichloroacetic acid, 175,177,
  178
triglycerides, 228,229,257
trigramin, 213
tropomyosin, 71,72
trypan blue exclusion, 174,176
trypsin/*EDTA*, 173
tryptophan codon, 242
tubulin, 73,74
tumor cell-endothelium adhesion,
  37,230
tumor necrosis factor/cachectin
  (*TNF*), 162-165,173,179,182
  TNF-α, 163,191-195,198
tunica intima, 11,12
tunica media, 13
turbulence, 46
tyrosine, 242
tyrosine kinase activity, 144,
  145

ultraviolet light-treated
  cholesterol, 2,3
uterus, 147
umbilical artery endothelium,
  188
urokinase (*u-PA*), 187-189,192

vascular cell phenotype
  modulation, 127-133
vascular endothelial growth
  factor (*VEGF*), 146
vascular growth, 38,40
vascular homeostasis, 161
vascular permeability, 115,173
vascular permeability factor
  (*VPF*), 146
vascular resistance, 153
vascular or vessel wall injury,
  37,227,228
  and repair, 127,130
vasculitides, 166
vasoactive factors, 153
vasoconstrictor, 17,37,44
vasoconstrictor and vasodilator
  prostaglandins, 153
vasodilator response, 35,37,43
vasomotor tone, 153-157
vaso-occlusive lesion, 7
vasospasm, 37,43,153,209
very low density lipoproteins
  (*VLDL*), 249
  lipolytic degradation of,
  249
vessel circumference, 37,41

vessel lumen, 14,17,37,44,45
vessel radius, 42,44
vessel relaxation, 153
vessel size, 40
vessel wall, 8,11,12,14-16,33-35,37,41,42,45,46,139,142,143,145,148,211,212,227-229
  integrity of, 127
  monocytosis of, 16,90
  necrosis of, 7
  and *PAI-1* activity, 187,192,198
  remodeling of, 37,43
  thickening, 209
vessel wall-blood interface, 67
vessel wall perturbation, 101
vimentin, 53,54,74
vinculin, 69,72,73,77,129
viscoelasticity, 115,122
vitamin *E*, 251
vitronectin, 189,210,213
  receptor, 230,231
von Willebrand factor, 16,140,173,209-214,230
  receptor, 210,211
v-sis gene, 140

wall stress, 101-103,105-107,109,110,112
  and elastin production, 106-108
wall tension or stretch, 38,41-44,46
  aortic, 44
  arterial responses to, 43,44
  biological responses to, 34
  local elevations of, 46
  pulmonary artery, 44
wall thickness, 40-42,44,101,102,107
warfarin, 174
Weibel-Palade bodies, 211,214
Western blots, 55
whole blood serum (*WBS*), 8
WKY rats, 11
wound healing, 60,94,131,187,192

xanthomas, 238
X chromosome, 12
X-irradiation, 2, 101

YIGSR-containing peptides, 129